V. A. Peña Rodríguez · R. C. Mercader · C. Pizzaro ·
C. A. Barrero Meneses · E. Passamani Caetano
Editors

LACAME 2010

Proceedings of the 12th Latin American Conference on the Applications of the Mössbauer Effect (LACAME 2010) held in Lima, Peru, 7–12 November 2010

Previously published in *Hyperfine Interactions*
Volumes 202 and 203, 2011

Springer

Editors
Víctor A. Peña Rodríguez
Profesor Principal
Universidad Nacional Mayor
de San Marcos
Facultad de Ciencias Físicas
Pabellón de Laboratorios
de Investigación de Física
Av. Venezuela s/n.
Ciudad Universitaria de San Marcos
Lima 1, Perú
rodiguez.vap@gmail.com

Roberto C. Mercader
Departamento de Fisica - IFLP
Facultad de Ciencias Exactas
Universidad Nacional de La Plata
C.C. 67, 115 & 49
1900 La Plata, Argentina
mercader@fisica.unlp.edu.ar

Carmen Pizarro
Facultad de Química y Biología
Universidad de Santiago de Chile
Av. L. B. O'Higgins 3363
7254758 Santiago, Chile
carmen.pizarro@usach.cl

César Augusto Barrero Meneses
Grupo de Estado Sólido
Instituto de Física
Universidad de Antioquia
Calle 67 No. 53 - 108
A.A.1226 Medellín, Colombia
barrero05@yahoo.com

Edson Passamani Caetano
Universidade Federal
do Espírito Santo (UFES)
Centro de Ciências Exatas,
Departamento de Física
Av. Fernando Ferrari, 514,
Campus de Goiabeiras
Goiabeiras 29075-910, Vitória ES, Brazil
edson@cce.ufes.br

ISBN 978-94-007-4300-7
Springer Dordrecht Heidelberg London New York

Library of Congress Control Number: 2012933972

Printed on acid-free paper

Springer is part of Springer Science+Business Media (www.springer.com)

Table of Contents

Hyperfine Interact (2011) 202:1–2
DOI 10.1007/s10751-011-0377-6

Twelfth Latin American conference on the applications of the Mössbauer effect, LACAME 2010, Lima, Peru, 7–12 November 2010

Chairpersons

Víctor A. Peña Rodríguez
Jorge A. Bravo Cabrejos

Local Organizing Committee

C. Landauro Saenz – UNMSM
Ch. Rojas Ayala – UNMSM
J. Quispe Marcatoma – UNMSM
A. Bustamante Domínguez – UNMSM
E. Torres Tapia – UNMSM
J. Rojas Tapia – IPEN

Latin American Committee

E. M. Baggio Saitovitch – CBPF Brazil
F. González Jiménez – UCV Venezuela
G. A. Pérez Alcázar – UV Colombia
J. A. Jaén – UP Panama
N. R. Furet Bridón – CNIC Cuba
N. Nava – IMP México
R. C. Mercader – UNLP Argentina
V. A. Peña Rodríguez – UNMSM Peru

Scientific Committee

E. M. Baggio Saitovitch – CBPF Brazil
R. C. Mercader – UNLP Argentina
F. H. Sánchez – UNLP Argentina
C. Saragovi – CNEA Argentina
C. Partiti – USP Brazil
C. Larica – UFES Brazil
E. C. Passamani – UFES Brazil
C. A. Barrero Meneses – UA Colombia
N. R. Furet Bridón – CNIC Cuba
J. A. Jaén – UP Panama
A. Bustamante Domínguez – UNMSM Peru
J. R. Gancedo – IQFR Spain

Editorial Committee

Acknowledgements

This Conference has been sponsored and organized by the Physical Faculty of the Universidad Nacional Mayor de San Marcos in Lima, Peru.

Financial support has been provided by the following organizations:

* Consejo Nacional de Ciencia y Tecnología (CONCYTEC), Lima.
* Centro Latinoamericano de Física (CLAF) – Rio de Janeiro
* Office of Naval Research Global (ONRG) – London
* IOP Publishing, Inc. – Philadelphia
* Springer Science, Business Media – The Netherlands
* Universidad Nacional de Ingeniería (UNI), Lima.
* Sociedad Peruana de Física (SOPERFI), Lima.

The Editorial Committee of the LACAME 2010 takes this opportunity to thank Prof. Roberto Mercader for his relevant contribution in the manuscript evaluation process. Prof. Mercader has exchanged/passed his experience with other members, making the entire evaluation process slightly easier than it would really be if he was not present.

Chairpersons

Víctor A. Peña Rodríguez
Jorge A. Bravo Cabrejos

Hyperfine Interact (2011) 202:3–4
DOI 10.1007/s10751-011-0376-7

Preface

Víctor A. Peña Rodríguez · Carmen Pizarro · César A. Barrero Meneses · Edson Passamani Caetano · Roberto C. Mercader

Published online: 30 September 2011

The Mössbauer effect, resonant emission and absorption of gamma rays, is a sophisticated tool to investigate basic and applied subjects of importance to solid state and materials science applied to a variety of fields. Partly as a result of the series of Latin American Conferences on the Applications of the Mössbauer Effect (LACAME), Mössbauer spectroscopy has become a widely used and important technique in many countries across Latin America, where the Mössbauer laboratories exhibit the highest increase in publications and development in the world.

Since 1988 the cycle of biennial LACAME conferences aims at encouraging the knowledge and scientific exchange between the research groups and applied science relevant to the needs of the Latin American countries. At the same time, it allows that young researchers, students and the societies in general, can meet with foreign visitors that come to the conferences and bring along their expertise and last developments in the different topics covered by the scientific programs.

LACAME 2010 was organized well in advance in a flawless way, which helped improve and increase the cooperation between laboratories across the region. A total of one hundred nineteen participants from fifteen countries attended the conference, which was developed in a warmhearted and pleasant atmosphere that

V. A. Peña Rodríguez
Lima, Peru

C. Pizarro
Santiago de Chile, Chile

C. A. Barrero Meneses
Medellín, Colombia

E. Passamani Caetano
Vitória, Brazil

R. C. Mercader (✉)
La Plata, Argentina
e-mail: mercader@fisica.unlp.edu.ar

cooperated toward the mentioned goals and encouraged the participants to keep on strengthening their links and developing the series of Latin American meetings.

Adding to the charming atmosphere of the 2010 LACAME, the attendants could enjoy the gorgeous surroundings that Lima provided. Lima is a modern and outstanding city, with many buildings set in a magnificent landscape; all painted and renewed to the very last detail, including the building where the University of San Marcos, the oldest university in the New World (founded in 1551) is currently sited. Most participants, not used to dwell in old countries themselves, experienced a unique and deeply emotional moment when the inaugural and ending ceremonies were held in a building kept as beautiful as four centuries ago. In addition, the plenary talks were delivered in the main hall of the University, where the position of the speakers was, like in the very old halls, to the side of the audience. This place added to the cultural enrichment of the participants. Lima is also a lively capital with exceptionally charming people that were happy to receive all the participants in their proudly well-kept city.

Compared to previous conferences, this one carried out in Lima exhibited a remarkable group of young participants who, coming from new laboratories from across the region, manifested a noticeable interest in the topics developed, presented a great number of contributions, and displayed a contagious optimism about the development of science in Latin America. Researchers engaged in heated scientific discussions and many showed interest in organizing the next LACAME conferences in their countries. A noticeable difference, which probably demonstrates the importance that research has gained in the region, is that, back in 1996, when LACAME 96 was also organized in Peru, very few attendants could afford paying their air tickets, a remarkable contrast with LACAME 2010. The current situation supports the impression that Mössbauer spectroscopy development in Latin America is gaining momentum and is here to enjoy a long life.

Hyperfine Interact (2011) 202:5–16
DOI 10.1007/s10751-011-0335-3

Radiological and hyperfine characterization of soils from the Northeastern region of the Province of Buenos Aires, Argentina

M. L. Montes · R. C. Mercader · M. A. Taylor · J. Runco · P. Imbellone · P. C. Rivas · J. Desimoni

Published online: 25 August 2011

Abstract The activity concentrations of both natural (^{238}U and ^{232}Th chains and ^{40}K) and anthropogenic (^{137}Cs) radionuclides down along the soil profile have been determined in soil samples collected from inland and coastal areas of the La Plata River, located in the Northeastern region of the Province of Buenos Aires, Argentina. These studies were complemented with ^{57}Fe Mössbauer spectroscopy characterization, pH, texture and organic carbon content measurements. From Mössbauer results, the sample compositions differ from one area to the other. Spectra from both soil samples are dominated by the Fe^{3+} paramagnetic signal. For soil samples from the coastal area, the α-Fe_2O_3 contribution is lower, Fe_3O_4 was not detected, and the relative areas of each spectral contribution are nearly constant with depth. For samples from the inland area, the Fe^{3+} paramagnetic fraction increases up to 82%, mainly at the expense of the magnetically ordered phase. The main observed activity originates from the decay of ^{40}K (540–750 Bq/kg), followed by ^{238}U (60–92 Bq/kg) and ^{232}Th (37–46 Bq/kg) chains. The activity of ^{235}U was in all the cases lower than the detection limit ($L_D = 0.02$ Bq/kg). The only determined anthropogenic nuclide was ^{137}Cs, arising from the fallout of the Southern Hemisphere nuclear weapon tests. Three of the observed differences in the depth distributions can be described by the dispersion-convection model. A correlation

M. L. Montes · R. C. Mercader · M. A. Taylor · J. Runco · J. Desimoni (✉)
Departamento de Física, Facultad de Ciencias Exactas, Universidad Nacional de La Plata,
Instituto de Física La Plata – CONICET, Buenos Aires, Argentina
e-mail: desimoni@fisica.unlp.edu.ar

P. Imbellone
Instituto de Geomorfología y Suelos, Universidad Nacional de La Plata,
Buenos Aires, Argentina

P. C. Rivas
Facultad de Ciencias Agrarias y Forestales, Universidad Nacional de La Plata,
Instituto de Física La Plata – CONICET, Buenos Aires, Argentina

between the natural nuclide activities and the Mössbauer relative fractions was found, whereas no correlation was found between the ^{137}Cs profile and the relative fraction of Fe_3O_4 or with other iron species.

Keywords Natural nuclides · Nuclear weapons fallout · ^{137}Cs · Soil · Mössbauer spectroscopy

1 Introduction

The assessment of the gamma radiation dose from natural sources is of particular importance because it embodies the largest contribution to the external dose regularly received by the population. This dose varies depending upon the concentration of the natural nuclides, ^{238}U, ^{232}Th, their daughter products, and ^{40}K, naturally present in soil and rocks, which in turn depend on the local geology of each region [1, 2]. However, human activities can strongly modify their natural concentrations due to the presence of residues or accumulation of elements caused by the release of effluents by the nuclear industry. In addition, artificial nuclides may be released to the environment. Among the class of anthropogenic nuclides, ^{137}Cs is the most prominent gamma emitter isotope in the Earth crust, originated by fission processes, and is considered as one of the most hazardous environmental contaminants due to human external exposure to the human body or as intake of contaminated food with ^{137}Cs distributed all along the human trophic chain [1, 2]. Consequently, the knowledge of the distribution of radionuclides in soils becomes of critical importance in order to allow predicting the impact on man and on the natural environment. Regarding ^{137}Cs, such predictions can be made only with models that realistically describe the underlying processes, which include the local physicochemical conditions, encompassing the very complex system that soil is. Indeed, ^{137}Cs transport in soil is strongly affected by the soil composition, particularly by its mineralogy, pH, and concentration of inorganic ions, redox potential and concentration of organic matter [3, 4]. For these reasons, the distribution profiles of ^{137}Cs are used to more broadly predict its dispersion in the soil, to estimate the plant root uptake and the transfer to animals. Therefore, it is necessary either to test existing models [5, 6] or to develop new ones to account for specific local conditions.

In South America, according to the United Nation Scientific Committee on the Effects of the Atomic Radiation (UNSCEAR) report [2], only the activity concentration of the natural nuclide ^{40}K in soils has been compiled for Argentina, with an activity concentration ranging from 559 to 773 Bq/kg. More recently, data of ^{226}Ra (73 Bq/kg) and ^{40}K (740 Bq/kg) of semi-natural grassland soils of the central part of the country (Province of San Luis) have been reported [7]. In the Brazilian State of Rio Grande do Norte [8], the average concentrations of ^{226}Ra (29.2 Bq/kg), ^{232}Th (47.8 Bq/kg) and ^{40}K (704 Bq/kg) in unperturbed soils have been reported.

Regarding the anthropogenic nuclides, only a few studies dealing with the determination of the activity of the ^{137}Cs gamma emitter, globally occurring in soil due to nuclear weapons tests, have been performed in the southern hemisphere. Important differences in the ^{137}Cs depth distribution were found in studied sites of the South-Central region of Brazil, ascribed to chemical, physical, mineralogical and biological differences in soils [9]. The spatial distribution and behavior of the ^{137}Cs in tropical, subtropical and equatorial unperturbed Brazilian soils have been investigated, for soil depth down to 40 cm [10]. The majority of the ^{137}Cs content was observed in the 10–15 cm upper layer, whereas minor quantities were detected down to 35 cm. Extremely low deposition densities were detected across the Amazon region; the southern area of Brazil exhibited considerable larger concentrations, confirming the UNSCEAR predictions [2, 11].

Total inventories and depth distributions of ^{137}Cs were determined at four sites in Chile (latitude, 38–40° S) in two agriculturally used soil-types of high milk-production regions. The inventory reported values that were always higher than those previously estimated activities for the Southern Hemisphere [2] depending on the annual rainfall. The differences in depth distributions and the high cesium mobility in coarser textured soils of low clay content were ascribed to be due to differences in edaphic parameters [12]. The depth distribution of ^{137}Cs in well-developed cropping land soil at 29 sites in different southern regions (latitude, 38–44° S) was also studied [13]. The depth distribution in most sites follows no systematic pattern in the upper few centimeters; below this depth an exponential behavior is observed. The penetration increased with decreasing clay content and increasing volume of coarse pores, being correlated with the mean annual rainfall rate of the sampled areas. The South Patagonia sheep-raising region (50–54° S and 68–74° W) was also studied, and the areal activity density was significantly correlated with the mean annual precipitation rate [14]. The convection velocity resulted rather high when compared with data for temperate regions of Chile [12], but the diffusion coefficient is closer to those obtained in the Antarctic region [13].

In Argentina, the ^{137}Cs reference activity profile was determined in the Pampa Ondulada region (33°50′ S, 59°52′ W) [15, 16]. The ^{137}Cs activity declined sharply from 4.71 Bq/kg in the first 10 cm to 0.93 Bq/kg at 30 cm. An average integrated activity value was slightly higher than the ^{137}Cs values of global distribution [17] and lower than the estimated values by UNSCEAR using the ^{90}Sr/^{137}Cs yield ratio [2]. Vertical migration of ^{137}Cs was studied in soils of natural and semi-natural grassland areas of San Luis Province (33°40′ S, 65°23′ W) [18]. The obtained diffusion coefficients and convection velocities were in the range reported in the bibliography [19]. Two other areas (33°40′10″ S, 65°23′27″ W and 33°39′56″ S, 65°23′16″ W), undisturbed for least in the last four decades, were studied by the same authors [7]. Despite that inventories were similar for both areas, the depth profile resulted different, i.e., the maxima of the distributions were placed at different depths. The diffusion coefficients estimated with a convection–diffusion model [5, 6] agreed well with previous data, while the convection velocities were one order of magnitude higher. The great penetration in these soils was ascribed to the high sand and low fine materials content.

Because of the lack of a more systematic information about natural and anthropogenic activities baselines, the controversies concerning the nuclide transport in soil and in view of the retaken nuclear program in Argentina [20, 21], we have started a long term project aiming at contributing to monitor the activity of gamma emitter nuclides, complemented with a physicochemical characterization of soils of the Province of Buenos Aires. This is an important agricultural and industrial area where nuclear power plants and research nuclear reactors are located, adding to the natural inventory on small amounts of activities of anthropogenic nuclides. Despite the long period of nuclear of activity in the region, no systematic study has been carried out until now to establish the actual conditions. We present here activities depth profiles of the gamma emitters nuclides of ^{238}U (^{226}Ra) and ^{232}Th (^{208}Tl, ^{212}Bi, ^{212}Pb and ^{228}Ac) chains, ^{40}K and ^{137}Cs of unperturbed soil. Two sampling sites are located in the inland area and two to coastal area of La Plata River. These studies have been complemented with a Mössbauer spectroscopy characterization, pH, texture and organic carbon content measurements.

2 Experimental

Undisturbed soils were sampled: (1) the costal (M1: 34°48.5′ S, 58°5.3′ W and M2: 34°54.1′ S, 57°55.1′ W) border of the La Plata river and (2) inland (C1: 34°54.5′ S, 58°8.4′ W and C2: 35°3.3′ S, 57°51.2′ W) areas, in the neighbourhood of La Plata city [22–24].

The samples for radiological analysis were collected with a spade 3 cm high and 30 cm wide, while the corresponding ones for Mössbauer spectroscopy were taken with small plastic tubes. After collection, the samples were dried at 363 K, crushed and sieved through a 2 mm mesh without previous separation. The sampling was performed at depths down to 50 cm from the surface.

For gamma spectroscopy, the soil sample was properly accommodated in a Marinelli-type box, sealed and kept for nearly 3 weeks before analysis, allowing it to achieve the secular equilibrium. Spectra were taken inside an EG&G Ortec low-background chamber in the range of 30 keV to 1.8 MeV, using a GMX10 gamma EG&G Ortec detector, with a standard electronic chain and 8192 channels multichannel. The energy calibration was performed with ^{60}Co, ^{133}Ba, ^{137}Cs and ^{152}Eu sources. The efficiency calibration was carried out using an admixture of known amounts of naturally occurring ^{176}Lu and ^{138}La long-lived isotopes, dispersed in a soil substrate [25]. The laboratory background was determined and peak corrections were performed on the spectrum. Spectra were recorded during four days and analyzed with a commercial computer program. The activity of the ^{238}U chain was obtained by considering the activity of the gamma emitter ^{226}Ra, and ^{232}Th chain activity was determined from ^{208}Tl, ^{212}Bi, ^{212}Pb and ^{228}Ac nuclides.

Mössbauer spectra were recorded at room temperature in a 512 channels conventional constant acceleration spectrometer with a ^{57}Co*Rh* source of approximately 5 mCi nominal activity. Velocity calibration was performed with a 12 μm thick α-Fe foil at room temperature. Isomer shifts are referred to this standard. The spectra were numerically analyzed using hyperfine field and quadrupole distributions [26].

The pH and the texture measurements were performed according to international protocols [27, 28]. The organic carbon content was evaluated using colorimetric titration, through CO oxidation with potassium dichromate [28].

3 Results and discussion

Figure 1 shows selected room temperature Mössbauer spectra recorded for the collected samples. Two main contributions to the spectral area are observed; magnetic sextets and a central set of doublets. To help assign the magnetically split sextets, AC magnetic susceptibility was measured at different temperatures for some samples. Most of the them displayed a thermal dependence (not shown) typical of the

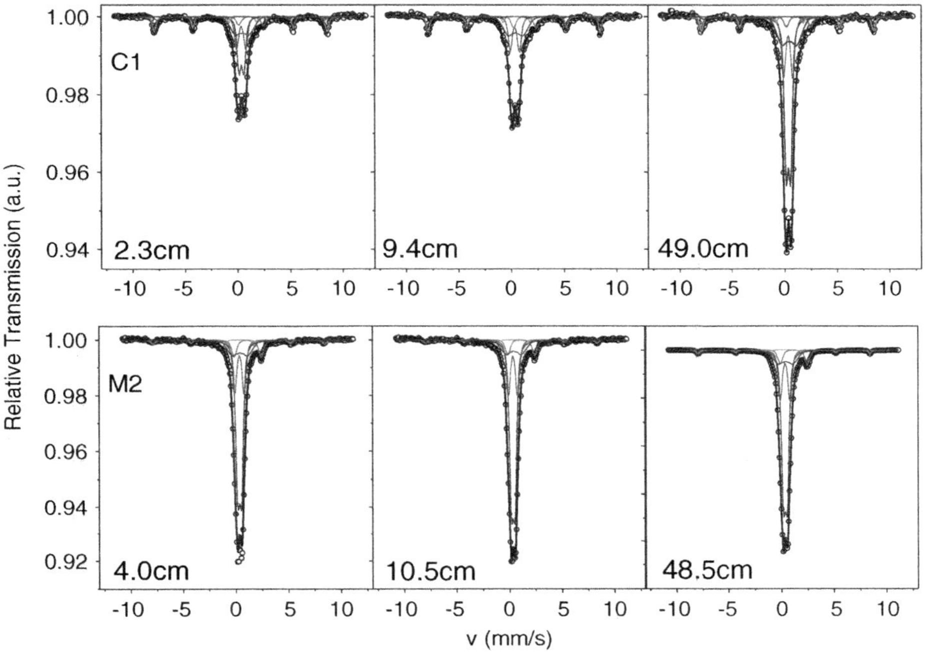

Fig. 1 Room temperature Mössbauer spectra of two different analyzed areas (*Top*: inland area and *bottom*: costal area) labeled with the depth. The *thick solid* and *thinner lines* are the result of the fitting and the corresponding hyperfine interactions components as described in the text

mixture of two magnetic phases, whose shape allowed assigning Verwey and Morin transitions, which are characteristic of magnetite and hematite, respectively. Accordingly, the Mössbauer sextets were fitted assuming three signals, one for hematite and two for the two sites due to magnetite ($I_{\alpha-Fe_2O_3}$: $B = 51.1$ T, $\varepsilon = -0.11$ mm/s, $\delta = 0.39$ mm/s; $I_{Fe_3O_4\,(I)}$: $B = 49.1$ T, $\varepsilon = 0.00$ mm/s, $\delta = 0.28$ mm/s; $I_{Fe_3O_4\,(II)}$: $B = 46.0$ T, $\varepsilon = 0.00$ mm/s, $\delta = 0.66$ mm/s). However, the relative populations varied from sample to sample, and some profiles did not support the existence of magnetite. This fact proved that the Mössbauer spectra taken at room temperature, do not allow determining accurately the magnetic species contributing to the spectral area.

The broad lines of the spectra produced by the fine particles of the soil samples made it reasonable to perform the fittings assuming hyperfine field distributions. To account for the total spectrum, superimposed to the above mentioned sextets, it was necessary to include quadrupole split doublets associated to paramagnetic Fe^{3+} and Fe^{2+} species ($I^{3+}_{Fe\,(I)}$: $\Delta = 0.46$ mm/s, $\delta = 0.36$ mm/s, $I^{3+}_{Fe\,(II)}$: $\Delta = 0.88$ mm/s, $\delta = 0.34$ mm/s; I^{2+}_{Fe}: $\Delta = 2.31$ mm/s, $\delta = 1.27$ mm/s) and a paramagnetic relaxation (I_{PR}: $B = 6.0$ T, $\varepsilon = 0.00$ mm/s, $\delta = 0.44$ mm/s) [29].

Because of the overlapping of different Fe species with similar parameters [29–32], e.g., hydroxides, finely divided Fe oxides in superparamagnetic relaxation regimes, or Fe ions partially substituted for other ions, like Al or Mn, the assignment of the Mössbauer parameters of the doublets to iron compounds is uncertain. Based

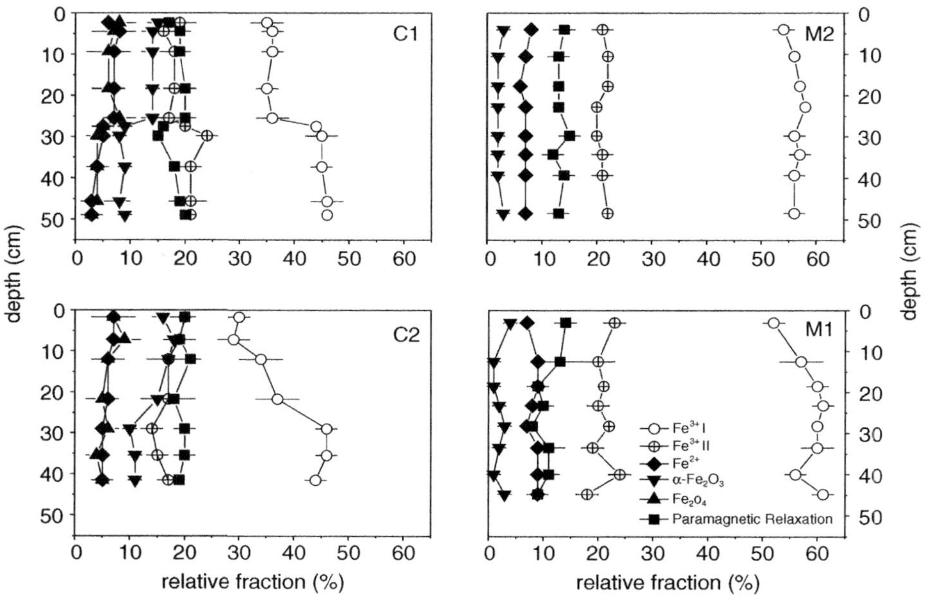

Fig. 2 Evolution of the Mössbauer relative subspectral areas with depth. *Left*: inland area and *right*: coastal area

on low-temperature previous results [33], the determined quadrupole interactions could be related to the Fe^{3+} and the Fe^{2+} interactions existing at the octahedral and tetrahedral sites in paramagnetic phases [29–32].

Figure 2 displays the evolution of the relative subspectral areas for ^{57}Fe Mössbauer spectra in different chemical environments with depth. It is observed that for all the samples and studied depths, the main component corresponds to the paramagnetic Fe^{3+} doublets. However, sample compositions differ from one soil location to the other. For the soils samples from the costal area, Fe_3O_4 was not detected, the α-Fe_2O_3 contribution is lower, and the relative fractions are nearly constant in depth. In the inland area, Fe^{3+} fraction increases up to 82%, mainly at the expense of the α-Fe_2O_3 spectral area.

Concerning the organic carbon content, no noticeable differences were observed in the depth distributions of the four studied soils, as observed in Fig. 3. No systematic trend is observed in the pH depth determinations (Fig. 3); the M1 and M2 top layers depict higher values. The texture determinations indicate that M2 and C1 soils are silty clay loam, C2 is silty loam and M1 belongs to silty clay and clay, showing M1 soil to contain the highest clay concentration.

The gamma spectra (not shown here) display the characteristic lines associated with nuclides belonging to the ^{238}U and ^{232}Th chains and that corresponding to ^{40}K. The only anthropogenic detected radionuclide was ^{137}Cs. The determined gamma activities of both natural and anthropogenic nuclides for all samples are presented in Fig. 4. The main observed activity is originated from the decay of the ^{40}K following

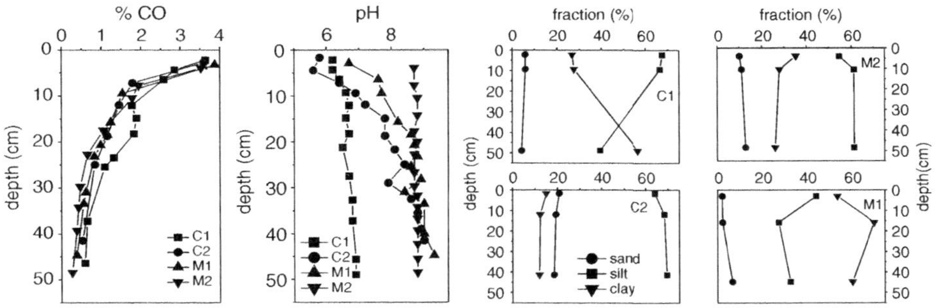

Fig. 3 Evolution in depth of organic carbon content (*left*), pH (*center*) and texture (*right*)

in importance those of ^{238}U and ^{232}Th chains. Activities of ^{235}U in all cases are lower than the detection limit ($L_D = 0.02$ Bq/kg).

The activity values of the ^{238}U and ^{232}Th chains lie in the intervals 60–92 and 37–46 Bq/kg, respectively. The activity of the ^{232}Th chain is closer for the C1 and C2 sites and a slightly increment is observed for the M1 and M2 sites. In the case of the ^{238}U, the activities result to some extent high when compared with data from Brazil and close to those reported for the central region of Argentina [7]. It is observed in Fig. 5 that soils without magnetite and lower hematite relative fraction (M1 and M2) present a higher Ra and Th activity concentrations, probably related to the geological origin of the soils. The ^{40}K activity profiles are quite different when comparing the monitored soils, ranging the surface activities values from 531 to 720 Bq/kg as observed in Fig. 4. In the C1 and C2 profiles, the activity increases with depth. The activity of the top layer samples are in good agreement with data reported by the UNSCEAR [2] and the depletion of the activity in the approximately first 20 cm of the C1 and C2 soils could be correlated with the uptake of K by the plant roots [2]. As observed in Fig. 5, the Fe^{3+}(I) relative fractions and the ^{40}K distribution have quite similar behavior. This correlation can be ascribed to the soil pedogenetic and edaphic properties [34], as well as to the presence of plant roots that use these ions as nutrients [2]. It is worth to mention that the differences in activities of natural nuclides of the upper soil layers produce variations in the exposure dose [2], indicating that the contribution to the dose in the coastal area is 40 % higher than in the inland region studied.

Even though the ^{137}Cs integrated activities of the profiles are similar in all soils (average value: 11 Bq/kg) and coincident with the reported average values around the world [2, 7, 12–14, 17, 18], compatible with the fallout of the nuclear weapon tests in the Southern Hemisphere [2], differences in the ^{137}Cs depth distributions are detected (Fig. 4). The profiles of C1 and M2 sites follow a Gaussian-type feature, typical of a convective–diffusive process [5, 6]. Both profiles were analyzed using the distribution proposed in [5]. The fitted diffusion coefficient (D) and convection velocity (v) result equal to 0.728 and 0.23 cm/year for the C1, while the values for M2 are 0.5 and 0.22 cm/year, respectively. The profile of C2 is quite different since a Gaussian-shape is established from 7 cm in depth. The analysis of the data in the frame of the theoretical models [5, 6] lead to diffusion coefficient and convection velocity values of 0.39 and 0.34 cm/year, respectively. The present set of transport

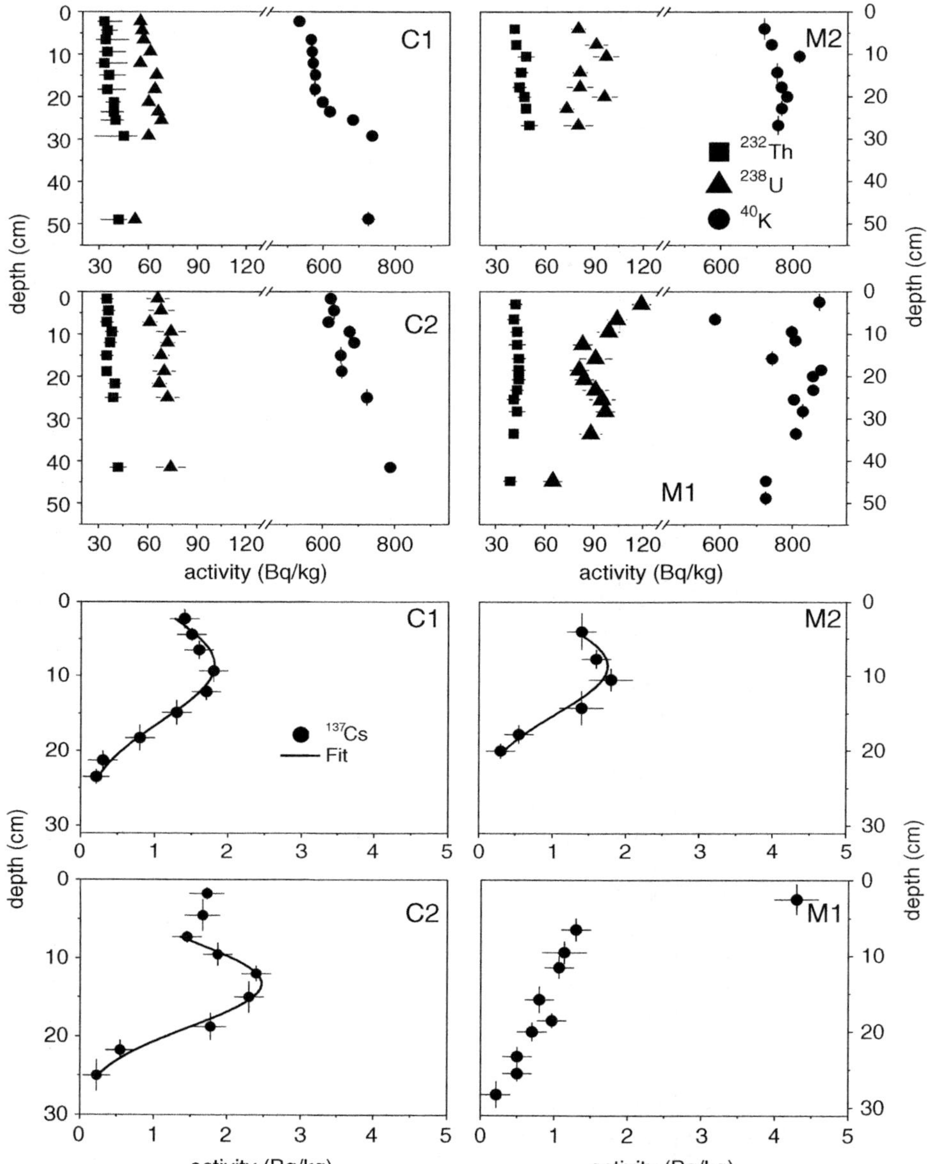

Fig. 4 ^{40}K, ^{238}U and ^{232}Th (*top*) and ^{137}Cs (*bottom*) activity profiles. (L_D is the detection limit and the *horizontal* and *vertical bars* represent the activity errors and analyzed soil depth, respectively)

parameter values agrees well with the South American values [7, 12–14, 18], as observed in Fig. 6, but is slightly larger than the average values reported by IAEA [19]. In the case of the M1 soil, placed at 5 km from the La Plata river coast, the activity values are higher at the surface and then suddenly decrease. Both facts, the

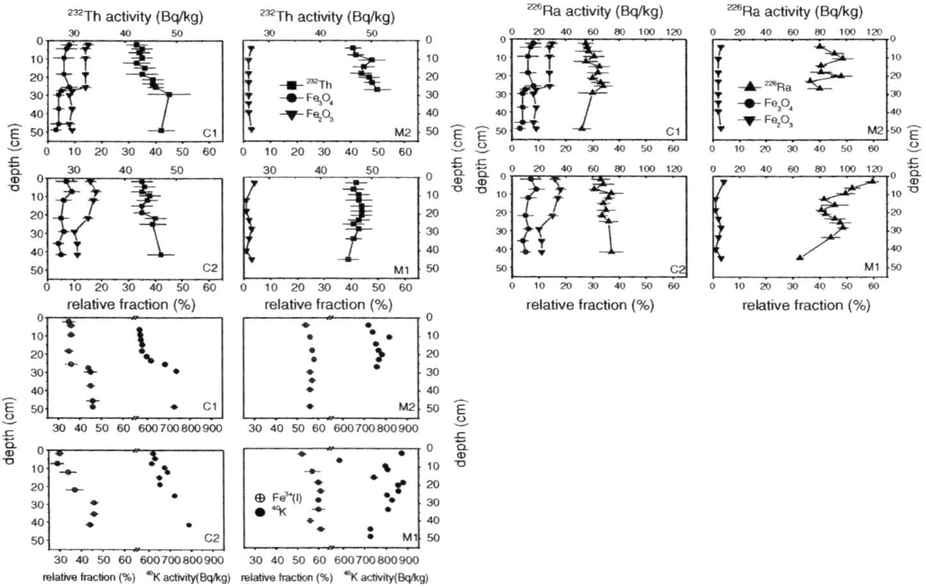

Fig. 5 Comparison between Mössbauer relative fractions and natural nuclide activity profiles

Fig. 6 Diffusion coefficient (D) and convection velocity (v) parameters obtained using the model of [5]. Present study (*open symbols*). IAEA (*solid line*) [19] and South American data (*filled symbols*) [7, 12–14, 18] are included

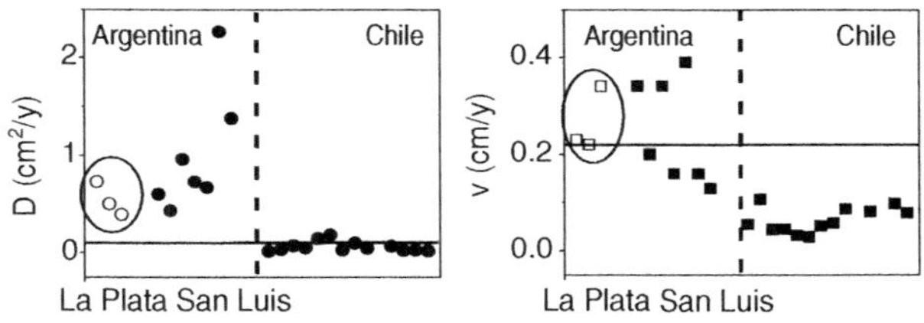

high values at the surface and the deviation from the models [5, 6], could be explained considering the fine texture (Fig. 3) and the flat relief of the region which induce water-logging, i.e., this area shows a low permeability of the underlying horizons, and the phreatic water affects the deepest horizons [34].

Finally, vis-à-vis the ^{137}Cs depth profile and its correlation with Fe_3O_4, controversial results on their sorption by Fe_3O_4 have been reported [35–37]. It was found that the percentage sorption of ^{137}Cs by magnetite is pH dependent, increasing by increasing pH of the suspension. The results have been explained in terms of the electrostatic interaction between the positively charged metal ions and the surface charge of the magnetite which becomes increasingly negative with increasing pH. By comparing the present results displayed in Figs. 2 and 4 it is observed that there is not an apparent correlation between the ^{137}Cs profile and the Mössbauer relative fraction of Fe_3O_4 as well as with the other iron species.

4 Conclusions

These results—for samples taken from the surface down to a depth of 50 cm—are the first ones to establish baseline activity for the natural and anthropogenic occurring gamma emitter radionuclides in unperturbed soils of inland and coastal river areas around La Plata city. The ^{238}U and ^{232}Th chains were detected, as well as the presence of ^{40}K. ^{232}Th and ^{238}U activity chains vary in depth and space, i.e., soils with low hematite content and absence of magnetite display slightly higher activities, probably related to the geological origin of the soils. The main contribution to the exposure dose arises from the ^{40}K activity, in agreement with previous reported data, and correlates with the Fe^{3+} content, probably relate to the pedogenesis and edaphic properties of the soils, and/or to the presence of roots. The variation in the top layer activities gives rise to higher exposure dose in the coastal region.

The only anthropogenic detected nuclide was ^{137}Cs, arising from the fallout of the nuclear weapons tests carried out in the Southern Hemisphere. Differences in depth distributions were detected; three of them follow the behavior associated to a convective–diffusive process. The determined diffusion coefficient and convection velocity are in the range of South American and worldwide determinations. The deviation from the convection–diffusion model observed in the depth profile of the La Plata River coastal soil could be explained considering the high clay content, the flat relief and the water dynamics of the region. These properties will be further included in the transport equation to simulate the complex migration processes of ^{137}Cs in these soils.

Finally, putting together the gamma and Mössbauer results, it is tempting to conclude that there is not an apparent correlation between the ^{137}Cs profile and the relative fraction of Fe_3O_4 or of other iron species.

Acknowledgements Research grants PIP 5283 and PIP 0230 from Consejo Nacional de Investigaciones Científicas y Técnicas (CONICET, Argentina) and BID 1728 OC/AR PICT 38047 from Agencia de Promoción Científica (ANCYT, Argentina) are gratefully recognized.

References

1. Cooper, J.R., Randle, K., Sokhi, R.S.: Radioactive Releases in the Environment, Ed. John Wiley & Sons Ltd. The Atrium, Southern Gate, Chichester, West Sussex, England (2003)
2. UNSCEAR 2008 Report Vol. 1: Sources of ionizing radiation. http://www.unscear.org/unscear/en/publications/2008_1.html (2008)
3. Cornell, R.M.: Adsorption of cesium on minerals: a review. J. Radioanal. Nucl. Chem. **171**, 483–500 (1993)
4. Sawhney, B.L.: Selective sorption and fixation of cations by clay minerals: a review. Clays Clay Miner. **20**, 93–100 (1972)
5. Bossen, P., Kirchner, G.: Modeling the vertical distribution of radionuclides in soils. Part I: the convection-dispersion equation revisited. J. Environ. Radioact. **73**, 127–150 (2004)
6. Almgren, S., Isaksson, M.: Vertical migration studies of ^{137}Cs from nuclear weapons fallout and the Chernobyl accident. J. Environ. Radioact. **91**, 90–102 (2006)
7. Juri Ayub, J., Velasco, R.H., Rizzotto, M., Quintana, E., Aguiar, J.: ^{40}K, ^{137}Cs and ^{226}Ra soil and plant content in semi-natural grasslands of Central Argentina. In: Paschoa, A.S. (ed.) The Natural Radiation Environment—8th International Symposium. American Institute of Physics 978-0-7354-0559 (2008)
8. Malanca, A., Gaidolfi, L., Pessina, V., Dallara G.: Distribution of ^{226}Ra, ^{232}Th, and ^{40}K in soils of Rio Grande do Norte (Brazil). J. Environ. Radioact. **30**, 55–67 (1996)

9. Correchel, V., Oliveira Santos Bacchi, O., Reichardt, K., Cereci de Maria I.: Random and systematic spatial variability of ^{137}Cs inventories at references sites in south-central Brazil. Sci. Agric. **62**, 173–178 (2005)
10. Handl, J., Sachase, R., Jakob, D., Michel, R., Evangelista, H., Gonçalves, A.C., de Freitas, A.C.: Accumulation of ^{137}Cs in Brazilian soils and its transfer to plants under different climatic conditions. J. Environ. Radioact. **99**, 271–287 (2008)
11. UNSCEAR 1982 Report: Ionizing radiation: sources and biological effects. http://www.unscear.org/unscear/en/publications/1982.html (1982)
12. Schuller, P., Ellies, A., Kirchner, G.: Vertical migration of fallout ^{137}Cs in agricultural soils from southern Chile. Sci. Total Environ. **193**, 197–205 (1997)
13. Schuller, P., Voigt, G., Handl, J., Ellies, A., Oliva, L.: Global weapons fallout ^{137}Cs in soils and trasnfer to vegetation in south-central Chile. J. Environ. Radioact. **62**, 181–193 (2002)
14. Schuller, P., Bunzl, K., Voigt, G., Ellies, A., Castillo, A.: Global fallout ^{137}Cs accumulation and vertical migration in selected soils fron south Patagonia, J. Environ. Radioact. **71**, 43–60 (2004)
15. Bujan, A., Santanatoglia, O.J., Chagas, C., Massobrio, M., Castiglioni, M., Yañez, M., Ciallella, H., Fernandez, J.: Preliminary study on the use of the ^{137}Cs method for soil erosion investigation in the pampean region of Argentina. Acta Geol. Hisp. **35**, 271–277 (2000)
16. Bujan, A., Santanatoglia, O.J., Chagas, C., Massobrio, M., Castiglioni, M., Yañez, M., Ciallella, H., Fernandez, J.: Soil erosion evaluation in a small basing though the use of ^{137}Cs technique. Soil Tillage Res. **69**, 127–137 (2003)
17. Garcia Agudo, E.: IAEA-TECDOC-1028 (1998)
18. Juri Ayub, J., Rizzotto, M., Toso, J., Velasco, H.: ^{137}Cs deposition and vertical migration in soils from Argentina. In: Proc. Int. Conf. Environ. Radioctivity (2007)
19. IAEA Technical Report Series no. 42. Handbook of Parameter values for the Prediction of Radionuclide Transfer in Terrestrial and Freshwater Environments (2010)
20. De Vido, J.M.: Discurso, Reactivación de la industria nuclear en Argentina (2006)
21. Carlos Rey, D.F: Reactivación del plan nuclear Argentino. www.pagina12.com.ar (2007)
22. Instituto de Geomorfología y Suelos. Facultad de ciencias Naturales y Museo, Universidad Nacional de la Plata. Centro de Investigaciones de Suelos y Aguas de Uso Agropecuario. Elaboración y Transferencia de Cartografía Temática e Implementación de un Sistema de Información Geográfica para el Planeamiento (Partido de Berisso) (2005)
23. Instituto de Geomorfología y Suelos. Facultad de ciencias Naturales y Museo, Universidad Nacional de la Plata. Centro de Investigaciones de Suelos y Aguas de Uso Agropecuario. Análisis ambiental del Partido de La Plata. Aportes a ordenamiento territorial (2006)
24. Instituto de Geomorfología y Suelos. Facultad de ciencias Naturales y Museo, Universidad Nacional de la Plata. Centro de Investigaciones de Suelos y Aguas de Uso Agropecuario. Suelos del Partido de Berasategui como base para el planeamiento ambiental y ordenamiento territorial (2006)
25. Perillo Isaac, M.C., Hurley, D., McDonald, R.J., Norman, E.B., Smith, A.R.: A natural calibration source for determining germanium detector efficiencies. Nucl. Instrum. Methods Phys. Res. A **397**, 310–316 (1997)
26. Lagarec, K., Rancourt, D.G.: Mössbauer Spectral Analysis Software for Windows, Version 1.0. University of Ottawa- Department of Physics (1998)
27. United Nations Econimic Comission for Europe Convention on long-range Transboundary Air Pollution, Manual on methods and criteria for harmonized sampling, assessment, monitoring and analysis of the effects of air pollution on forests Part IIIa Sampling and Analysis of Soil (2006)
28. Claver, R.: Procedure in Sedimentary Petrology. Wiley Interscience, New York (1971)
29. Murad, E.: Clays and clay minerals: what can Mössbauer spectroscopy do to help understand them? Hypefine Interact. **117**, 39–70 (1998)
30. Fabris, J.D., Coey, J.M.D.: Topics em Ciencia do Solo II, Soc. Brasileira de Ciencia do Solo 47–102 (2002)
31. Paduani, C., Samudio Pérez, C.A., Gobbi, D., Ardisson, J.D.: Mineralogical characterization of iron-rich clayey soils from the middle plateau in the Southern of Brazil. Appl. Clay Sci. **42**, 559–562 (2009)
32. MacCannon, C.A.: Mössbauer Spectroscopy of Minerals in Mineral Physics & Crystallography: Handbook of Physical Constants, pp. 332–347. American Geophysical Union, Washington DC (1995)
33. Montes, M.L., Taylor, M.A., Mercader, R.C., Siver, F.R., Desimoni J.: Hyperfine and radiological characterization of soils of the province of Buenos Aires, Argentina. J. Phys.: Conf. Series **217**, 012058 (2010). doi:10.1088/1742-6596/217/1/012058

34. Imbellone, P.A., Guichon, B.A., Giménez, J.E.: Latinamercian J. sedimentology and basin analysis **16**, 3–18 (2009)
35. Singh, B.K., Jain, A., Kumar, S., Tomar, B.S., Tomar, R. Manchanda, V.K., Ramanathan, S.: Role of magnetite and humic acid in radionuclide migration in the environment. J. Contam. Hydrol. **106**, 144–149 (2009)
36. Catallette, H., Dumonceau, J., Ollar, P.: Sorption of cesium, barium and europium on magnetite. J. Contam. Hydrol **35**, 151–159 (1998)
37. Marnier, N., Fromage, F.: Sorption of Cs(I) on magnetite in the presence of silicates. J. Colloid. Interf. Sci. **223**, 83–88 (2000)

Hyperfine Interact (2011) 202:17–24
DOI 10.1007/s10751-011-0336-2

Loading channels of SiO_2 mesoporous matrices with Fe oxides

N. A. Fellenz · L. A. Cano · J. F. Bengoa · I. de Souza Azevedo · R. C. Mercader · S. G. Marchetti

Published online: 25 August 2011

Abstract After embedding hematite nanowires (15 wt.% Fe) into a MCM-41 hard template, we have explored alternative routes to induce the structural transformations that lead from hematite to maghemite and magnetite embedded nanowires. The impregnation media (ethanol or water) and the calcination atmosphere (air and NO/He) on the hematite nanowires production play a significant role at the time of reducing and re-oxidizing the embedded hematite nanoparticles. The solids were characterized by X-ray diffraction, nitrogen adsorption, and Mössbauer spectroscopy. The results indicate that the effect of the solvent on the structural properties of the iron species is more important than the calcination atmosphere. The best conditions for iron magnetic nanowires not to get outside of the MCM-41 channels over the treatments are reached using water as the solvent and air as the calcination atmosphere. When ethanol is the solvent used over the preparation step, the end iron oxides are in the form of nanotubes spread out on the amorphous silica walls of the matrix.

Keywords Mesoporous silicates · MCM-41 · Embedded nanowires synthesis · Mössbauer spectroscopy

N. A. Fellenz · L. A. Cano · J. F. Bengoa · S. G. Marchetti
CINDECA, CONICET, CICPBA, Universidad Nacional de La Plata,
Facultad de Ciencias Exactas, Calle 47, No 257,
1900 La Plata, Argentina

I. de Souza Azevedo
Centro Brasileiro de Pesquisas Físicas,
Rua Dr. Xavier Sigaud 150, Rio de Janeiro 22290-180, Brazil

R. C. Mercader (✉)
Departamento de Física, IFLP-CONICET,
Facultad de Ciencias Exactas, Universidad Nacional de La Plata,
115 y 49, 1900 La Plata, Argentina
e-mail: mercader@fisica.unlp.edu.ar

1 Introduction

The applications of regular arrays of nanowires and nanotubes have stimulated lately quite a few studies. Some time ago we set about preparing these solids, starting with the embedding of hematite nanoparticles into the regular array of mesoporous matrices [1]. More recently, we have investigated different conditions of preparation to enhance some special properties of the solids, namely, higher metal loadings or larger length-to-diameter ratios of the iron species [2].

Because of their magnetic properties, magnetite or maghemite-loaded solids display more promising properties than hematite, e.g., for biomedical applications or high-density magnetic recording media. Therefore, once we reached a certain degree of expertise in the preparation of hematite-loaded solids, we have used them as precursors for producing the magnetic species of iron, maghemite and magnetite [3, 4], and also to synthesize catalysts of active Fe^0 nanoparticles supported inside the channels of the MCM-41 matrix to be used in the Fischer–Tropsch hydrocarbon synthesis.

The aim of the present work is to study alternative routes to induce the structural transformations that lead from hematite to maghemite and magnetite embedded nanowires. We report the synthesis of three samples prepared in different conditions by incipient wetness impregnation of MCM-41 with iron nitrate to obtain an iron nominal concentration of 15 wt.%.

2 Experimental

The MCM-41 matrix was prepared according to the methodology proposed by Ryoo and Kim [5]. Treating MCM-41 in the way described in [2], three solids were obtained with iron concentrations of about 15 wt %. From now on we will use the following notation: Fe/MCM-41(aq-air), Fe/MCM-41(alc-air), and Fe/MCM-41(alc-NO/He) according to these treatments:

Fe/MCM-41 (aq-air): incipient wetness impregnation with $Fe(NO_3)_3{\cdot}9H_2O$ aqueous solution. The sample was dried at room temperature in atmospheric air, and calcined in dry air stream (100 cm^3/min) from 298 to 723 K at 1 K/min and kept at 723 K for 4 h,

Fe/MCM-41 (alc-air): incipient wetness impregnation with $Fe(NO_3)_3{\cdot}9H_2O$ in ethanol solution. The sample was dried at room temperature in atmospheric air, calcined in the same way than Fe/MCM-41 (aq-air),

Fe/MCM-41 (alc-NO/He): impregnated in the same way as Fe/MCM-41 (alc-air). The sample was dried at room temperature in atmospheric air, calcined in a dry mixture of 1% (*V*/*V*) of NO in He (100 cm^3/min) from 298 to 723 K at 1 K/min and kept at 723 K for 4 h following the methodology proposed by Sietsma et al. [6].

After this stage, the three samples were subjected to the following treatment. Firstly, they were reduced in a flowing (20 cm^3/min) mixture of 10% *V*/*V* of H_2/N_2, heating from 298 to 698 K at 10 K/min, kept at 698 K for 2 hours and quenched to 298 K thus obtaining the solids that we call (R)Fe/MCM-41(alc-air), (R)Fe/MCM-41(aq-air), and (R)Fe/MCM-41(alc-NO/He). Finally, they were reoxidized in air at 353 K for 24 h so producing the samples called (O)Fe/MCM-41(alc-air), (O)Fe/MCM-41(aq-air), and (O)Fe/MCM-41(alc-NO/He).

The resulting solids were characterized by X-ray diffraction (XRD), Nitrogen adsorption (BET), and Mössbauer spectroscopy.

All XRD patterns were measured using a standard automated powder x-ray diffraction system Philips PW 1710 with diffracted-beam graphite monochromator using Cu$K\alpha$ radiation ($\lambda = 1.5406$ Å). Nitrogen adsorption isotherms were recorded at the temperature of liquid nitrogen (77 K) using a Micromeritics ASAP 2020 apparatus.

The Mössbauer spectra at 298 and 30 K were taken in transmission geometry with a constant acceleration spectrometer. A source of ^{57}Co in Rh matrix of nominally 50 mCi was used. The velocity calibration was performed with a metallic 12 μm-thick α-Fe foil at room temperature. All isomer shifts (δ) are referred to this standard. The temperature between 30 and 298 K was varied using a Displex DE-202 Closed Cycle Cryogenic System. The measurement at liquid helium was taken in an Oxford Instruments cryostat. The Mössbauer spectra were fitted using the Recoil program [7]. At 30 K the Blume and Tjon's dynamic line shape site analysis option was used [7]. Lorentzian lines with equal linewidths were considered for each spectrum component in a complete superparamagnetic regime.

3 Results and discussion

The structural properties typical of an ordered mesoporous solid MCM-41, used as template, were confirmed by XRD and by their textural properties measured by N_2 adsorption. The textural properties obtained were characteristic of this mesoporous material: specific surface area (S_g): 864 m^2/g, pore diameter (d_p): 2.8 nm and pore volume (V_p): 0.86 cm^3/g. The XRD measurements (not shown here) enabled us to verify that the impregnation, calcinations, reduction and re-oxidation treatments did not modify the ordered mesoporous network of MCM-41. However, a decrease of 35–43% in S_g and 42–55% in V_p were measured after iron loading. Instead, no changes in d_p were detected. These results are coherent with a partial pore filling with iron species.

As already reported in [2], superparamagnetic α-Fe_2O_3 with different blocking temperatures was detected in all solids, indicating diverse magnetic shape anisotropy constant for the three samples, being the largest one found for the Fe/MCM-41 (aq-air) sample.

Figure 1 displays the Mössbauer spectra of the reduced hematite embedded nanoparticles, that we have called (R)Fe/MCM-41 samples. At room temperature all the spectra display central signals that are Fe^{3+} and Fe^{2+} doublets. Their parameters belong to Fe_3O_4 nanoparticles in a superparamagnetic magnetic regime or to Fe^{3+} and Fe^{2+} ions that have migrated into the mesoporous channels walls over the heat treatments.

Table 1 displays the values for the parameters of the spectra at 30 K obtained after fittings assuming the parameter values of the Fe_3O_4 sites reported by Berry et al. [8] for low-temperature spectra. The leftmost spectrum, (aq-air), which is the one displaying the largest fraction of blocked nanoparticles, was fitted using the five-component model. The other two, (alc-air) and (alc-NO/He), in which the

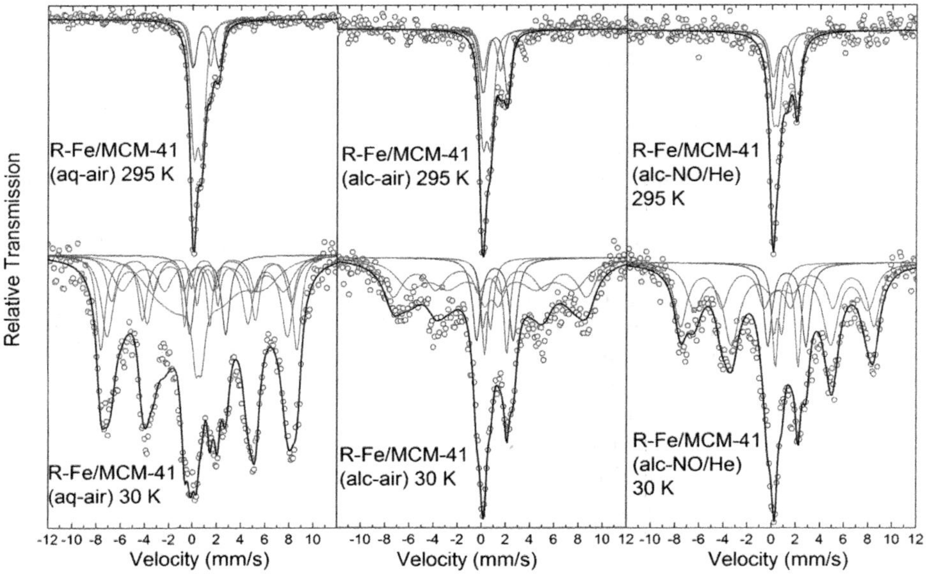

Fig. 1 Mössbauer spectra of the (R)Fe/MCM-41 samples taken at 295 and at 30 K. The *thick solid lines* are the results of the fittings described in the text. The *thin solid lines* are the components ascribed to superparamagnetic Fe_3O_4 and to Fe^{2+} and Fe^{3+} ions diffused into the MCM-41 walls

nanoparticles are mainly in a superparamagnetic regime, were fitted with parameters obtained by a weighted average over the three Fe^{3+} sites and the two Fe^{2+} sites according to the populations reported in [8]. The values of the hyperfine magnetic fields H obtained for these samples are lower than those reported for bulk magnetite. These reduced values are likely originated in the small size of the particles that undergo collective magnetic excitations that decrease the H value from that of the bulk [9]. Similar hyperfine parameters to those obtained in this work for the Fe^{2+} doublets with low relative area have been assigned by previous works to Fe^{2+} ions diffused into octahedral and tetrahedral sites of the silica lattice [10–12]. In the same way, a paramagnetic fraction of Fe^{3+} ions at low temperature was assigned to ions diffused into the silica lattice in the Fe/SBA-15 system [12]. The walls of this type of support are also made exclusively of SiO_2. Therefore, we have tentatively assigned both doublets at low temperature to Fe^{2+} and Fe^{3+} ions diffused into the MCM-41 walls. The first ones get oxidized to Fe^{3+} ions after the re-oxidizing step.

The analysis of the low-temperature spectra reveals that the same differences in the blocking temperatures that were detected with the unreduced samples [2] remain after the reduction treatment. Again, the higher magnetic shape anisotropy constant belongs to the (R)Fe/MCM-41(aq-air) sample.

Comparing the different magnetic behavior of the samples, the larger nanoparticles would have a higher magnetic blocking temperature. Hence, we can conclude that the use of water as a solvent in the impregnating solution leads to the agglomeration of the iron species inside the MCM-41 channels. Instead, irrespective of the calcination atmosphere used, the use of ethanol as the solvent favors the spreading out of the iron species on the walls of the MCM-41 channels.

Table 1 Mössbauer parameters of reduced (R)Fe/MCM-41 samples measured at 30 K

Species	Parameters	(aq-air)	(alc-air)	(alc-NO/He)
Site A of Fe_3O_4	H (T)	47.4 ± 0.7	–	–
	δ (mm/s)	0.37 (*)	–	–
	2ε (mm/s)	−0.02 (*)	–	–
	A_R (%)	23 ± 4	–	–
Site B_1 of Fe_3O_4	H (T)	50.5 ± 0.4	–	–
	δ (mm/s)	0.49 (*)	–	–
	2ε (mm/s)	0 (*)	–	–
	A_R (%)	18 ± 4	–	–
Site B_2 of Fe_3O_4	H (T)	46.8 ± 0.9	–	–
	δ (mm/s)	0.83 (*)	–	–
	2ε (mm/s)	−0.27 (*)	–	–
	A_R (%)	11 ± 3	–	–
Site B_3 of Fe_3O_4	H (T)	44.2 (*)	–	–
	δ (mm/s)	1.03 (*)	–	–
	2ε (mm/s)	−0.41 (*)	–	–
	A_R (%)	16 ± 4	–	–
Site B_4 of Fe_3O_4	H (T)	32 ± 2	–	–
	δ (mm/s)	0.96 (*)	–	–
	2ε (mm/s)	0.89 (*)	–	–
	A_R (%)	25 ± 3	–	–
Fe^{3+} sites of Fe_3O_4	H (T)	–	50 ± 1	49.9 ± 0.6
	δ (mm/s)	–	0.7 ± 0.1	0.47 ± 0.09
	2ε (mm/s)	–	0 (*)	0 (*)
	A_R (%)	–	33 ± 6	34 ± 4
Fe“2.5+” sites of Fe_3O_4	H (T)	–	44 ± 2	44.9 ± 0.7
	δ (mm/s)	–	0.8 ± 0.2	0.77 ± 0.08
	2ε (mm/s)	–	0 (*)	0 (*)
	A_R (%)	–	36 ± 7	38 ± 4
Fe_3O_4(sp)	δ (mm/s)	–	1.17 ± 0.05	1.24 ± 0.04
	Δ (mm/s)	–	1.9 ± 0.1	1.88 ± 0.08
	A_R (%)	–	11 ± 2	12 ± 2
Fe“3+” diffused into the MCM-41 walls	δ (mm/s)	–	0.39 ± 0.08	0.50 ± 0.08
	Δ (mm/s)	–	0.7 ± 0.1	0.68 (*)
	A_R (%)	–	8 ± 2	7 ± 1
Fe“2+” diffused into the MCM-41 walls	δ (mm/s)	1.26 ± 0.06	1.11 ± 0.07	1.26 ± 0.06
	Δ (mm/s)	3.0 ± 0.1	3.0 ± 0.1	3.1 ± 0.1
	A_R (%)	7 ± 1	12 ± 2	12 ± 2

H: hyperfine magnetic field in Tesla; δ: isomer shift (referred to α-Fe at 298 K); Δ: quadrupole splitting; 2ε: quadrupole shift; A_R: relative spectral area; (sp): superparamagnetic. (*) parameters held fixed while fitting

This situation has been depicted in the rough sketches of Fig. 2. Although we have not performed STEM-HAADF or HRTEM analysis to these samples, we have found in previous works [1] that when the impregnating solution is of Fe salts in an aqueous solution, the hematite particles embedded in the channels are not uniformly distributed and have the shape of beads whose diameter is limited by the channel transverse dimension and whose length is slightly larger than the diameter. For the case reported in [1], the hematite nanoparticles had a diameter of 3 nm and an average length of 3.8 nm.

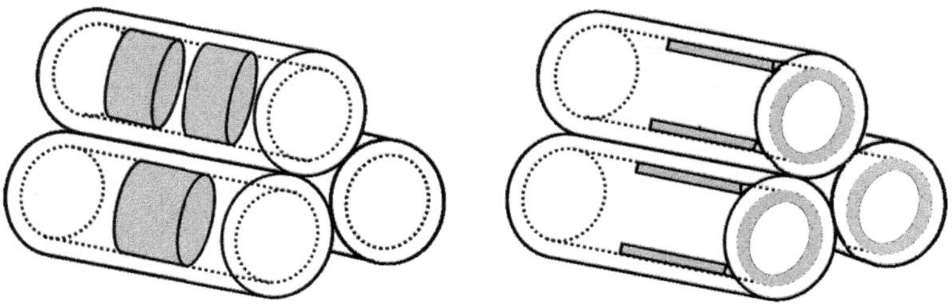

Fig. 2 Schematic representation of the MCM-41 channels (*drawn as cylinders*) where the gray shading indicates the agglomerated state of the iron species for water solvent (*left*) and of hollow nanotubes in contact with the walls for the ethanol solvent (*right*)

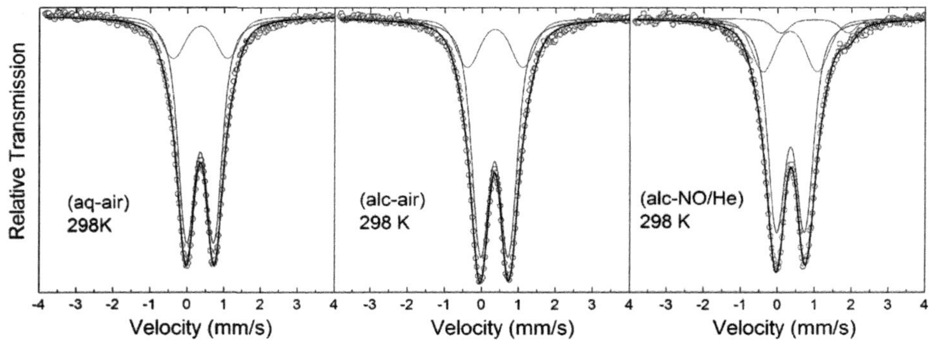

Fig. 3 Mössbauer spectra at room temperature of the (O)Fe/MCM-41 samples. The thick solid line is the fitted spectrum made up of the components shown as thin solid lines

Figure 3 shows the Mössbauer spectra of the re-oxidized samples at room temperature. Only a broad central signal can be observed for the three samples. The spectra can be fitted assuming two doublets of isomer shift values belonging to Fe^{3+} species. The two doublets of different quadrupole splitting can be ascribed to superparamagnetic maghemite and to Fe^{3+} originated in the Fe^{2+} ions that had migrated into the SiO_2 matrix walls. For the rightmost spectrum, however, a third doublet of Fe^{2+} was required, with a spectral area of $\approx$5% which, again, we assign to the Fe^{2+} ions diffused into the amorphous SiO_2 walls. The resulting parameters are shown in Table 2.

We interpret these results as an indication that for (O)Fe/MCM-41(aq-air) and (O)Fe/MCM-41(alc-air), after the re-oxidation step, all the iron species have been converted into superparamagnetic γ-Fe_2O_3 and Fe^{3+} diffused into the MCM-41 walls. However, for reasons that would need further studies, for (O)Fe/MCM-41(alc-NO/He) a small fraction ($\approx$5%) of the Fe^{2+} ions, which had already diffused into the amorphous SiO_2 walls, was not re-oxidized and remained unchanged. Because the samples are attracted to a hand magnet, the only materials that can have a ferromagnetic or ferrimagnetic behavior are the magnetic oxides of Fe, namely, maghemite or magnetite. To sort out the nature of the oxide species, we have taken Mössbauer spectra at liquid helium temperature.

Table 2 Mössbauer parameters of re-oxidized (R)Fe/MCM-41 samples measured at 298 K

Species	Parameters	(aq-air)	(alc-air)	(alc-NO/He)
γ-Fe_2O_3(sp)	δ (mm/s)	0.35 ± 0.02	0.35 ± 0.01	0.35 ± 0.01
	Δ (mm/s)	0.74 ± 0.01	0.76 ± 0.01	0.77 ± 0.02
	A_R (%)	80 ± 2	79 ± 2	72 ± 4
Fe^{3+}	δ (mm/s)	0.35 ± 0.01	0.36 ± 0.01	0.34 ± 0.02
	Δ (mm/s)	1.48 ± 0.08	1.51 ± 0.07	1.5 ± 0.1
	A_R (%)	20 ± 2	21 ± 2	23 ± 4
Fe^{2+}	δ (mm/s)	–	–	0.99 ± 0.08
	Δ (mm/s)	–	–	1.8 ± 0.1
	A_R (%)	–	–	5 ± 1

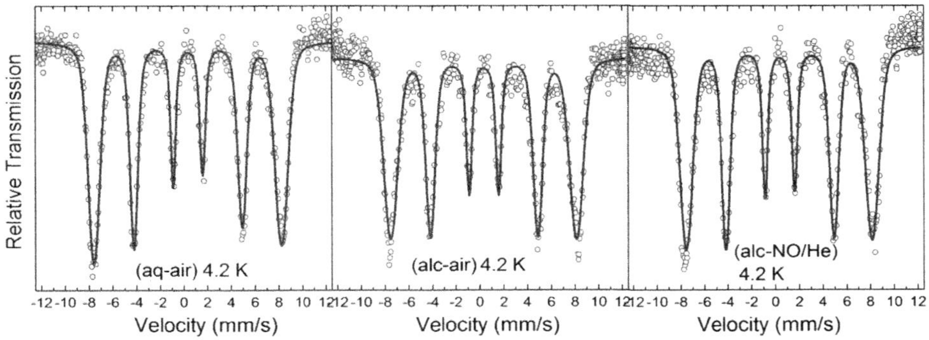

Fig. 4 Mössbauer spectra of the re-oxidized samples (O)Fe/MCM-41 at liquid helium temperature. The solid lines are the results after the fittings described in the text

Table 3 Mössbauer parameters of re-oxidized (O)Fe/MCM-41 samples measured at 4.2 K

Species	Parameters	(aq-air)	(alc-air)	(alc-NO/He)
γ-Fe_2O_3	H (T)	49.06 ± 0.09	48.5 ± 0.1	48.7 ± 0.1
	δ (mm/s)	0.34 ± 0.01	0.32 ± 0.01	0.34 ± 0.01
	2ε (mm/s)	-0.04 ± 0.02	-0.04 ± 0.03	-0.08 ± 0.02
	A_R (%)	100	100	100

Figure 4 exhibits the Mössbauer spectra of the three re-oxidized samples at 4.2 K. The spectra show that the central signals display at this temperature completely split sextets. Because of the small size of the nanoparticles that constitute the iron species in the sample, it is expected that the different sites of the ^{57}Fe probes experience diverse hyperfine fields. Therefore, a distribution of hyperfine fields was used to fit the spectra. The parameters belong to maghemite, as can be seen in Table 3. The signals of low intensity belonging to the paramagnetic species (Fe^{3+} and Fe^{2+} that we have tentatively assigned to ions diffused into the MCM-41 walls) can not be distinguished from the background because of the poor statistics of these spectra.

4 Conclusions

We have demonstrated that in Fe/MCM-41 (15 wt.%) reduced system, the effect of the solvent on the structural properties of the iron species is more important than the calcination atmosphere. Considering that in all samples the maximum diameter of the iron oxide particles is determined by the channel diameter of the MCM-41 matrix, the higher blocking temperature in (R)Fe/MCM-41(aq-air) would indicate a higher shape anisotropy magnetic constant. Therefore, the best conditions for iron magnetic nanowires to get embedded inside the MCM-41 hard template are achieved using water as the solvent and air as the calcination atmosphere. If, for a particular application, nanotubes are desired instead of nanowires, ethanol is the more appropriate solvent.

Acknowledgements The authors acknowledge support of this work by Consejo Nacional de Investigaciones Científicas y Técnicas, Comisión de Investigaciones Científicas de la Provincia de Buenos Aires, Universidad Nacional de La Plata and Agencia Nacional de Promoción Científica y Tecnológica (PICT N° 38337).

References

1. Cagnoli, M.V., Gallegos, N.G., Bengoa, J.F., Alvarez, A.M., Moreno, M.S., Roig, A., Marchetti, S.G., Mercader, R.C.: AIP Conference Proceedings **765**, 13–22 (2005)
2. Cano, L.A., Bengoa, J.F., Stewart, S.J., Mercader, R.C., Marchetti, S.G.: Hyperfine Interact. **195**, 93–98 (2010)
3. Bengoa, J.F., Cagnoli, M.V., Gallegos, N.G., Alvarez, A.M., Mogni, L.V., Moreno, M.S., Marchetti, S.G.: Microporous Mesoporous Mater. **84**, 153–160 (2005)
4. Moreno, M.S., Weyland, M., Midgley, P.A., Bengoa, J.F., Cagnoli, M.V., Gallegos, N.G., Alvarez A.M., Marchetti, S.G.: Micron **37**, 52–56 (2006)
5. Ryoo, R., Kim, J.M.: J. Chem. Soc., Chem. Commun. 711 (1995)
6. Sietsma, J.R.A., den Breejen, J.P., de Jongh, P.E., van Dillen, A.J., Bitter, J.H., de Jong, K.P.: Stud. Surf. Sci. Catal. **167**, 55–60 (2007)
7. Lagarec, K., Rancourt, D.G.: Mössbauer Spectral Analysis Software, Version 1.0. Department of Physics, University of Otawa (1998)
8. Berry, F.J, Skinner, S., Thomas, M.F.: J. Phys., Condens. Matter **10**, 215–220 (1998)
9. Mørup, S.: J. Magn. Magn. Mater. **37**, 39–50 (1983)
10. Clausen, B.S, Topsøe, H., Mørup, S.: Appl. Catal. **48**, 327 (1989)
11. Cano, L.A., Cagnoli, M.V., Fellenz, N.A., Bengoa, J.F., Gallegos, N.G., Alvarez, A.M., Marchetti, S.G.: Appl. Catal A: Gral **379**, 105–110 (2010)
12. Cano, L.A., Cagnoli, M.V., Bengoa, J.F., Alvarez, A.M., Marchetti, S.G.: J. Catal. **278**, 310–320 (2011)

Hyperfine Interact (2011) 202:25–38
DOI 10.1007/s10751-011-0337-1

Application of Mössbauer spectroscopy to the study of tannins inhibition of iron and steel corrosion

Juan A. Jaén · J. De Obaldía · M. V. Rodríguez

Published online: 30 August 2011

Abstract The inhibitory effect of tannins was investigated using, among others, potentiodynamic polarizations and Mössbauer spectroscopy. These techniques confirmed that the nature, pH and concentration of tannic solution are of upmost importance in the inhibitory properties of the solutions. It is observed that at low tannin concentration or pH, both, hydrolizable and condensed tannins, effectively inhibit iron corrosion, due to the redox properties of tannins. At pH ≈ 0, Mössbauer spectra of the frozen aqueous solutions of iron(III) with the tannin solutions showed that iron is in the form of a monomeric species $[Fe(H_2O)_6]^{3+}$, without coordination with the functional hydroxyl groups of the tannins. The suspended material consisted of amorphous ferric oxide and oxyhydroxides, though with quebracho tannin partly resulted in complex formation and in an iron (II) species from a redox process. Other tannins, such as chestnut hydrolysable tannins, do not complex iron at this low pH. Tannins react at high concentrations or pH (3 and 5) to form insoluble blue–black amorphous complexes of mono-and bis-type tannate complexes, with a relative amount of the bis-ferric tannate generally increasing with pH. Some Fe^{2+} in the form of hydrated polymeric ferrous tannate could be obtained. At pH 7, a partially hydrolyzed ferric tannate complex was also formed. The latter two phases do not provide corrosion protection. Tannin solutions at natural pH react with electrodeposited iron films (approx. 6 μm) to obtain products consisting only on the catecholate mono-complex of ferric tannate. Some aspects of the mechanism of tannins protection against corrosion are discussed.

J. A. Jaén (✉)
Depto. de Química Física, Universidad de Panamá, CITEN, Lab. N° 105,
Edificio de Laboratorios Científicos-VIP, Panama, Panama
e-mail: jjaen@ancon.up.ac.pa

J. De Obaldía · M. V. Rodríguez
Escuela de Química, Facultad de Ciencias Naturales, Exactas y Tecnología,
Universidad de Panamá, Panama, Panama

Keywords Tannins · Corrosion inhibition · Mössbauer spectroscopy · Potentiodynamic polarization curves

1 Introduction

There have been several recent studies concerning the use of tannins as corrosion inhibitors in aqueous media [1, 2], rust converters [3], pigments in paint coatings [4], pretreatment of reinforcement steel in concrete [5], chemical cleaning agents for removing iron deposits [6] and cement paste coloring in concretes [7] within many other applications.

Tannins are a complex mixture of non-toxic, biodegradable polyphenolic compounds, extracted from plant material. They are able to form chelates with iron and other metallic cations, including rust-phase components, due to the vicinity of hydroxyl groups in ortho position in the polyphenolic moieties [8]. In the reaction between tannic acid in high concentrations and other polyphenols with iron (and other corrosion products) insoluble amorphous complexes of ferric tannates or polyphenolates are obtained [9–11]. At low concentrations hydrolysis reactions are favored resulting in the formation of iron oxides and oxihydroxides [10].

The corrosion inhibiting properties of tannins are usually attributed to the formation of the highly insoluble blue–black ferric tannate, though there have been contradictory results in the use of tannins for clean and rusted steel, particularly because of the diversity of conditions prevailing in the different corrosion studies [1]. Several studies [12–14] have shown that condensed tannins are good cathodic inhibitors with a performance that depends on the concentration. Other condensed tannins, like mimosa tannin, are good mixed inhibitors of low-carbon steel in acidic solutions [13]. Inhibition occurs via adsorption of ferric tannate on the metal surface. Also, it has been found that the efficiency of inhibition tannins decreases with increasing pH [15]. According to several authors [16, 17], they are more effective when used in conjunction with phosphoric acid. Nevertheless, other authors [18–20] concluded that the use of tannins or their addition into formulations does not improve the anticorrosion properties.

On the other hand, the nature of ferric and ferrous-tannates may have an influence of the corrosion rate with iron. Thus, it is of interest the study of the mechanism of formation of the ferric tannates, and other possible compounds, from the interaction of tannins with ferric ions.

This report has dual purpose; first to further establish the influence of tannin concentration on the effectiveness as corrosion inhibitors using potentiodynamic polarization (Tafel plots), and next to investigate the effect of pH on the nature and composition of the complexes formed by Mössbauer spectroscopy in order to ascertain some aspects of the mechanism of tannins protection against iron and steel corrosion.

2 Experimental

Electrochemical tests were carried out in a three-electrode electrochemical cell consisting of a mild steel rod embedded in an epoxy polyester resin with an exposed area of 1.0 cm^2. All potentials are referred to this electrode. The cell was connected to a Radiometer Copenhagen DEA 332-Digital Electrochemical Analyzer controlled using the software Volta Master 2. After polishing with 220 and 600 emery papers, the electrode was washed thoroughly with doubly distilled water and cleaned with ethanol. The electrode was cathodically polarized for 5 min at −800 mV relative to the open-circuit potential. Then the potentiodynamic curve was recorded from -1000 mV to 1000 mV versus with a sweep rate of 1 mV s^{-1}. The scanning was always from cathodic to anodic direction. Corrosion current densities (I_{corr}) were calculated by extrapolating the Tafel lines to *Ecorr*. The standard solution consisted of 0.1 mol/kg Na_2SO_4 and the test solutions contained 0.1% (w/V), 1% (w/V) and 5% (w/V) of the tannin. Tannic acid–penta-*m*-digalloylglucose—$C_{76}H_{52}O_{46}$, M.M. 1701.18 g/mol Merck, chestnut (*Castanae* sp.), and quebracho (*Schinopsis sp.*) were used. The pH of the inhibiting solution was not adjusted for the electrochemical measurements.

10 ml of the a 5% aqueous tannin solution were mixed with 10 ml 10% AR grade ferric nitrate $Fe(NO_3)_3{\cdot}9H_2O$ (Fluka) solution. The mixture was mechanically stirred and left to react at room temperature for 1 day. The pH of the solution dropped overnight to ~1 and then was adjusted to 3, 5 and 7 using KOH solution. The blue black reaction products were separated by centrifugation. The resultant powder was washed with bidistilled water and allowed to dry under vacuum.

For the experiments at pH ~ 0, a 15% enriched in ^{57}Fe 10% solution in nitric acid was used. Mixing of the solutions resulted in dark, scarce and slender suspended precipitate, mainly in reaction to chestnut. After standing at room temperature for 1 day, the mixture was vacuum filtered with a 0.45 μm paper. The supernatant was carefully decanted, and approx. 3 cm^3 were placed in a sample holder, and then rapidly frozen in liquid nitrogen. Then it was mounted in a previously cooled at ~80 K Mössbauer cryostat.

The precipitates were characterized by means of Fourier transform infrared (FT-IR) and Mössbauer spectroscopy (MS). IR spectra were obtained using 150 mg of KBr and 1.5 mg of the dry powder. The spectra were recorded from 400 cm^{-1} to 4000 cm^{-1} using a Nicolet Avatar 360 FTIR spectrometer. The resolution of the spectra was 4 cm^{-1}. Mössbauer spectra (MS) were recorded using a conventional spectrometer of constant acceleration with a $^{57}Co(Rh)$ source of 10 mCi. A closed-cycle cryostat (CCS 850 Janis) was employed for low temperature measurements. The Mössbauer data were evaluated with the Recoil software (University of Ottawa, Canada) using a Lorentzian-based spectral fitting routine for the doublets and Voigt base fitting for broadened sextets. All isomer shifts are given referred to α-Fe.

Also electrodeposited iron films of approximately 5 μm to 6 μm were let to interact with the tannin solutions for 10 min and 1 h. The tannate films formed were examined by transmission Mössbauer analysis.

3 Results

Potentiodynamic polarization curves for the tannic acid and quebracho tannin at different concentrations are shown in Figs. 1 and 2. The respective kinetic parameters are given in Table 1. It can be clearly seen that the presence of 0.1% tannic acid resulted in a marked shift in the cathodic branches of the polarization curves. The inhibition efficiency (% I.E.) of this solution is calculated to be around 89%.

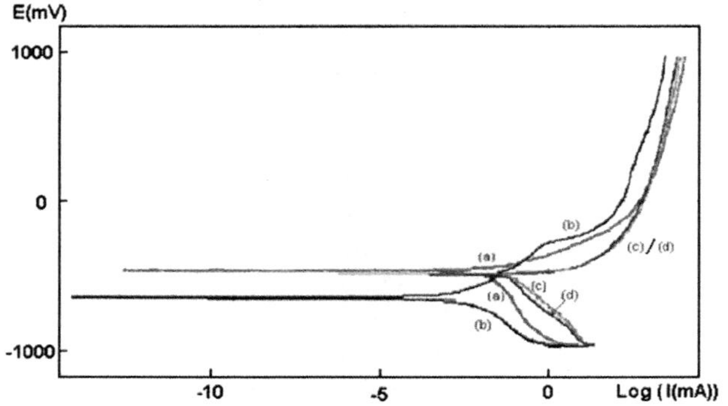

Fig. 1 Polarization curve of mild steel in tannic acid solutions at concentrations of: **a** 0%, **b** 0.1%, **c** 1% and **d** 5%

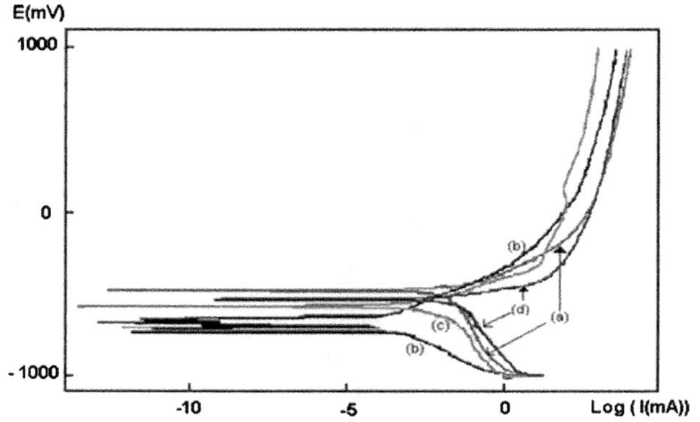

Fig. 2 Polarization curve of mild steel in quebracho tannins solutions at concentrations of: **a** 0%, **b** 0.1%, **c** 1% and **d** 5%

Table 1 Kinetic parameters derived from potentiodynamic polarization plots of mild steel immersed in aerated solution of 0.1 M Na_2SO_4 at 298 K

Sample	Concentration(%)	E_{corr}(mV/SCE)	I_{corr}(mA cm^{-2})	b_c(mV)	b_a(mV)
Blank	–	−480	0.18	263	39
Tannic acid	0.1	−645	0.02	84	47
	1	−498	0.24	181	19
	5	−501	0.28	87	15
Quebracho tannin	0.1	−670	0.01	100	86
	1	−568	0.06	284	71
	5	−531	0.12	274	33

Nevertheless, using 1% and 5% tannic acid solution increased the corrosion current density, accelerating the corrosion process. The shape of the anodic portion of the polarization curve and the Tafel slope of 47 mV should be taken as indicative of inhibition of the anodic process due to adsorption of substances at low current densities and desorption from the surface at high current densities, when the metal dissolution exceed adsorption velocity [21]. When examining the electrode at the end of the polarization, there was a dark film on the electrode, due to the well-known formation of the iron tannates complexes.

Quebracho solutions of low concentration, 0.1%, in 0.1 m Na_2SO_4 had the largest decrease in the corrosion current and there is a significant shift of the corrosion potential towards more negative values. At higher concentrations, there is obtained a displacement of the potential for corrosion to more negative values. The quebracho solutions inhibit both processes cathodic and anodic, but in higher proportion the cathodic reactions.

The trend in the curves of polarization is that the lower concentration of quebracho, the greater is the resistance to corrosion. The results are summarized in Table 1. It is clear that 0.1%, 1% and 5% quebracho solutions are good inhibitors of steel corrosion, and that the higher efficiency is obtained with the solution of 0.1% quebracho.

The chemical compositions of the precipitates formed in the interaction of chestnut and quebracho tannins with Fe^{III} nitrate aqueous solutions at various pH (0, 3, 5 and 7) were determined by means of Mössbauer spectroscopy. The Mössbauer spectra consisted of quadrupole doublets of high-spin iron (III) and iron (II), as can be seen from Figs. 3 and 4. The ferric doublet with lower quadrupole splitting is attributed to ferric mono-tannate complex, while the largest quadrupole splitting corresponds to bis-type tannate complexes of ferric tannate, the parameters of which are presented in Tables 2 and 3. The identification of compounds was done according to the report of Gust and Suwalski [8]. They worked at different pH and reported the formation of mono and bis complexes as a result of a reaction between rust-phase components and oak tannins in aqueous solution. These hyperfine parameters are similar to those obtained by us when reacting several plant extracts of Panama with ferrous and ferric salts [22] and for other tannates and polyphenolates [10, 11, 23, 24]. It should be noted that the ferric tannates have Mössbauer parameters similar to those of oxides or oxyhydroxides of iron (III) amorphous or small particle size, but the FT-IR spectra show no evidence of the presence of these compounds.

Samples obtained at pH 7 and measured at 20 K, show a magnetically split sextet with broadened lines (see Figs. 3 and 4), that was interpreted as a partially hydrolysed tannate complex due to the absence of evidence for oxides or hydroxides in the FT-IR spectra, as can be seen in Fig. 5. This component is formed at the expense of the bis tannate complexes.

The experiments at low temperature show the contribution of one (or more) species of iron (II), which can easily pass unnoticed at room temperature due to low quantities and the low value of the recoil free factor of the specimen. As a matter of fact, the lower absorption for the precipitated tannates observed in the room temperature Mössbauer spectra, gives an indication of the polymeric nature of the precipitates. As there was no iron (II) present in the stock solution of iron nitrate, this finding is a direct evidence of the reductive capacity of tannic solutions which is able to reduce iron (III) to form species of iron (II). The Mössbauer parameters at room temperature are within the range observed in other polyphenolic complex of iron (II) (e.g. gallates, pirogallates and plant extracts) [11, 23, 24]. Mössbauer parameters at 80 K are very close to the blue complex of catechol-iron (II) determined from freeze-dried solutions of pH 5.4, and the predominant green complex at pH of 2.9 [25]. These Mössbauer parameters refer to a kind of Fe^{2+} in a distorted octahedral environment, in which ferrous ion is bound by dative bonds to phenolic oxygen atoms forming a complex of iron tannate. Differences in the isomer shift and quadrupole splitting are due to structural distortions for the reason that the amount of water of

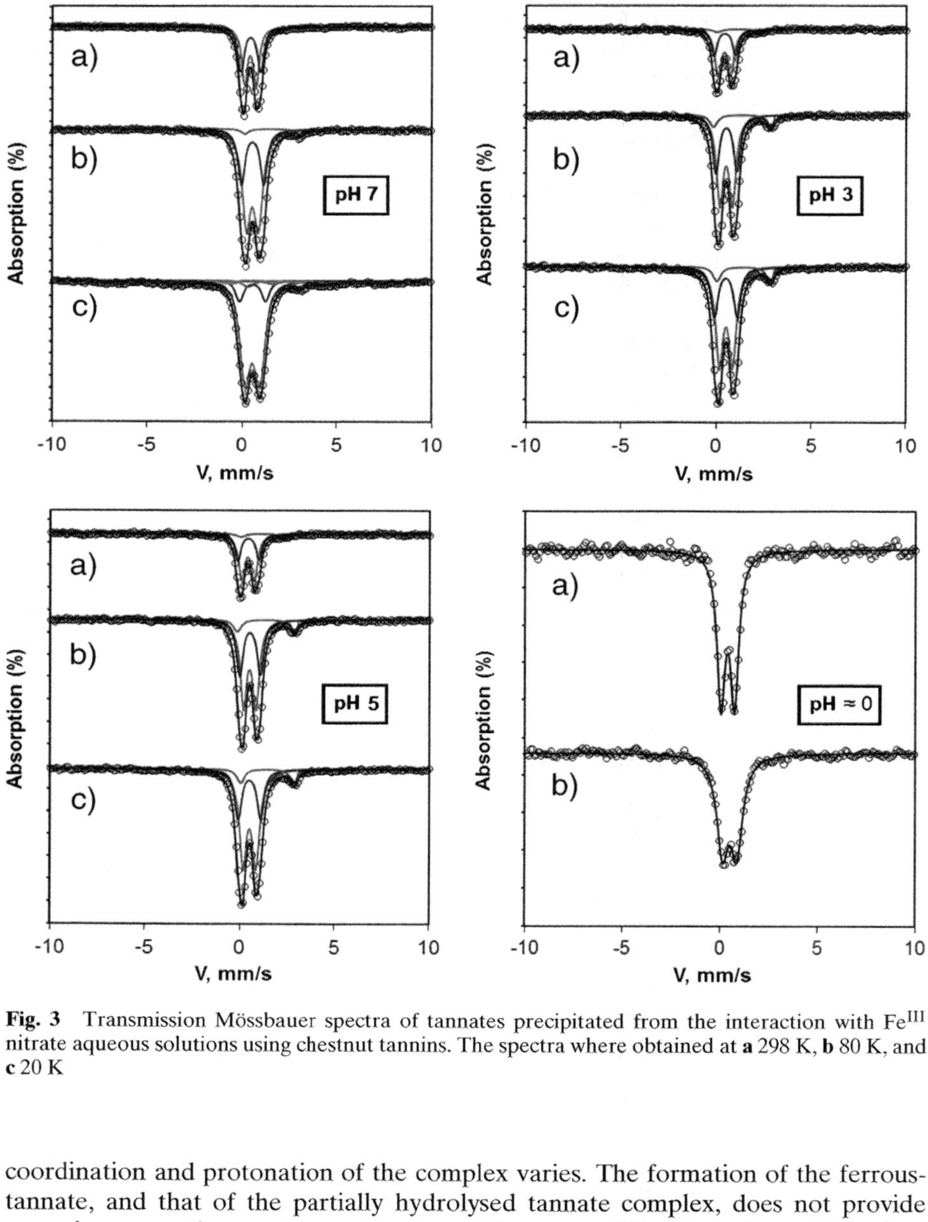

Fig. 3 Transmission Mössbauer spectra of tannates precipitated from the interaction with Fe^{III} nitrate aqueous solutions using chestnut tannins. The spectra where obtained at **a** 298 K, **b** 80 K, and **c** 20 K

coordination and protonation of the complex varies. The formation of the ferrous-tannate, and that of the partially hydrolysed tannate complex, does not provide corrosion protection; on the contrary, provide a way of leaching of metal ions into the solution [12].

The Mössbauer spectrum at room temperature and 80 K of the reaction product separated from the interaction with chestnut with the ferric nitrate solution at pH ≈ 0 gave a spectrum consisting of a quadrupole doublet as shown in Fig. 3. The fitting with one doublet resulted in Mössbauer parameters, given in Table 2, in the typical range of oxides and amorphous iron oxyhydroxides such as ferrihydrite and feroxyhite [26] and slightly larger quadrupole splittings than other iron oxides such

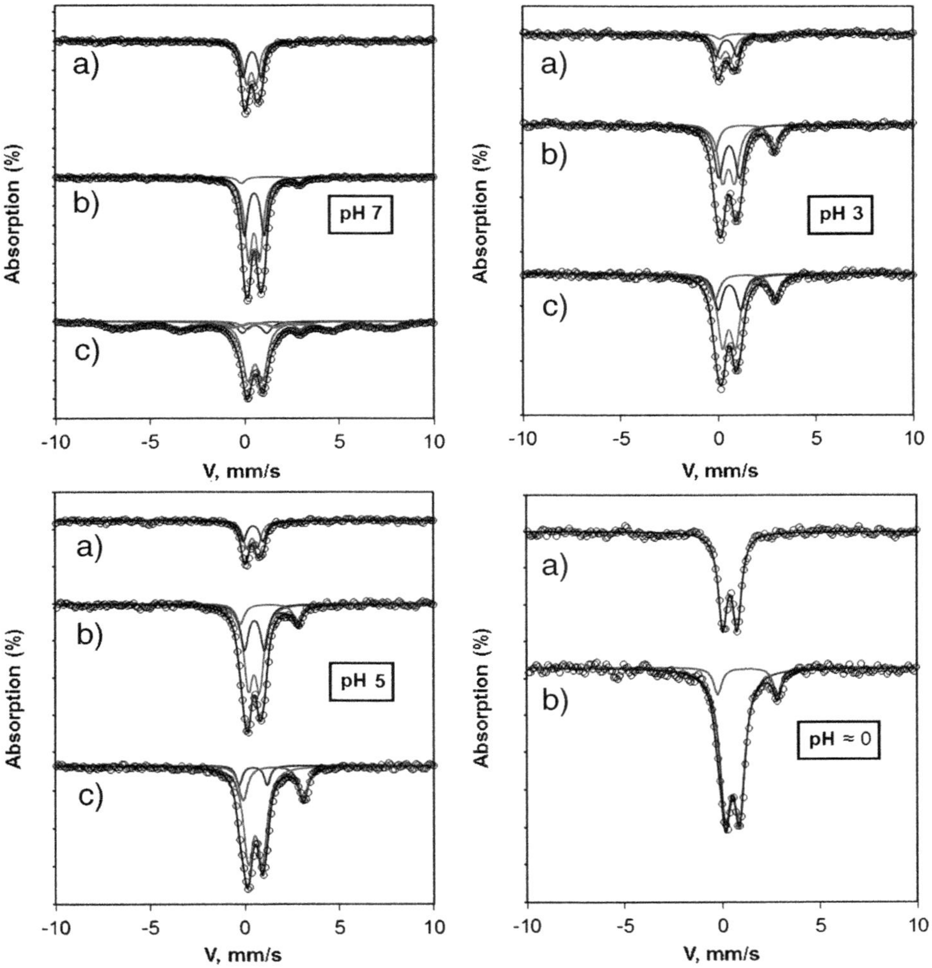

Fig. 4 Transmission Mössbauer spectra of tannates precipitated from the interaction with Fe^{III} nitrate aqueous solutions using quebracho tannins. The spectra where obtained at **a** 298 K, **b** 80 K, and **c** 20 K

as lepidocrocite and superparamagnetic goethite [27]. These parameters match the gels of amorphous iron oxides [28] and ferrihydrite, which can be superparamagnetic to temperatures down to 23 K [29]. Try to consider the spectrum as the result of the superposition of two or more doublets did not improve fitting. FT-IR spectra of the reaction products of the samples at pH 3 to 7 have all the distinguishing features of the formation of iron tannate complexes [24]. For the sample of the precipitate obtained at pH $\approx$ 0 with chestnut tannins, the FT-IR spectrum profile (Fig. 5a) is very similar to that of natural δ-FeOOH, but without the remarkable Fe-OH bending band at 1110 cm^{-1} of this phase [30, 31]. In the synthetic variety

Table 2 Mössbauer parameters of the spectra of the precipitates formed in Fe(III) nitrate aqueous solutions with chestnut tannins at various pH

pH	T, K	δ, mm/s	Δ, mm/s	A_r, %
0	298	0.38 ± 0.01	0.69 ± 0.01	100
	80	0.50 ± 0.01	0.72 ± 0.02	100
3	298	0.41 ± 0.01	0.70 ± 0.07	61.8
		0.46 ± 0.02	1.13 ± 0.08	38.2
	80	0.50 ± 0.08	0.71 ± 0.17	57.4
		0.49 ± 0.01	1.20 ± 0.03	32.4
		1.48 ± 0.43	2.66 ± 0.87	10.2
	20	0.55 ± 0.02	0.74 ± 0.05	51.0
		0.53 ± 0.01	1.22 ± 0.08	36.0
		1.53 ± 0.11	2.86 ± 0.21	12.8
5	298	0.41 ± 0.02	0.67 ± 0.04	61.2
		0.43 ± 0.01	1.12 ± 0.04	32.6
		1.16 ± 0.31	2.25 ± 0.62	6.2
	80	0.52 ± 0.01	0.66 ± 0.03	55.7
		0.53 ± 0.01	1.10 ± 0.04	35.0
		1.35 ± 0.03	3.00 ± 0.06	9.3
	20	0.52 ± 0.01	0.68 ± 0.03	60.4
		0.51 ± 0.01	1.19 ± 0.04	29.9
		1.45 ± 0.04	2.81 ± 0.08	9.7
7	298	0.42 ± 0.01	0.62 ± 0.02	53.3
		0.44 ± 0.02	1.05 ± 0.03	46.7
	80	0.54 ± 0.01	0.63 ± 0.04	65.1
		0.55 ± 0.01	1.18 ± 0.04	31.5
		1.57 ± 0.11	2.88 ± 0.21	3.4
	20	0.54 ± 0.01	0.67 ± 0.05	56.0
		0.55 ± 0.01	1.24 ± 0.05	25.9
		1.76 ± 0.04	2.62 ± 0.08	3.5
		0.39 ± 0.13[a]	–	14.5

δ chemical shift relative to α-Fe, Δ quadrupole splitting, A_r Relative area
[a]Relaxed sextet with H = 43.5 ± 0.3 T

of feroxihite δ'-FeOOH, this band is very weak, reducing and even disappearing with the decrease of the crystallinity of the sample [32, 33]. Giving the fact that the color of this material was dark brown, completely different from the characteristic blue–black of the iron tannates, and the lack of evidence of formation of these compounds in the infrared spectra lead to consider this phase as an amorphous ferric oxyhydroxide. The blackish material from the interaction with quebracho tannin at pH ≈ 0, also depict the high spin doublet of Fe^{3+}, and in this case we attribute it to a sum of different species, including partial formation of iron tannate complexes, oxides or oxyhydroxides of iron (III) amorphous. Infrared spectroscopy didn't help differentiating the phase composition of this material. The IR spectrum shows a general profile with some few similarities to those of the precipitates synthesized in the same way at higher pH values. The most notable difference is the very strong absorption peak around 1380 cm^{-1} due to nitrate adsorbed [34] on the precipitate of quebracho ferric tannate and other amorphous ferric phases. As shown in Fig. 4, this sample also give an iron (II) doublet, with parameters similar to those of ferrous tannates, but also to those reported for the aqua complex of iron (II), hinting to

Table 3 Mössbauer parameters of the spectra of the precipitates formed in Fe^{III} nitrate aqueous solutions with quebracho tannins at various pH

pH	T, K	δ,mm/s	Δ,mm/s	A_r, %
0	298	0.40 ± 0.01	0.73 ± 0.02	100
	80	0.50 ± 0.01	0.72 ± 0.02	87
		1.31 ± 0.04	3.09 ± 0.08	13
3	298	0.40 ± 0.11	0.60 ± 0.22	46.3
		0.42 ± 0.04	1.07 ± 0.20	42.3
		1.40 ± 0.45	2.70 ± 0.90	11.4
	80	0.52 ± 0.02	0.61 ± 0.12	40.2
		0.56 ± 0.02	1.06 ± 0.13	40.3
		1.36 ± 0.02	3.09 ± 0.05	19.5
	20	0.53 ± 0.14	0.67 ± 0.13	53.5
		0.56 ± 0.03	1.15 ± 0.17	25.5
		1.38 ± 0.03	3.14 ± 0.06	21.0
5	298	0.40 ± 0.02	0.59 ± 0.11	56.6
		0.46 ± 0.03	1.04 ± 0.10	43.4
	80	0.49 ± 0.01	0.59 ± 0.12	56.0
		0.52 ± 0.02	1.05 ± 0.15	30.1
		1.29 ± 0.04	3.03 ± 0.07	13.9
	20	0.54 ± 0.01	0.74 ± 0.07	66.9
		0.44 ± 0.02	1.48 ± 0.13	9.0
		1.49 ± 0.06	3.16 ± 0.12	24.1
7	298	0.35 ± 0.01	0.52 ± 0.05	53.0
		0.39 ± 0.01	0.93 ± 0.05	47.0
	80	0.48 ± 0.01	0.59 ± 0.03	55.2
		0.49 ± 0.01	1.04 ± 0.01	39.4
		1.34 ± 0.04	3.05 ± 0.21	5.4
	20	0.52 ± 0.02	0.60 ± 0.09	10.0
		0.50 ± 0.01	1.01 ± 0.19	47.1
		1.50 ± 0.10	2.81 ± 0.21	6.6
		0.50 ± 0.04[a]	–	36.3

δ chemical shift relative to α-Fe, Δ quadrupole splitting, A_r Relative area
[a]Relaxed sextet with H = 43.5 ± 0.3 T

the possibility that hydrated iron (II) form in the course of the gradual reduction of iron (III) in this acid solution. Ionic equilibrium calculations [12] indicate that the formation of mono-ferric tannate complexes begins at pH values higher than 1, so that is not expected complex formation at lower pH values.

The Mössbauer spectra of rapidly frozen solutions of 10% ferric nitrate and other supernatants product of their interaction with tannins from chestnut and quebracho, all at pH ≈ 0 are shown in Fig. 6. The spectrum in (a) shows a very broad relaxed line shape, as expected for magnetic relaxation of monomeric species $[Fe(H_2O)_6]^{3+}$ at these concentrations and pH values [35], because the spin–spin interactions and spin-lattice are weak due to low concentrations of Fe^{3+} and low temperatures (80 K), respectively.

Mössbauer spectra of both supernatants obtained from the interaction of nitrate solutions of iron (III) with tannins from chestnut (Fig. 6b) and quebracho (Fig. 6c) show the relaxed magnetic profile of the mononuclear species, indicating that Fe^{3+} ions in solution are not as tannate complexes. There was no evidence of other species,

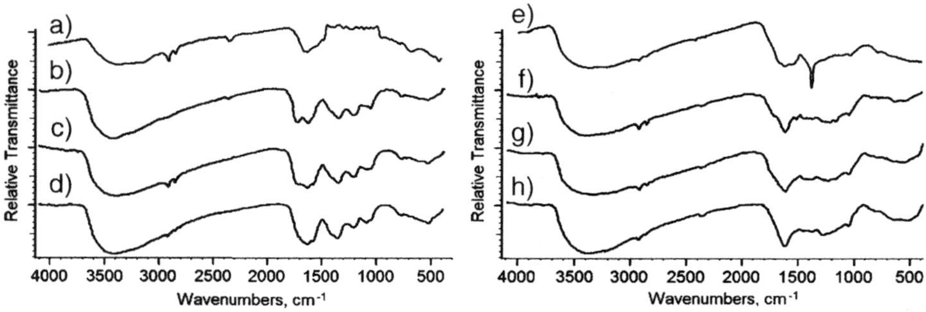

Fig. 5 Fourier transformed infrared spectra of the products from the interaction of Fe^{III} nitrate aqueous solutions and chestnut tannins at pH (*a*) ≈ 0, (*b*) 3, (*c*) 5, (*d*) 7; and quebracho tannins at pH (*e*) ≈ 0, (*f*) 3, (*g*) 5 and (*h*) 7

Fig. 6 Mössbauer spectra at 80 K and pH ≈ 0 of the frozen solution of **a** $[Fe(NO_3)_3{\cdot}9H_2O]$ enriched 15% in ^{57}Fe, and the supernatants resulting of its interaction with **b** chestnut **c** quebracho tannins

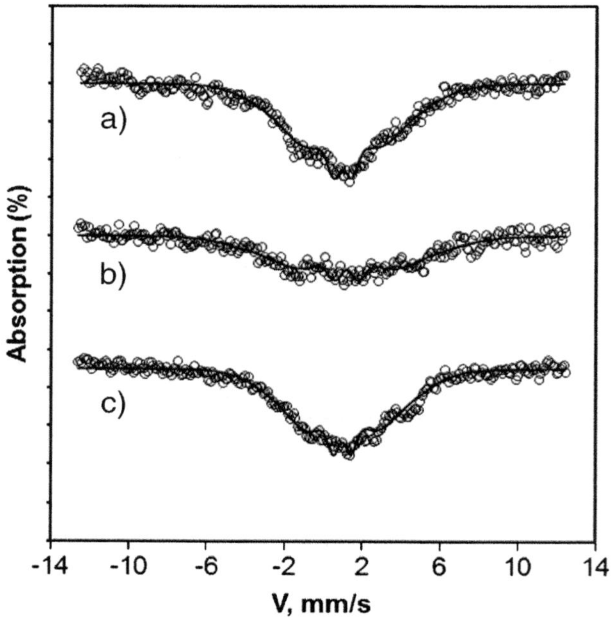

e.g. species with a hydroxy-iron bridge and crystalline hydrates $[Fe(NO_3)_3{\cdot}nH_2O]$. Frozen solutions of iron (III) nitrate, at pH values (0.2–2) and 77 K, give Mössbauer spectrum with two doubles, one of $\Delta = 1.67$ mm/s and other $\Delta = 0.38$ mm/s, respectively [36]. It follows from this that the Fe^{3+} in the solution under consideration is more likely to retain water molecules in its coordination sphere, without interacting with the hydroxyl functional groups of the tannins, forming species monomer $[Fe(H_2O)_6]^{3+}$.

We also let electrodeposited iron films to interact with the tannin solutions (Table 4), obtaining products consisting only of mono-type complexes, as previously observed in a reaction on a rusted surface covered with oak tannin giving mainly

Table 4 Room temperature Mössbauer parameters of the doublet from the spectra of samples formed in the interaction of the tannin solution with electrodeposited iron films of ~ 6 μm

Sample	δ (mm/s)	Δ (mm/s)	Linewidth (mm/s)	A_r (%)
Tannic acid—10 min	0.417	0.669	0.448	55.8
Tannic acid—1 h	0.396	0.669	0.442	58.7
Quebracho tannin—10 min	0.400	0.654	0.428	70.3
Tannic acid—1 h	0.365	0.737	0.440	48.7

mono-type complexes [8]. It is inferred that physical adsorption occurs through complexation of surface iron forming ferrous-tannates which can be easily oxidized into mono-chatecol-ferric-tannates, a blue–black deposit, when in contact with oxygen.

4 Discussion

The inhibitory capacity of tannins is affected by pH and by the relative amounts of tannins and iron in the solution. The relative concentrations of tannins determine the prevalence of the reaction products of hydrolysis or metal complexes. In less concentrated solutions of 0.1% to 1%, the polarization curves show a region of passivation, with neutral species of tannin spontaneously adsorb onto the surface. Mössbauer spectroscopy clearly shows the formation of iron (II) species from a redox process, due to the high reducing power of tannins [10].

It has been reported that quebracho tannins are good cathodic inhibitors with a performance that depends on the concentration [12–14]. Also, the efficiency of inhibition of quebracho, mangrove and mimosa tannins, as well as that of the hydrolysable tannin chestnut, decreases with increasing pH. At low pH values, Martinez et al. [13] found that mimosa tannins behave as a mixed-type inhibitor, with an adsorption mechanism that can be approximated by Tempkin and Frumkin isotherms at low pH. The free energy of adsorption suggests a mechanism based on the chemical adsorption of tannin molecules onto the metal surface. This adsorption occurs through the sharing of an oxygen pair of electrons (OH^- moiety in the aromatic rings of tannins) onto the iron surface. This mechanism has been further established using semi empirical and density functional theory for mangrove tannins [15]. Thus, at very acidic pH, 0 and 0.5, the inhibitory properties of studied tannins solutions should be attributed to the chemisorption of molecules on the surface of metals. The reduction of any specie is made by the oxidation of the tannin to its quinonoid form, in agreement with the Mössbauer observation of no iron tannate complex formation. Any incipient oxyhydroxide, in the presence of low tannins (and other polyphenols) concentration (0.1%), can be further reduced to compounds like magnetite.

Our results, and those of several other researchers [13, 15], show that at higher pH (and high tannin concentrations), there is a diminishing of the inhibiting capabilities of the tannins. At these higher pH values, tannins interact directly on the Fe^{3+} ions available to form ferric-tannates. Tannins can also complex with Fe^{2+} ions to form ferrous-tannates which can be easily oxidized into ferric-tannates in the presence of oxygen. Moreover, in the presence of large amounts of tannins (and other the polyphenols), ferric-tannates are also formed in the conversion of iron oxides and oxyhydroxides [10, 23]. With the help of Mössbauer spectroscopy it has been shown that γ-FeOOH is a most reactive phase and that reactivity of other aged iron oxides and oxyhydroxides with tannins depends on their nature, and the size and degree of crystallinity. Mössbauer experiments, in addition to the polarization curves, on the other hand, show that solutions of 2% and 5% of tannins anodic dissolution is followed by the formation of a visible dark blue ferric tannate, that comes from the metal surface towards the solution during potentiodynamic polarizations. Also, the interaction with the surface of iron metal gives blue ferric tannate compounds. The adsorption of these iron-tannate complexes obeys a Freundlich type isotherm, with values of free energy of adsorption that suggest a physical type adsorption mechanism [13, 14]. The interactions between molecules adsorbed and those of the solution cause desorption from the surface of the substrate, thereby the inhibiting efficiency decreases. Furthermore, the cracked morphology and the loosely adherence of the deposits results in the possibility of wetting of the surface, lessening inhibitory capacity and even stimulating the overall corrosion reaction. At neutral or basic pH, in addition to ferrous-tannates, partially hydrolyzed tannate complexes can form not favoring the formation of protective films on the surface.

5 Conclusion

The different nature, pH and concentration of tannic solution seem to be of upmost importance in the inhibitory properties of the solutions. The inhibitive properties of tannins at low concentration or pH are due to the chemisorption of the tannin molecules onto the surface of iron, steel or rusted steel. Under this conditions, corrosion protection results from the redox properties of tannins, though they may form small amounts of sparingly soluble deposits (i.e. quebracho tannates) reinforcing their protective effect.

At higher concentration and pH (pH 3 to 5), there is a decrease in efficiency in the inhibition behavior of the tannins. Tannins react at pH of 3, 5 and 7 to form mono-and bis-type tannate complexes. At natural pH, they react with iron to obtain products consisting only on the catecholate mono-complex of ferric tannate. These insoluble blue–black amorphous complexes physically adsorbed to grow a layer that provides some protection, but with a lot of cracks and low adherence, losing easily their protective effect, even increasing metal dissolution.

At pH 7 (and even higher) a partially hydrolyzed ferric tannate complex and some Fe^{2+} in the form of hydrated polymeric ferrous tannate are obtained. These phases do not provide corrosion protection.

Acknowledgements The authors are thankful to Secretaría Nacional de Ciencia, Tecnología e Innovación (Panamá), for financial help through the project Contrato N° 9-SENACYT, P-06-0017-A—PNUD Panamá, 2005.

References

1. Lahodny-Šark, O., Kapor, F.: Corrosion inhibition of carbon steel in the near neutral media by blends of tannin and calcium gluconate. Mater. Corros. **53**, 264–268 (2002)
2. Rahim, A.A., Rocca, E., Steinmetz, J., Jain Kassim, M.: Inhibitive action of mangrove tannins and phosphoric acid on pre-rusted steel via electrochemical methods. Corros. Sci. **50**, 1546–1550 (2008)

3. Ocampo, L.M, Margarit, I.C.P., Mattos, O.R., Córdoba-de-Torresi, S.I., Fragata, F.L.: Performance of rust converter based in phosphoric and tannic acids. Corros. Sci. **46**, 1515–1525 (2004)
4. Hornus Sack, S., Romagnoli, R., Vetere, V.F., Elsner, C.I., Pardini, O., Almalvy, J.I., Di Sarli, A.R.: Evaluation of steel/primer based on chestnut tannin/paint film systems by EIS. J. Coat. Technol. **74**(3), 63–69 (2002)
5. Singh, DDN., Yadav, S.: Role of tannic acid based rust converter on formation of passive film on zinc rich coating exposed in simulated concrete pore solution. Surf. Coat. Technol. **202**, 1526–1542 (2008)
6. Al-Mayouf, A.M.: Inhibitors for chemical cleaning of iron with tannic acid. Desalination **121**, 173–182 (1999)
7. Chinchón, S., García, J., López-Atalaya, M., Linares, A., Vera, R.: Cement paste colouring in concretes. Cem. Concr. Res. **34**, 1987–1991 (2004)
8. Gust, J., Suwalski, J.: Use of Mössbauer spectroscopy to study reaction products of polyphenols and iron compounds. Corrosion **50**(5), 355–365 (1994)
9. Iglesias, J., García de Saldaña, E., Jaén, J.A.: On the tannic acid interaction with metallic iron. Hyperfine Interact. **134**, 109–114 (2001)
10. Jaén, J.A., Araúz, E.Y., Iglesias, J., Delgado, Y.: Reactivity of tannic acid with common corrosion products and its influences on the hydrolysis of iron in alkaline solutions. Hyperfine Interact. **148–149**, 199–209 (2003)
11. Jaén, J.A., González, L., Vargas, A., y Olave, G.: Gallic acid, ellagic acid and pyrogallol reaction with metallic iron. Hyperfine Interact. **148–149**, 227–235 (2003)
12. Martinez, S., Štern, I.: Ferric-tannate formation and anticorrosive properties of mimosa tannin in acid solutions. Chem. Biochem. Eng. Q. **13**(4), 191–199 (1999)
13. Martinez, S., Štern, I.: Inhibitory mechanism of low-carbon steel corrosion by mimosa tannin in sulphuric acid solutions. J. Appl. Electrochem. **31**(9), 973–978 (2001)
14. Martinez, S., Štern, I.: Thermodynamic characterization of metal dissolution and inhibitor adsorption processes in the steel/mimosa tannin/sulfuric acid. Appl. Surf. Sci. **199**(1–4), 83–89 (2002)
15. Rahim, A.A. Rocca, E., Steinmentz, J. Kassim, M.J., Adnan, R., Sani Ibrahim, M.: Mangrove Tannins and their flavanoid monomers as alternative steel corrosion inhibitors in acidic medium. Corros. Sci. **49**, 402–417 (2007)
16. Pardini, O.R., Amalvy, J.I., Di Sarli, A.R., Romagnoli, R., Vetere, V.F.: Formulation and testing of a waterborne primer containing chestnut tannin. J. Coat. Technol. **73**(913), 99–106 (2001)
17. Matamala, G., Smeltzer, W., Droguett, G.: Use of tannin anticorrosive reaction primer to improve traditional coating systems. Corrosion **50**, 270–275 (1994)
18. Morcillo, M., Feliu, S., Simancas, J., Bastidas, J.M., Galván, J.C., Feliu Jr., S., Almeida, E.M.: Corrosion of rusted steel in aqueous solutions of tannic acid. Corrosion **48**(12), 1032–1039 (1992)
19. Barrero, C.A., Ocampo, L.M., Arroyave, C.E.: Possible improvements in the action of some rust converters. Corros. Sci. **43**, 1003–1018 (2001)
20. Ocampo, L.M., Margarit, I.C.P., Mattos, O.R., Córdoba-Torresi, S.I., Fragata, F.L.: Performance of rust converter based in phosphoric and tannic acids. Corros. Sci. **46**, 1515–1525 (2004)
21. Caballero, L., Felloni, L., Trabanelli, G., Pulidori, F.: The anodic dissolution of the behaviour of some corrosion inhibitors investigated by the potentiodynamic method. Electrochim. Acta **9**(5), 485–494 (1964)
22. Jaén, J.A., García de Saldaña, E., Hernández, C.: Characterization of reaction products of iron and iron salts and aqueous plant extracts. Hyperfine Interact. **122**, 139–145 (1999)
23. Jaén Osorio, J.A., Adames, I.M.: Evidencias espectroscópicas de la reacción de acomplejamiento y reducción de algunos productos de corrosión por el ácido gálico y el pirogalol. Rev. Cubana de Química (ISSN 0258-5595) XIX(1) (2007)
24. Jaén, J. A., Navarro, C.: Mössbauer and infrared spectroscopy as a diagnostic tool for the characterization of ferric tannates. Hyperfine Interact. **192**(1), 61–67 (2009)
25. Hider, R.C., Howlin, B., Miller, J.R., Rahim Mohd-Nor, A., Silver, J.: Model compounds for microbial iron-transport compounds. IV: further solution chemistry and Mössbauer studies on iron (II) and iron (III) catechol complexes. Inorg. Chim. Acta **80**, 51–56 (1983)
26. Murad, E., Johnston, J.H.: Iron oxides and oxyhydroxides. In: Long, G.J. (ed.) Mössbauer Applied to Inorganic Chemistry, vol. 2, pp. 507–582. Plenum Pub Co, New York–London (1987)
27. Vértes, A., Czakó-Nagy, I.: Mössbauer spectroscopy and its application to corrosion studies. Electrochim. Acta. **34**(6), 721–758 (1989)
28. Coey, J.M.D., Readman, P.W.: Characterisation and magnetic properties of natural ferric gel. Earth Planet. Sci. Lett. **21**(1), 45–51 (1973)

29. Jambor J.L., Dutrizac, J.E.: Occurrence and constitution of natural and synthetic ferrihydrite, a wide spread iron oxyhydroxide. Chem. Rev. **98**, 2549–2585 (1998)
30. Raman, A., Kuban, B., Razvan, A.: The application of infrared spectroscopy to the study of atmospheric rust systems. Corr. Sci. **32**(12), 1295–1306 (1991)
31. Cornell, R.M., Schwertmann, U.: The Iron Oxides, Structure, Properties, Reactions, Occurrences and Uses, 2nd edn. Wiley-VCH (2003)
32. Carlson, L., Schwertmann, U.: Natural occurrence of feroxyhite (δ'-FeOOH). Clay Miner. **28**(4), 272–280 (1980)
33. Kahani, S.A., Jafari, M.: A new method for preparation of magnetite from iron oxyhydroxide or iron oxide and ferrous salt in aqueous solution. J. Magn. Magn. Mater. **321**, 1951–1954 (2009)
34. Miller, F.A., Wilkins, C.H.: Infrared spectra and characteristics frequencies of inorganic ions. Anal. Chem. **24**(8), 1253–1294 (1952)
35. Vértes, A., Parak, F.: A study of the relationship between the spin relaxation and certain chemical properties of paramagnetic iron (III)-salt solutions by Mössbauer spectroscopy. J. Chem. Soc. Dalton Trans. 2062–2068 (1972). doi:10.1039/DT9720002062
36. Chavez, F.A.B., Garg, V.K.: Mössbauer frozen solution studies of Fe(III) nitrate and Fe(III) perchlorate. J. Inorg. Nucl. Chem. **37**, 2283–2285 (1975)

Hyperfine Interact (2011) 202:39–46
DOI 10.1007/s10751-011-0358-9

Complementary ^{57}Fe and ^{119}Sn Mössbauer study of mechanochemical redox reaction

V. Šepelák · M. J. Nasr Isfahani · M. Myndyk · M. Ghafari · A. Feldhoff · K. D. Becker

Published online: 25 August 2011

Abstract A non-thermal route to oxide nanocomposites by mechanochemical redox reactions in the α-Fe_2O_3 + SnO system is presented. The important impact of the work, from the methodology point of view, is the combined application of ^{57}Fe and ^{119}Sn Mössbauer spectroscopies to the study of the mechanically induced redox processes, resulting in the reduction of α-Fe_2O_3 and the oxidation of SnO. In addition, the obtained product was characterized by X-ray diffraction and transmission electron microscopy.

Keywords Mechanochemistry · Redox reaction · Mössbauer spectroscopy

V. Šepelák is on leave from Slovak Academy of Sciences, 04353 Košice, Slovakia.

V. Šepelák (✉) · M. Ghafari
Institute of Nanotechnology, Karlsruhe Institute of Technology,
Herrmann-von-Helmholtz-Platz 1, 76344 Eggenstein-Leopoldshafen, Germany
e-mail: vladimir.sepelak@kit.edu

M. J. Nasr Isfahani
Department of Physics, Lenjan Branch, Islamic Azad University, Isfahan, Iran

M. Myndyk
Department of Inorganic Chemistry, Dresden University of Technology,
Mommsenstrasse 6, 01062 Dresden, Germany

A. Feldhoff
Institute of Physical Chemistry and Electrochemistry, Leibniz University Hannover,
Callinstrasse 33A, 30167 Hannover, Germany

K. D. Becker
Institute of Physical and Theoretical Chemistry, Braunschweig University of Technology,
Hans-Sommer-Strasse 10, 38106 Braunschweig, Germany

1 Introduction

The energy needed for the activation of chemical reactions is usually provided by heat, light, or electrical potential. Correspondingly, terms such as *thermochemistry*, *photochemistry*, or *electrochemistry* are generally common in the chemistry literature. A fundamentally different way of initiating or accelerating a chemical reaction is the use of mechanical force. It should be noted that the idea of inducing a reaction by exerting mechanical action dates back to the end of ninetieth century, when Carey Lea [1] observed the decomposition of silver and mercuric halides by trituration in a porcelain mortar, although the same compounds are known to melt or sublime undecomposed when heated. Mechanically induced chemistry (*mechanochemistry*) thus has a long history and it continues to be of high importance [2–5].

In addition to mechanosynthesis [6, 7] and mechanochemical decomposition of compounds [8, 9], chemical reduction-oxidation (redox) processes induced by mechanical treatment represent an important class of solid-state reactions [10]. Many examples have been reported where reduction of simple binary oxides by a more reactive metal occurs during high-energy milling [11–17]. In addition to simple binary oxides, there have also been reports on the mechanochemical reduction of hydroxides [18], sulphides [19], chlorides [20], suphates [21] and ternary complex oxides such as titanates [22], wolframates [23], ferrites [10, 24–27]. Depending upon the value of the adiabatic temperature of the redox reaction and milling conditions, the reaction may occur either in a steady-state manner during milling or by unstable thermal combustion of the reactants [28].

Whereas the mechanochemical reduction of simple binary oxides by a reactive metal is discussed extensively in the literature, the mechanically induced redox reactions involving two metal oxides, e.g., $M1O$ and $M2_2O_3$ with different valences of metal cations $M1$ and $M2$, are treated only to a limited extent. This is due to the fact that mechanical treatment of the $M1O + M2_2O_3$ mixtures mostly leads, instead of redox processes, to the formation of a complex oxide. The latter is demonstrated, for example, by numerous mechanosyntheses of ferrites MFe_2O_4 from the mixture of MO and Fe_2O_3 ($M =$ Zn, Mn, Mg, Ni, Ca) [29–33].

In this article, we will report on the mechanically induced evolution of the α-Fe_2O_3 + SnO mixtures submitted to high-energy milling. The important impact of the present work, from the methodology point of view, is the first combined application of ^{57}Fe and ^{119}Sn Mössbauer spectroscopies to the study of the mechanically induced redox reaction. It is obvious that the two spectroscopies are remarkably complementary in the present case. It is shown that the mechanochemical redox process provides an easy route to nanocomposites. In addition, the obtained product is characterized by comprehensive techniques including X-ray diffraction (XRD) and electron microscopy.

2 Experimental

The α-Fe_2O_3 + SnO mixtures (5 g) were milled for various times (up to 20 h) in a SPEX 8000 shaker mill (Spex CertiPrep Inc., USA) at room temperature. A grinding chamber and balls made of hardened steel were used. The ball-to-powder weight ratio was 7:1. Milling experiments were performed in air. ^{57}Fe and ^{119}Sn

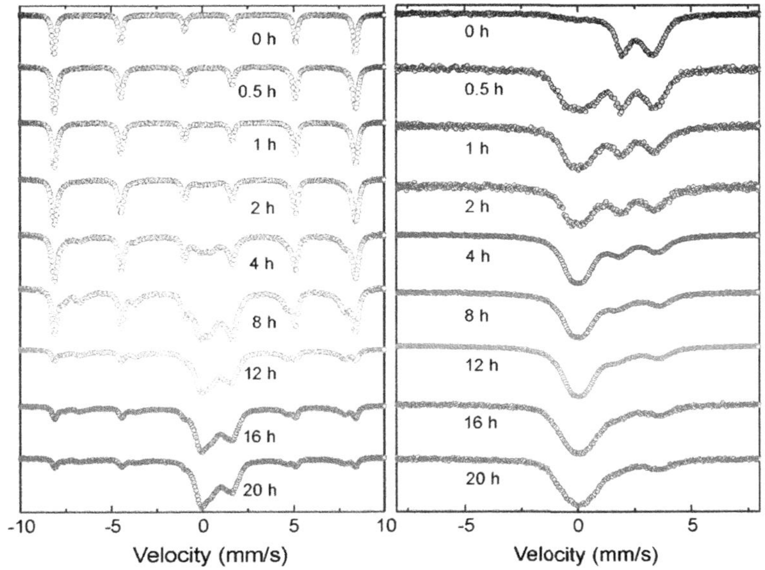

Fig. 1 ^{57}Fe (*left*) and ^{119}Sn (*right*) Mössbauer spectra of the α-Fe_2O_3 + SnO mixtures milled for various times. The milling times are shown in the figure

Mössbauer spectroscopic measurements were carried out in transmission mode at room temperature using the ^{57}Co/Rh and ^{119}Sn/$CaSnO_3$ γ-ray sources, respectively. The isomer shifts of the presented ^{57}Fe and ^{119}Sn Mössbauer spectra are referred against ^{57}Fe in Rh and ^{119}Sn in $CaSnO_3$, respectively. Recoil spectral analysis software [34] was used for the quantitative evaluation of the Mössbauer spectra. The XRD patterns were collected using a PW 1820 powder diffractometer (Philips, Netherlands) with Cu Kα radiation. The JCPDS PDF database [35] was utilized for phase identification using the STOE software. The morphology of powders was studied using a combined field-emission (scanning) transmission electron microscope (S)TEM (JOEL JEM-2100F). Prior to TEM investigations, powders were crushed in a mortar, dispersed in ethanol, and fixed on a copper-supported carbon grid.

3 Results and discussion

The mechanically induced evolution of the α-Fe_2O_3 + SnO mixtures submitted to high-energy milling was folowed by ^{57}Fe and ^{119}Sn Mössbauer spectroscopies. Figure 1 shows the room-temperature ^{57}Fe and ^{119}Sn Mössbauer spectra of the mixtures milled for various times up to 20 h. ^{57}Fe Mössbauer spectrum of the starting α-Fe_2O_3 + SnO mixture (Fig. 1, left) shows a sextet with a magnetic hyperfine field of 51.6(4) T corresponding to α-Fe_2O_3. ^{119}Sn Mössbauer spectrum of this sample (Fig. 1, right) is dominated by a doublet characteristic of Sn^{2+} cations in SnO (IS = 2.48(5) mm/s, QS = 2.18(4) mm/s). With increasing milling time, the sextet typical of α-Fe_2O_3 becomes asymmetric toward the inside of each line, slowly collapses, and

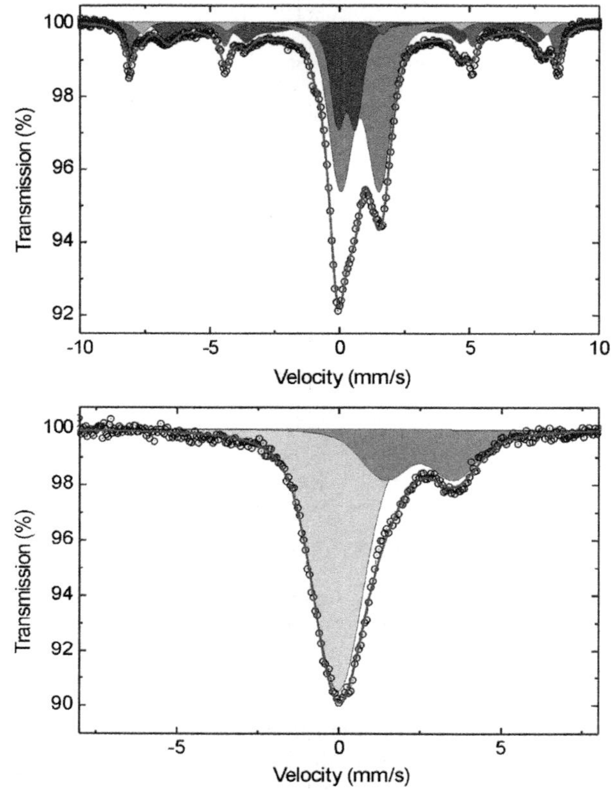

Fig. 2 ^{57}Fe (*top*) and ^{119}Sn (*bottom*) Mössbauer spectrum of the α-Fe_2O_3 + SnO mixture milled for 20 h. The subspectra corresponding to FeO, superparamagnetic Fe_3O_4, Fe(A) ions in Fe_3O_4, Fe[B] ions in Fe_3O_4, and α-Fe_2O_3 are denoted by *green*, *dark blue*, *light blue*, *violet*, and *red* colors, respectively. The subspectra due to Sn^{2+} and Sn^{4+} cations are denoted by *orange* and *yellow* colors, respectively

is gradually replaced by a central doublet. Qualitative changes are also observed in ^{119}Sn Mössbauer spectra of the milled samples; after only 30 min of mechanical treatment, clear evidence is observed of a new spectral component in the central part of the spectrum, which corresponds to octahedrally coordinated Sn^{4+} cations in SnO_2 (IS = −0.02(4) mm/s, QS = 0.92(7) mm/s). With prolonged milling, the intensity of this spectral component increases at the expense of the intensity of the doublet belonging to SnO. Thus, ^{119}Sn Mössbauer spectroscopy indicates that mechanical treatment of the α-Fe_2O_3 + SnO mixture leads to the oxidation of Sn^{2+} cations to Sn^{4+}.

Mössbauer spectra of the α-Fe_2O_3 + SnO mixture milled for 20 h were analyzed in detail. The fitting of the ^{57}Fe Mössbauer spectrum of this sample (Fig. 2, top) reveals that it contains the Fe^{2+}-containing wüstite phase (FeO) (47.2%), magnetite (Fe_3O_4) in magnetically ordered and superparamagnetic states (40.6%), and the rest of hematite (α-Fe_2O_3) (12.2%). Thus, ^{57}Fe Mössbauer spectroscopy indicates that mechanical treatment of the α-Fe_2O_3 + SnO mixture leads to the reduction of Fe^{3+} to Fe^{2+} cations, i.e., the following mechanochemical reduction reaction related to the iron-containing phases takes place: α-$Fe_2O_3 \rightarrow Fe_3O_4 \rightarrow$ FeO. The formation of magnetite (phase containing both Fe^{3+} and Fe^{2+} cations) as an intermediate product is clearly visible in the spectra of the samples milled for times longer than 4 h (see Fig. 1, left). The appearance of magnetite in superparamagnetic (19.1%)

Table 1 Hyperfine parameters obtained by fitting the room-temperature ^{57}Fe Mössbauer spectrum of the α-Fe_2O_3 + SnO mixture milled for 20 h

Spectral component due to	IS (mm/s)	QS (mm/s)	*B* (T)	*I* (%)
FeO	0.77(4)	1.49(9)	–	47.2(1)
Superparamagnetic Fe_3O_4	0.46(5)	0.69(1)	–	19.1(5)
Fe(A) ions in Fe_3O_4	0.15(1)	0	48.3(3)	8.6(4)
Fe[B] ions in Fe_3O_4	0.52(3)	0	37.2(3)	12.9(7)
α-Fe_2O_3	0.24(2)	−0.10(2)	51.0(8)	12.2(4)

A Lorentzian line width $\Gamma = 0.25(2)$ mm/s resulted from the fit of the spectrum
IS isomer shift, *QS* quadrupole splitting, *B* average magnetic hyperfine field, *I* relative intensity

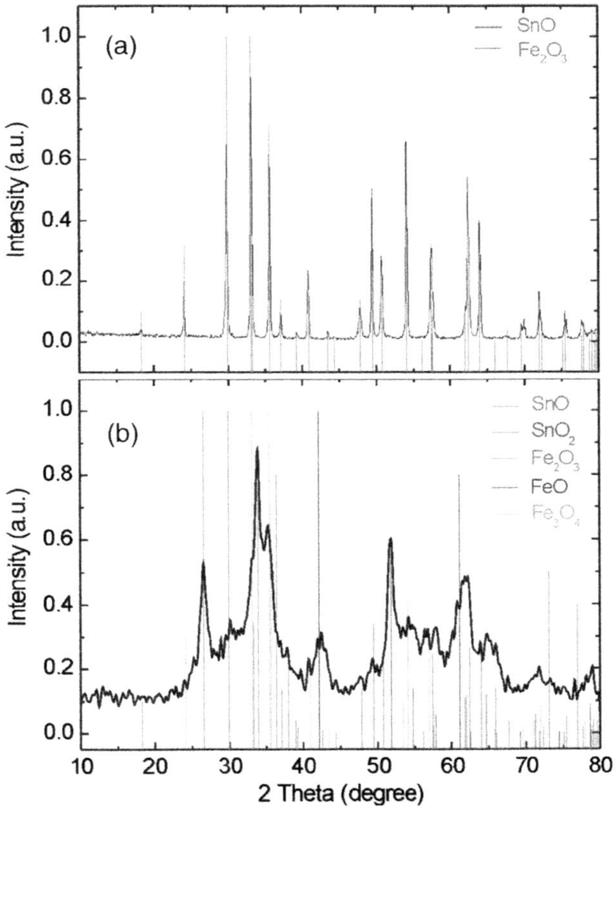

Fig. 3 XRD patterns of **a** the initial α-Fe_2O_3 + SnO mixture and **b** the nanocomposite prepared by mechanochemical redox reaction

and ferrimagnetic (21.5%) states corresponds to the formation of its smaller and larger crystals, respectively. The hyperfine parameters of the corresponding iron-containing phases in the milled sample are listed in Table 1. The fitting of the ^{119}Sn Mössbauer spectrum of the mixture milled for 20 h (Fig. 2, bottom) reveals that the relative intensity of the new spectral component corresponding to Sn^{4+} cations is about 74.1%.

Fig. 4 TEM image demonstrating the nanoscale nature of the Sn-oxide/Fe-oxide composite prepared by mechanochemical redox reaction

To determine the phase evolution of the α-Fe_2O_3 + SnO mixture during high-energy milling in greater detail and to support the findings of Mössbauer spectroscopy, the mechanochemical redox reaction was also followed by XRD. Figure 3 compares the XRD patterns of the initial and milled mixtures. In agreement with the above-presented results of spectroscopic measurements, XRD reveals the presence of two phases, SnO (JCPDS PDF 06–0395) and α-Fe_2O_3 (JCPDS PDF 33–0664) [35], in the initial mixture. This is in contrast to the mixture milled for 20 h, which is found to contain the following five phases: SnO (JCPDS PDF 06–0395), α-Fe_2O_3 (JCPDS PDF 33–0664), Fe_3O_4 (JCPDS PDF 19–0629), FeO (JCPDS PDF 46–1312), and SnO_2 (JCPDS PDF 41–1445) [35]. Taking into account the results of both Mössbauer and XRD measurements, it can be stated that milling of the α-Fe_2O_3 + SnO mixture initiates the redox reaction, resulting in the formation of composites containing FeO, Fe_3O_4, α-Fe_2O_3, SnO_2, and SnO phases. Note that in ideal case, the redox reaction between α-Fe_2O_3 and SnO proceeds in the following manner: $\alpha\text{-}Fe_2O_3 + SnO \rightarrow 2\ FeO + SnO_2$. The broad shape of the diffraction peaks in the XRD pattern of the milled sample indicates its nanoscale nature, i.e., the *nanocomposite* is formed by the present mechanochemical redox reaction. The average particle size of the nanocomposite was found to be about 10 nm, see Fig. 4.

Although several mechanisms have been proposed for the reduction processes occurring during high-energy milling [36–38], to our best knowledge, no conclusive explanation for mechanically induced redox reactions has been given yet. It has been suggested that the mechanochemical reduction process takes place at the surface of the milled oxide [36, 37]. According to these articles, oxygen bonds on the cleaved oxide surfaces are broken during the mechanical treatment leading to the reduction of the sample and to a release of oxygen to the vial's internal volume. In this case, the rupture of oxide surface layers and surface stress are assumed to provide the driving force for the reduction. It has also been proposed that the oxygen partial pressure in the vial atmosphere is the critical parameter responsible for the mechanochemical reduction. Zdujić et al. [38] presumed that adsorption of oxygen on atomically clean surfaces created by particle fracture may cause a decrease of the oxygen partial pressure to some critical value below which the given phase becomes locally unstable and transforms to a stable one. The transformation occurs at the moment of impact by the formation of high-energy localized sites of short lifetime (sometimes called "hot spots" or "thermal spikes"). To conclude, the identification of the factors representing the main driving force for mechanically induced redox processes, the elucidation of the microscopic mechanism(s), and the determination of rate-determining steps of mechanochemical reduction represent major challenges for fundamental research and require further efforts.

4 Conclusions

The novel Sn-oxide/Fe-oxide nanocomposite was synthesized by the mechanochemical redox reaction between α-Fe_2O_3 and SnO. The reduction of α-Fe_2O_3 and the oxidation of SnO induced by high-energy milling is documented by complementary ^{57}Fe and ^{119}Sn Mössbauer spectroscopic studies. The event of mechanically induced redox reactions presents novel opportunities for the non-thermal manipulation of materials and provides a promising field for future fundamental as well as applied research.

Acknowledgements The present work was supported by the DFG in the framework of the Priority Program "Crystalline Nonequilibrium Phases" (SPP 1415). Partial support by the VEGA (2/0174/11) and the Iranian Nanotechnology Initiative is gratefully acknowledged.

References

1. Carey Lea, M.: Am. J. Sci. **43**, 527 (1892)
2. Avvakumov, E., Senna, M., Kosova, N.: Soft Mechanochemical Synthesis: A Basis for New Chemical Technologies. Kluwer, Boston (2001)
3. Boldyrev, V.V.: Russ. Chem. Rev. **75**, 177 (2006)
4. Delogu, F., Mulas, G.: Experimental and Theoretical Studies in Modern Mechanochemistry. Transworld Research Network, Kerala (2010)
5. Šepelák, V., Bergmann, I., Indris, S., Feldhoff, A., Hahn, H., Becker, K.D., Grey, C.P., Heitjans, P.: J. Mater. Chem. **21**, 8332 (2011). doi:10.1039/c0jm03721d
6. Düvel, A., Wilkening, M., Uecker, R., Wegner, S., Šepelák, V., Heitjans, P.: Phys. Chem. Chem. Phys. **12**, 11251 (2010)
7. Šepelák, V., Becker, K.D., Bergmann, I., Suzuki, S., Indris, S., Feldhoff, A., Heitjans, P., Grey, C.P.: Chem. Mater. **21**, 2518 (2009)
8. Justin Joseyphus, R., Narayanasamy, A., Sivakumar, N., Guyot, M., Krishnan, R., Ponpandian, N., Chattopadhyay, K.: J. Magn. Magn. Mater. **272–276**, 2257 (2004)
9. Wei, Y., Yan, J., Lu, S., Li, X.: J. Environ. Sci. **21**, 1761 (2009)
10. Šepelák, V., Menzel, M., Becker, K.D., Krumeich, F.: J. Phys. Chem. B **106**, 6672 (2002)
11. Takacs, L.: Nanostruct. Mater. **2**, 241 (1993)
12. Basset, D., Matteazzi, P., Miani, F., Le Caër, G.: Hyperfine Interact. **95**, 2235 (1994)
13. Tokomitsu, K.: Solid State Ionics **101–103**, 25 (1997)
14. Kerr, A., Welham, N.J., Willis, P.E.: Nanostruct. Mater. **11**, 233 (1999)
15. Goya, G.F., Rechenberg, H.R.: J. Phys. Condens. Matter **12**, 10579 (2000)
16. Grigoryeva, T.F., Novakova, A.A., Kiseleva, T.Y., Barinova, A.P., Ancharov, A.I., Talako, T.L., Vorsina, I.A., Becker, K.D., Šepelák, V., Tsybulya, S.V., Bulavchenko, O.A., Lyakhov, N.Z.: J. Phys.: Conf. Ser. **144**, 012076 (2009)
17. Kano, J., Kobayashi, E., Tongamp, W., Miyagi, S., Saito, F.: J. Alloy. Compd. **484**, 422 (2009)

18. Iwasaki, T., Sato, N., Kosaka, K., Watano, S., Yanagida, T., Kawai, T.: J. Alloy. Compd. **509**, L34 (2011)
19. Baláž, P., Takacs, L., Godoèíková, E., Škorvánek, I., Kováè, J., Choi W.S.: J. Alloy. Compd. **434–435**, 773 (2007)
20. Ding, J., Tsuzuki, T., McCormick, P.G., Street, R.: J. Magn. Magn. Mater. **162**, 271 (1996)
21. Setoudeh, N., Welham, N.J.: Int. J. Miner. Process. **98**, 214 (2011). doi:10.1016/j.minpro.2010.12.002
22. Welham, N.J.: J. Alloy. Compd. **274**, 303 (1998)
23. Welham, N.J.: Mater. Sci. Eng. A **248**, 230 (1998)
24. Takacs, L.: Mater. Lett. **13**, 119 (1992)
25. Goya, G.F., Rechenberg, H.R., Jiang, J.Z.: J. Appl. Phys. **84**, 1101 (1998)
26. Shi, Y., Ding, J.: J. Appl. Phys. **90**, 4078 (2001)
27. Menzel, M., Šepelák, V., Becker, K.D.: Solid State Ionics **141–142**, 663 (2001)
28. Yang, H., McCormick, P.G.: J. Solid State Chem. **107**, 258 (1993)
29. Šepelák, V., Steinike, U., Uecker, D.C., Wißmann, S., Becker, K.D.: J. Solid State Chem. **135**, 52 (1998)
30. Muroi, M., Street, R., McCormick, P.G., Amighian, J.: Phys. Rev. B **63**, 184414 (2001)
31. Šepelák, V., Feldhoff, A., Heitjans, P., Krumeich, F., Menzel, D., Litterst, F.J., Bergmann, I., Becker, K.D.: Chem. Mater. **18**, 3057 (2006)
32. Šepelák, V., Bergmann, I., Feldhoff, A., Heitjans, P., Krumeich, F., Menzel, D., Litterst, F.J., Campbell, S.J., Becker, K.D.: J. Phys. Chem. C **111**, 5026 (2007)
33. Berchmans, L.J., Myndyk, M., Da Silva, K.L., Feldhoff, A., Šubrt, J., Heitjans, P., Becker, K.D., Šepelák, V.: J. Alloy. Compd. **500**, 68 (2010)
34. Lagarec, K., Rancourt, D.G.: Recoil - Mössbauer Spectral Analysis Software for Windows. University of Ottawa, Ottawa, ON (1998)
35. Joint Committee on Powder Diffraction Standards (JCPDS) Powder Diffraction File (PDF): International Centre for Diffraction Data, Newton Square, PA (2004)
36. Kaczmarek, W.A., Ninham, B.W.: IEEE Trans. Magn. **30**, 732 (1994)
37. Campbell, S.J., Kaczmarek, W.A., Wang, G.M.: Nanostruct. Mater. **6**, 735 (1995)
38. Zdujić, M., Jovalekić, Ć., Karanović, L., Mitrić, M.: Mater. Sci. Eng. A **262**, 204 (1999)

Hyperfine Interact (2011) 202:47–55
DOI 10.1007/s10751-011-0359-8

Mössbauer spectroscopic study of iron–nickel nitrides thin films prepared by ion beam assisted deposition

P. Prieto · M. Monti · J. de la Figuera ·
J. M. Sanz · J. F. Marco

Published online: 30 August 2011

Abstract In this paper we have shown the feasibility of ion beam assisted deposition to produce iron–nickel nitride thin films of various compositions. We have carried out their compositional and structural characterization by means of Rutherford backscattering spectroscopy (RBS), X-ray diffraction (XRD), X-ray photoelectron spectroscopy (XPS) and integral conversion electron Mössbauer spectroscopy (ICEMS). The results have shown that Ni is incorporated within the films and that the temperature of the substrate has a large influence both in the phase composition obtained as well as in the degree of crystallinity of the thin films produced.

Keywords Iron–nickel nitrides · Thin films · DIBS · ICEMS ·
Surface characterization

1 Introduction

The Fe-N phase diagram is quite complex. Depending on temperature and nitrogen concentration many different phases can be formed with different structures and physical properties [1]. For example, nitrogen-poor phases are usually ferromagnetic at room temperature while nitrogen-rich phases are paramagnetic. We have also recently shown that small differences in the composition of these materials have a large influence on their chemical stability in aggressive atmospheres [2]. In particular, nanocrystalline nitrogen-poor phases of iron nitride are good candidates for high frequency applications in which high saturation magnetization, low coercivity and

P. Prieto · J. M. Sanz
Departamento de Física Aplicada C-12I, Universidad Autónoma de Madrid,
Cantoblanco, 28049 Madrid, Spain

M. Monti · J. de la Figuera · J. F. Marco (✉)
Instituto de Química-Física "Rocasolano", CSIC, Serrano 119, 28006 Madrid, Spain
e-mail: jfmarco@iqfr.csic.es

Table 1 Sample denomination and main deposition parameters of iron–nickel nitride films: substrate temperature during deposition, $\%N_2^+$ ions and current density and energy of the assisting and sputtering beam

Sample	T (°C)	Assistance			Sputtering		Composition (RBS)
		$\%N_2^+$ ions	Current density ($mA\ cm^{-2}$)	Energy (eV)	Total current (mA)	Energy (eV)	
S1	20	33	0.046	55	10	500	$Fe_{0.74}Ni_{0.04}N_{0.22}$
S2	20	17	0.043	52	10	500	$Fe_{0.76}Ni_{0.09}N_{0.14}$
S3	20	33	0.046	55	10	500	$Fe_{0.60}Ni_{0.25}N_{0.14}$
S4	300	33	0.046	55	10	500	$Fe_{0.75}Ni_{0.04}N_{0.20}$
S4	300	33	0.046	55	10	500	$Fe_{0.70}Ni_{0.08}N_{0.21}$
S6	300	50	0.048	59	10	500	$Fe_{0.61}Ni_{0.06}N_{0.32}$

The sample compositions, as determined by RBS, are also given

high permeability are needed. The incorporation of interstitial nitrogen in iron can result in substantial increase of the resistivity and anisotropy field, retaining relatively larger magnetic moment of iron atoms. However, iron nitride films also show thermal stability problems that can be solved by the incorporation of a third element (e.g. a transition metal) in the Fe-N system. Ni is a good candidate because it can be incorporated up to relatively high concentrations maintaining a good combination of magnetic properties [3]. In fact, we have found that γ'-FeNiN thin films show important properties for high-frequency applications such as resonance ferromagnetic frequency in the gigahertz range, high anisotropy field, and well defined in-plane magnetic uniaxial anisotropy [4].

We have shown recently the feasibility of Ion Beam Assisted Deposition (IBAD) to prepare thin films of iron nitrides covering the full compositional range [5]. In the present paper we will also show the viability of this technique to produce thin films of iron–nickel nitrides. Under the experimental conditions used, many of the different phases produced are of nanocrystalline character and, therefore, conventional phase analysis techniques, such as X-ray diffraction, are not able to determine precisely the phase composition of the thin films. Thus, techniques as integral conversion electron Mössbauer spectroscopy (ICEMS) are very appropriate to elucidate the phase composition of the thin films deposited. We will show here several examples where the use of ICEMS has been of fundamental importance in the characterization of these materials.

2 Experimental

Iron–nickel nitride thin films were deposited on Si (100) using a Dual Ion Beam Sputtering (DIBS) system. The films were obtained by Ar^+ sputtering of a pure iron (99.99%) sputtering target with some nickel chips (99.99% purity) on its surface and under simultaneous bombardment with a controlled mixture of low energy argon and nitrogen ions, which assist the deposition. The composition of FeNiN films was adjusted by varying both the amount of Ni chips mounted on the iron target and the deposition parameters. Table 1 summarizes the deposition conditions for all the films characterized in this work. Two different temperatures of the substrate were used during deposition.

Fig. 1 RBS data recorded from sample S3

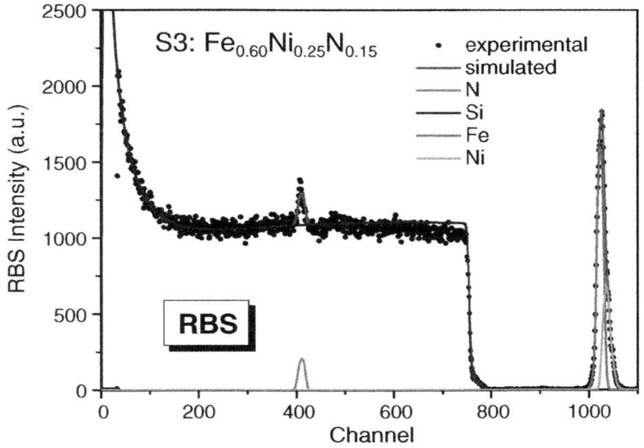

The as-deposited samples were characterised by Rutherford backscattering spectroscopy (RBS), X-ray diffraction (XRD), X-ray photoelectron spectroscopy (XPS) and integral conversion electron Mössbauer spectroscopy (ICEMS).

Rutherford backscattering spectroscopy, using a 5 MV tandem accelerator, was used to determine the average composition of the films. An incident beam of 3.7 MeV $^4He^+$ ions was used and the backscattered ions at scattering angles of 165° were detected to determine by resonant RBS the composition of the films with an accuracy better than 5%. The distribution and quantification of the elements was determined by simulation with the SIMNRA code using published $^{14}N(\alpha, \alpha)^{14}N$ elastic-scattering cross sections [6].

The structure of the different films was analyzed by XRD at a grazing incidence of 1° with respect to the surface plane, using Cu Kα radiation ($\lambda = 0.15408$ nm) in a Siemens 5000 spectrometer.

XPS data were recorded in a PHI 3027 spectrometer equipped with a double-pass cylindrical mirror analyzer and a non-monochromatic Mg Kα radiation source ($h\nu =$ 1253.6 eV). The pass energy of the spectrometer was set at 50 eV and the binding energies are accurate within 0.1 eV.

Integral conversion electron Mössbauer spectroscopy (ICEMS) data were recorded at room temperature in a conventional constant acceleration spectrometer using a $^{57}Co(Rh)$ source and a parallel plate avalanche counter. All the isomer shifts were referred to the centroid of the spectrum of α-Fe at room temperature.

3 Results

Figure 1 shows a typical RBS spectrum recorded from one of the samples (S3) measured with $^4He^+$ particles of 3.7 MeV. The spectrum shows peaks arising from the presence of N, Si, Ar, Fe and Ni. The Si signal arises from the substrate whilst Ar (<1 at.%) is incorporated during the sputtering process. The quantitative analysis of the different spectra leads to the average compositions given in Table 1. The RBS results indicate that Ni is incorporated in the deposited films, a fact which is also confirmed by the XPS results. Figure 2 shows the Ni 2p and Fe2p XPS spectra from

Fig. 2 Ni 2p and Fe2p XPS spectra for samples S1, S2 and S3 with different Ni content. The % Ni at. in Fe_xNi_{1-x} have been also included

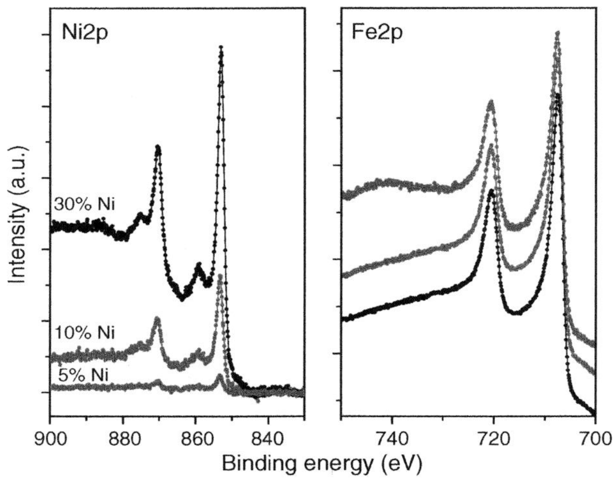

samples S1, S2 and S3 after sputtering with Ar^+ ions to remove any oxide layer which could be present at the surface. For comparison, the spectra have been normalized to the Fe2p intensity and due to the overlap of the Ni2p spectrum with the Fe2s and N KLL spectra, the Ni2p spectra in Fig. 2 have been obtained after substration of the Fe2s+NKLL spectra of FeN (without Ni) grown at the same deposition conditions. Note the increase in intensity of the Ni2p spectra with increasing Ni concentration in the Fe+Ni target. Both techniques, RBS and XPS, give values of 5, 10 and 30% Ni at., at the Fe_xNi_{1-x}.

The XRD data (Fig. 3) are composed in general by very broad lines which is indicative of the nanocrystalline character of the samples. This is particularly evident in the case of the films grown with the temperature of the substrate kept at 20°C (Fig. 3 top). In this case the line width of the Bragg peaks allows estimating using the Scherrer relationship an average grain size lower than 6 nm for samples S1, S2 and S3. However, the broadness of the diffraction peaks makes the identification of the different phases present difficult. In the case of samples grown at 300°C (Fig. 3 bottom), we were able to associate some of the diffraction lines with those characteristic of some iron–nickel nitride compounds, mainly a γ'-$(Fe_{1-y}Ni_y)_4N$ phase with diffraction peaks associated with the (111) and (220) planes for sample S4 and with the (110), (111), (112) and (300) planes of ε-$(Fe_{1-y}Ni_y)_3N$ phase for sample S6.

The ICEMS spectra recorded from the films which were deposited with the substrate temperature kept at 20°C (Fig. 4) showed also broad lines. This fact is consistent with the nanocrystalline character of the samples as evidenced by XRD. The spectra were fitted using distributions of hyperfine parameters since this kind of fit is much more appropriate in the presence of nanocrystalline phases.

The spectrum recorded from sample S1 is shown in Fig. 4 together with its corresponding hyperfine magnetic field distribution. We used a fit with a linear correlation between the isomer shift and the hyperfine magnetic field such that the isomer shifts corresponding to the sextets with higher magnetic fields were close to zero and those in the range 15–20 T had an isomer shift in the range 0.25–0.15 mm s^{-1}. In the present

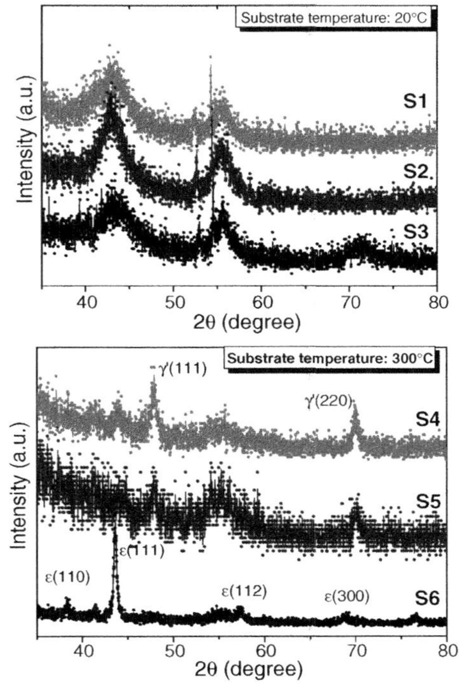

Fig. 3 XRD data recorded from the different samples studied in this work

case, the hyperfine field distribution presented several maxima: the one appearing at 31.9 T and having a corresponding isomer shift of ca. 0.07 mm s^{-1}, could correspond to a martensitic α'-(Fe_{1-y} Ni_y)N type-phase [7, 8]. Since the distribution extends continuously from 39.6 T to the next maximum at 24.2 T with evident contributions in the range 27.5–31.9 T (there is no sharp decrease between these two maxima as it occurs in some other samples, see below) the occurrence of a phase of the type α''-($Fe_{1-y}Ni_y)_{16}N_2$ [7–9] cannot be discarded. The hyperfine magnetic field distribution presents, as we have mentioned, another intense maximum at around 24.2 T (corresponding isomer shift, 0.14 mm s^{-1}). This could be compatible with the presence of an ε-($Fe_{1-y}Ni_y)_{3.00}$N type-phase [10]. Finally, the distribution shows other minor relative maxima at 14.3 T and 7.7 T which are compatible with the occurrence of an ε-($Fe_{1-y}Ni_y)_{3-x}$N type-phase [10]. From the present evaluation of the spectra it is difficult to ascertain the presence of a ferromagnetic γ'-($Fe_{1-y}Ni_y)_4$N contribution. In any case, if it were present, this phase would have a low intensity since we have not observed any clear maximum in the distribution at ca. 21–22 T as it would have otherwise been expected. In any case, the interpretation of the spectra results difficult because of the concomitant occurrence of small particle size, presence of different phases and existence of possible local inhomogeneities around the different iron sites.

The spectrum recorded from sample S2 is also broad (Fig. 4), although the corresponding hyperfine magnetic distribution has a different shape than that observed in the previous sample. In the present case, such distribution is dominated by a single maximum at 33.0 T (corresponding isomer shift 0.06 mm s^{-1}) what is compatible with

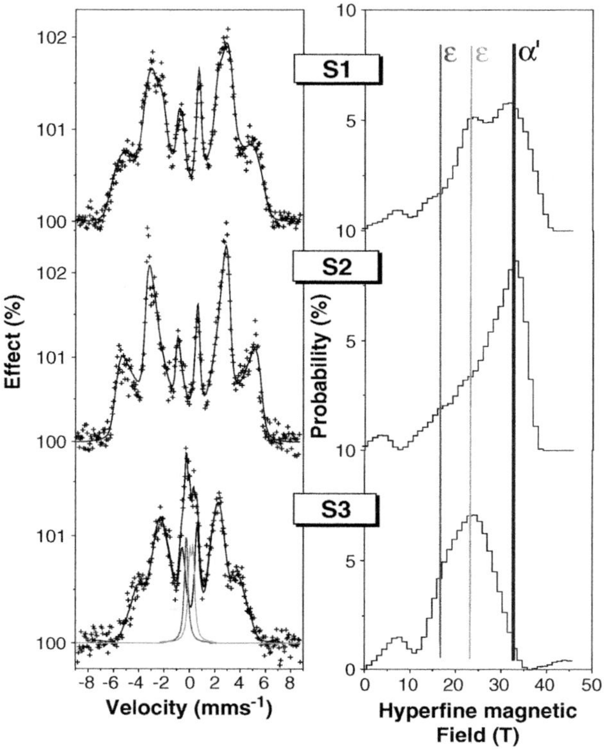

Fig. 4 ICEMS spectra and corresponding hyperfine field magnetic distributions corresponding to the samples grown with the temperature of the substrate kept at 20°C

the presence of an α'-(Fe_{1-y} Ni_y)N type-phase [7–9]. In this sample the distribution ends more abruptly on the high hyperfine magnetic field side than in the preceding case what precludes both the existence of γ'-$(Fe_{1-y}Ni_y)_4N$ and α'-$(Fe_{1-y}Ni_y)_{16}N_2$ type-phases. The presence of a small shoulder at ca. 22.0 T could be compatible with the occurrence of a ε-$(Fe_{1-y}Ni_y)_{3.00}N$ type-phase [10]. The results suggest that this sample is mainly composed of a nanocrystalline α'-(Fe_{1-y} Ni_y)N phase with a small contribution ε-$(Fe_{1-y}Ni_y)_{3.00}N$ of N type-phase.

The spectrum recorded from the S3 sample shows again very broad lines. Differently from the other samples, the spectrum showed, besides the broad magnetic component, discrete lines in the centre of the spectrum which could correspond to paramagnetic Fe-N phases. The spectrum was then fitted to a magnetic hyperfine distribution, a discrete singlet and a discrete doublet. The resulting hyperfine distribution was quite broad, asymmetrically ending at ca. 35 T and showing a main maximum at 23.5 T, a minor maximum at 7.7 T, and a shoulder at 17.6 T. The results could be compatible with the major presence of ε-$(Fe_{1-y}Ni_y)_{3.00}N$ type-phases, although a minor α'-(Fe_{1-y} Ni_y)N type-contribution cannot be discarded to account for the asymmetry of the distribution at larger magnetic fields. The parameters of the singlet ($\delta = -0.07$ mms^{-1}) and doublet ($\delta = 0.33$ mm s^{-1} and $\Delta = 0.28$ mm s^{-1}) are compatible with the presence (14%) of a paramagnetic γ-$(Fe_{1-y}Ni_y)N$ type-phase [10].

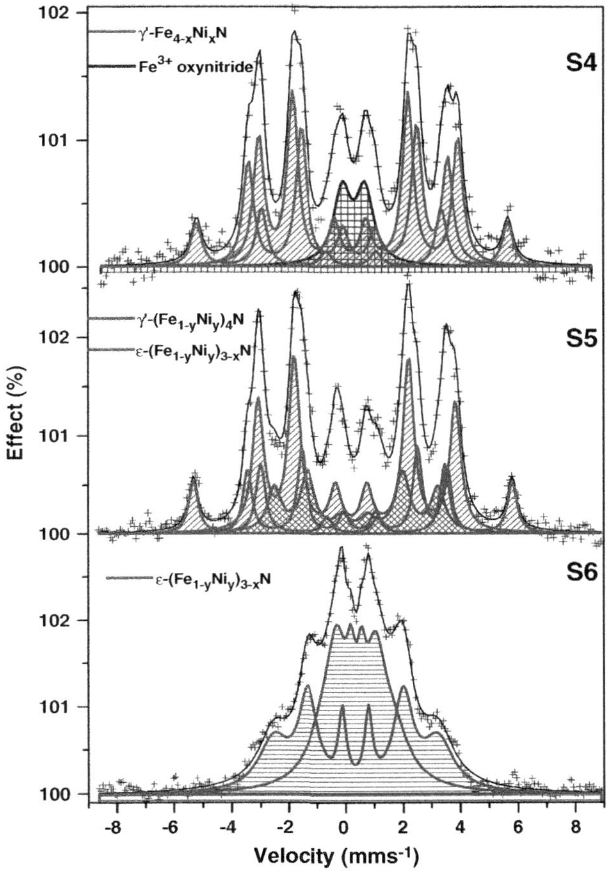

Fig. 5 ICEMS spectra recorded from the samples grown with the temperature of the substrate kept at 300°C

The spectra recorded from the samples which were prepared with the substrate temperature kept at 300°C had a much more defined shape. So, the spectrum corresponding to sample S4 (Fig. 5) is characteristic, except for a paramagnetic contribution in the centre of the spectrum which might correspond to an iron oxynitride compound [9], of a γ'-$(Fe_{1-y}Ni_y)_4N$ phase in good agreement with its diffraction pattern. The Mössbauer parameters obtained from the fit of the spectrum are collected in Table 2. γ'-Fe_4N has a cubic structure which contains iron in two non-equivalent crystallographic sites: the FeI atoms at the corners of the cubic unit cell and the FeII atoms occupying the face-center positions of the cube [9–11]. Because of the combined magnetic and electric interactions occurring in such structure and due to the different orientation of the internal magnetic field with the axis of the electric field gradient at the FeII sites, there are two magnetically unequivalent FeII sites in this material. Thus, the spectrum is composed by three different magnetic subspectra: one due to the FeI sites and two arising from the FeII (A and B) sites with a corresponding area ratio 1:2:1. It has been reported [11] that the substitution of Fe by Ni in γ'-Fe_4N occurs preferentially at the corner sites of the cubic structure and that due to this substitution the hyperfine magnetic field at these sites increases

Table 2 Mössbauer parameters obtained from the fit of the samples prepared with the temperature of the substrate kept at 300°C

Sample	δ (mm s^{-1})	2ε or Δ (mm s^{-1})	H(T)	Area (%)	Area$_{total}$(%)		Assignment
S4	0.23	0.03	33.6	14	88	Site I	γ'-$(Fe_{1-y}Ni_y)_4N$
	0.31	0.27	21.4	41		Site IIA	
	0.28	−0.44	21.7	33		Site IIB	
	0.35	0.78	–	12	12		Fe(Ni)-oxynitride
S5	0.24	0.01	34.5	14	72	Site I	γ'-$(Fe_{1-y}Ni_y)_4N$
	0.29	0.22	21.4	35		Site IIA	
	0.28	−0.44	21.7	23		Site IIB	
	0.35	0.04	17.5	20	28		$\varepsilon-(Fe_{1-y}Ni_y)_{3-x}N$
	0.35	0.25	5.0	8			
S6	0.34	0.002	17.8	47	63		$\varepsilon-(Fe_{1-y}Ni_y)_{3-x}N$
	0.41	0.08	7.1	53	31		

with increasing Ni content [11, 12]. It has been also reported that the magnitude of the quadrupole shift increases at the Fe II sites with increasing Ni substitution [11]. The results collected in Table 2 indicate that the ratio of the corresponding spectral areas Fe I sites/Fe II sites is smaller than the 0.33 ratio observed in γ'-Fe_4N, what is consistent with the literature findings. However, there is no clear increase observed in both the magnitude of the 2ε values or the hyperfine magnetic field, at least for this amount of Ni substitution. Taking together, the results do not allow to deduce immediately that a such substitution is taking place, even more considering the low amount of substitution (the changes observed in the magnitude of 2ε and H are more evident for larger Ni concentrations [11]) and the presence of the paramagnetic phase whose nature is uncertain.

In the case of sample S5 the results of the fit of the spectrum to discrete sextets (Table 2, Fig. 5) indicate that this sample is composed mainly by a γ'-$(FeNi)_4N$ phase (72%) and by a minor ε-$(Fe_{1-y}Ni_y)_{3-x}N$ type-phase (28%).

The spectrum recorded from sample S6 (Fig. 5) is totally characteristic of a ε-$Fe_{3-x}N$ phase [10], being composed by two broad superimposing magnetic sextets and in good agreement with its XRD diffraction pattern. The results of the fit of the spectrum to two discrete sextets are collected in Table 2. In terms of both the hyperfine magnetic fields and spectral areas of the different components the spectrum is very similar to that reported for a ε-$Fe_{2.40}N$ type-phase [10]. The existence of magnetic ordering contrasts with that recently reported [13] for a sample of composition ε-$Fe_{2.6}Ni_{0.4}N$ (which corresponds approximately to a 15% Ni substitution, as in the case reported here) synthesised by citrate routes which has showed to be paramagnetic at room temperature.

In summary, we have shown in the present paper that DIBS can be used to prepare Fe-Ni nitride thin films of various compositions which, by controlling the temperature of the substrate, can be produced with a marked nanocrystalline character. The results have also shown, however, the difficulty of preparing single-phase films when the temperature of the substrate is kept at 20°C. However, when the substrate temperature reaches 300°C it is possible to obtain almost pure γ' and ε-FeNiN phases. We have also shown the ability of ICEMS to characterize the different phases existing in the films in situations where X-ray diffraction has been unable to give an unequivocal identification.

Acknowledgements We acknowledge financial support from the Spanish Ministry of Science and Innovation through projects MAT2008-05763/NAN and MAT2009-14578-C03-01.

References

1. du Marchie van Voorthuysen, E.H., Chechenin, N.C., Boerma, D.O.: Metall. Mater. Trans. A **33A**, 2594 (2002)
2. Marco, J.F., Herranz, T., Gracia, M., Gancedo, J.R., Moutinho, F., Prieto, P., Sanz, J.M.: Surf. Interface Anal. **42**, 616 (2010)
3. Wang, X.F., Wu, P., Li, X.L., Li, Z.Q., Bai, H.L., Jiang, E.Y.: J. Magn. Magn. Mater **312**, 147 (2007)
4. Prieto, P., Pirota, K.R., Sanz, J.M., Jimenez, E., Camarero, J., Maccherozzi, F., Panaccione, G.: Appl. Phys. Lett. **90**, 032505 (2007)
5. Prieto, P., Marco J.F., Sanz, J.M.: Surf. Interface Anal. **40**, 781 (2008)
6. Feng, Y., Zhuying, Z., Gurquing Z., Fujia, Y.: Nucl. Instrum. Methods Phys. Res. **B94**, 11 (1994)
7. Jirasková, Y., Havlíček, S., Schneeweiss, O., Peřina, V., Blawert, C.: J. Magn. Magn. Mater. **234**, 477 (2001)
8. Kopcewicz, M., Jagielski, J., Turos, A., Williamson, D.L.: J. Appl. Phys. **71**, 4217–4226 (1992)
9. Borsa D.M., Boerma, D.O.: Hyperfine Interact. **151–152**, 31 (2003)
10. Schaaf, P.: Hyperfine Interact. **111**, 113 (1998)
11. Diao, X.G., Scorzelli, R.B., Rechenberg, H.R.: J. Magn. Magn. Mater. **218**, 81 (2001)
12. Panda, R.N., Balaji G., Gajbhiye, N.S.: Hyperfine Interact. **141–142**, 187 (2002)
13. Bajbhiye, N.S., Ningthoujam, R.S., Bhattacharyya, S.: Hyperfine Interact. **164**, 17 (2005)

Hyperfine Interact (2011) 202:57–61
DOI 10.1007/s10751-011-0338-0

Spin re-orientation in $FeCr_2S_4$

J. Engelke · F. J. Litterst · A. Krimmel · A. Loidl · F. E. Wagner · G. M. Kalvius · V. Tsurkan

Published online: 25 August 2011

Abstract The spinel $FeCr_2S_4$has been studied intensely for its peculiar magnetic and local structural changes which are sensitively influenced by the Jahn–Teller properties of Fe^{2+}in tetrahedral sulfur coordination. Recent muon spin rotation data give strong evidence that the commonly assumed collinear magnetic structure of this compound is only found between the Curie temperature $T_C = 165$ K and 50 K. For lower temperatures a helical structure has been proposed. We present new Mössbauer spectroscopic data taken on the same sample as used for muon spin rotation. Also the hyperfine spectra revealing non-equivalent iron sites support the appearance of a spin re-orientation around 50 K which may be related to the onset of short-range orbital order. Below 20 K severe dynamic broadenings are found which may indicate orbital fluctuations. Orbital order occurs around 11 K accompanied by severe changes in the crystalline electric field ground state as traced from quadrupole interaction.

LACAME 2010, Lima

J. Engelke · F. J. Litterst (✉)
Inst. Physik der Kondensierten Materie,
Technische Universität Braunschweig, 38106 Braunschweig, Germany
e-mail: j.litterst@tu-bs.de

A. Krimmel · A. Loidl · V. Tsurkan
Experimentalphysik V, Universität Augsburg,
86135 Augsburg, Germany

F. E. Wagner · G. M. Kalvius
Physics-Department, Technische Universität München,
85747 Garching, Germany

V. Tsurkan
Inst. Applied Physics, Acad. Sciences Moldova,
2028 Chisinau, Republic of Moldova

Keywords Mössbauer spectroscopy · Spin re-orientation · Orbital order · Jahn–Teller effect · Muon spin rotation

1 Introduction

The thiospinel $FeCr_2S_4$ has been studied since very early days of Mössbauer spectroscopy [1–3]. Magnetic order has been found below about 160 K [4]. Ferromagnetic coupling occurs between the B-site Cr. The A-site Fe^{2+} couples antiferromagnetically to the Cr sublattice resulting in ferrimagnetic order. $FeCr_2S_4$ has been re-investigated several times via Mössbauer spectroscopy for its peculiar behavior of hyperfine interactions which appear to be highly sensitive to sample quality also giving reason for controversial discussion [5–11]. From a more modern point of view the reason for this sensitivity can be sought in the competition between spin and orbital degrees of freedom of the Jahn–Teller ion Fe^{2+} in tetrahedral sulfur coordination. This has recently caused new interest in this spinel. Structural, caloric and magnetic studies give ample evidence for orbital order occurring below about 10 K [12].

Muon spin rotation experiments [13] clearly pick up the magnetic ordering around 165 K, yet with some partial order extending up to about 175 K. A single spontaneous rotation signal is indicative for the collinear ferromagnetic regime. Below about 50 K, however, other rotation frequencies are found which have been attributed to the appearance of an incommensurate modulated spin structure, probably of helical type. Notably there occurs a drop in magnetization in the same temperature range [12], yet no significant change in entropy is derived from heat capacity [12]. There is some indication from TEM for a structural change [14] and an anomaly in ultrasound velocity both occurring around 60 K [15].

Here we report on Mössbauer data recorded on the same sample material as used for muon spin rotation and also heat capacity in order to preclude any problems with varying sample quality. Some preliminary data have been included in [13]. We will concentrate on a gross interpretation of the magnetic hyperfine interaction, a detailed discussion related to the behavior of the electric field gradients and the spin dynamics close to the magnetic and orbital transitions will be presented elsewhere.

2 Experimental

The sample material has been prepared via a solid state reaction. For details of preparation and analytical details, see [12]. The ^{57}Fe Mössbauer absorption experiments have been performed in a standard spectrometer with sinusoidal velocity sweep using a ^{57}Co in Rh source at room temperature. The absorbers were kept under a static Helium atmosphere in a flow cryostat (Cryovac). Absorber temperature was controlled with a stability of about ± 0.1 K.

Spectra were least-squares fitted using a general Hamiltonian allowing the treatment of electric quadrupole and magnetic hyperfine interactions of comparable strengths and variable Euler angles.

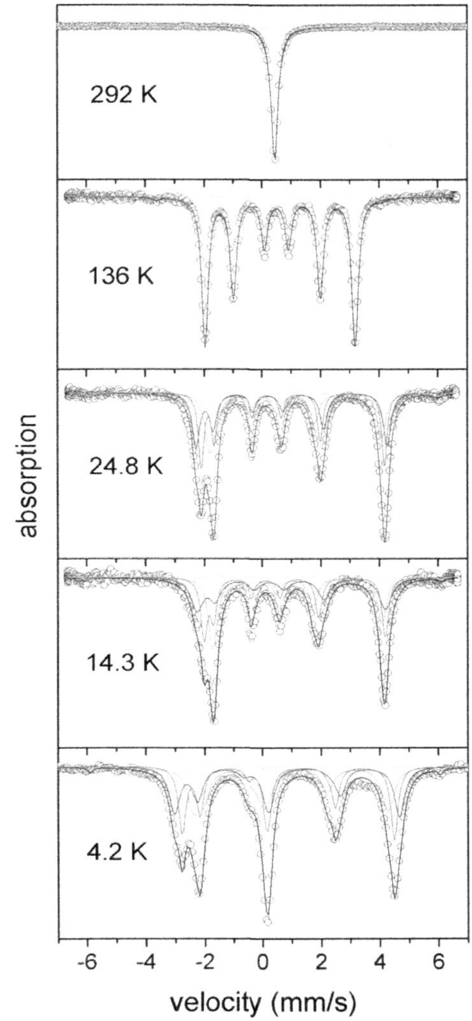

Fig. 1 ^{57}Fe absorption spectra of $FeCr_2S_4$ at various temperatures. Note that several sub-spectra are needed for the fit below about 50 K. The quadrupole doublet of very low intensity seen at room temperature and the corresponding weak magnetic pattern at low temperatures are related to a very small deviation from ideal stoichiometry

3 Results and discussion

In Fig. 1 we show a series of absorption spectra characteristic for the different temperature regimes. The spectra basically resemble in shape those reported earlier. Above $T_C = 165$ K we have a dominant single Lorentzian resonance line as expected for the ideal A-site tetrahedral sulfur coordination of Fe^{2+}, accompanied by a less than 5% of spectral area impurity signal, probably due to slight local deviations from stoichiometry [7]. In contrast to the results from muon spin rotation where partial long range order is found already above T_C, we find no indication for this from our Fe hyperfine spectra, i.e. this must rather be related to the Cr and not the Fe sublattice. Below T_C we find a 6-line pattern typical for the collinear ferromagnetic structure, yet a small quadrupole interaction due to an axially symmetric electric

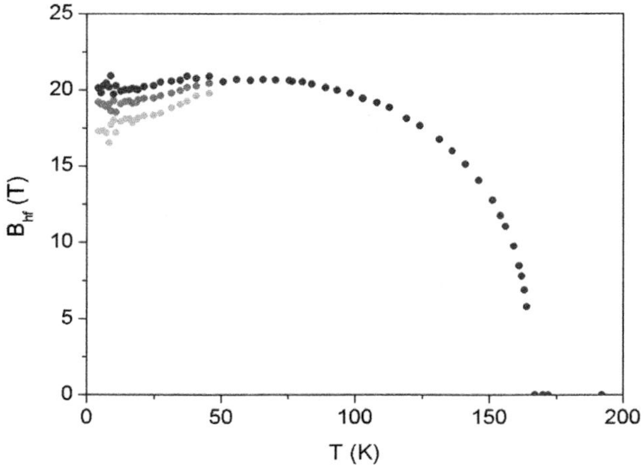

Fig. 2 Temperature dependence of hyperfine magnetic field B_{hf}

field gradient appears which is increasing with increasing magnetic hyperfine field as already described earlier [3]. The observation of a quadrupole interaction at iron below the Curie temperature despite neutron data prove cubic structure has been discussed controversially [1, 2, 5–11]. As first proposed by [2] this quadrupole interaction can be related to a magnetically induced electric field gradient. This may occur for the electronic ground state (E) of Fe^{2+} in tetrahedral crystalline electric field for certain orientations of the locally acting molecular magnetic field.

The variation of the magnetic hyperfine field B_{hf} with temperature is shown in Fig. 2. Its increase directly below T_C is typical for a second order magnetic phase transition and then follows a completely smooth sublattice magnetization curve down to about 50 K. A kink in B_{hf} around 130 K as indicated in [13] could not be reproduced.

Between about 50 K and 20 K the spectra can no more be described with a single magnetic site. We only could manage consistent fits using three sets of hyperfine interactions (see Fig. 2 for B_{hf}) having equal spectral weights standing for three magnetically and also structurally non-equivalent sites. The local symmetry is no more axially symmetric and an asymmetry parameter is necessary for describing the field gradient tensor which is in agreement with the TEM data of [14]. The non-equivalent magnetic sites found in our data give further support for the conclusions drawn from muon spin rotation regarding a change in spin structure from a collinear to a modulated one. Whether this is indeed helical with accompanying changes of the angles of B_{hf} vs. the electrical field gradient cannot reliably be resolved. We relate the spin re-orientation with the onset of short-range orbital ordering which is also indicated from X-ray data [12].

The gradual decrease of B_{hf} below 50 K is caused by continuous changes in the crystalline electric field. So there is no abrupt phase transition but a smooth change as also seen from heat capacity data [12].

Below 20 K we find strong line broadenings. This occurs in parallel to a maximum in mean-square displacement seen in X-ray data [12] and can be related to dynamical Jahn–Teller distortions. Finally below 10 K a dramatic change in quadrupole interaction becomes visible with a change of sign and re-orientation of the electric field gradient. Its major axis is now pointing perpendicular to the magnetic hyperfine field.

This has been described qualitatively already by [5]. We now managed quantitative fits of good quality, yet still three sub-spectra due to non-equivalent sites are necessary. The change of sign of electric field gradient indicates that the sequence of the lowest populated crystalline electric field states at Fe becomes modified upon orbital ordering.

From the temperature dependence of the turning of electrical field gradient we can define the orbital ordering transition for this sample with $T_{oo} = 10.6$ K in good agreement with specific heat data [12].

4 Conclusions

From our ^{57}Fe Mössbauer data we have found a sharp magnetic transition at $T_C = 165$ K indicating that the partial long-range order above T_C found by muon spin rotation [13] has to be related to Cr. A smooth variation of B_{hf} is traced from T_C down to about 50 K. There is a single magnetic Fe site consistent with a collinear ferrimagnetic structure. Below we find magnetically inequivalent Fe sites supporting spin re-orientation as proposed from muon data [13]. The sites are also structurally inequivalent which is in agreement with earlier TEM data [14]. This is interpreted with the onset of short-range Jahn–Teller distortions at Fe [12]. Orbital ordering occurs around 11 K accompanied by severe changes in the crystalline electric field ground state as traced from quadrupole interaction.

References

1. Hoy, G.R., Chandra, S.: J. Chem. Phys. **47**, 961 (1967)
2. Eibschutz, M., Shtrikman, S., Tenenbaum, Y.: Phys. Lett., A. **24** 563 (1967)
3. Hoy, G.R., Singh, K.P.: Phys. Rev. **172**, 514 (1968)
4. Shirane, G., Cox, D.E., Pickart, S.J.: J. Appl. Phys. **35**, 954 (1964)
5. Spender, M.R., Morrish, A.H.: Solid State Commun. **11**, 1417 (1972)
6. van Diepen, A.M., van Stapele, R.P.: Solid State Commun. **13**, 1651 (1973)
7. Lotgering, F.K., van Diepen, A.M., Olijhoek, J.F.: Solid State Commun. **17**, 1149 (1975)
8. Brossard, L., Dorman, J.L., Goldstein, L., Gibart, P., Renaudin, P.: Phys. Rev., B **20**, 2933 (1979)
9. Feiner, L.F., van Stapele, R.P.: Phys. Rev., B **22**, 2585 (1980)
10. Feiner, L.F.: J. Phys. C Solid State Phys. **14**, 1955 (1981)
11. Feiner, L.F.: J. Phys. C: Solid State Phys. **15**, 1495, 1515 (1982)
12. Tsurkan, V., Zaharko, O., Schrettle, F., Kant, Ch., Deisenhofer, J., Krug von Nidda, H.-A., Felea, V., Lemmens, P., Groza, J.R., Quach, D.V., Gozzo, F., Loidl, A.: Phys. Rev., B **81**, 184426 (2010)
13. Kalvius, G.M., Krimmel, A., Hartmann, O., Wäppling, R., Wagner, F.E., Litterst, F.J., Tsurkan, V., Loidl, A.: J. Phys. Condens. Matter. **22**, 052205 (2010)
14. Tsurkan, V., Barau, M., Szymczak, R., Szymczak, H., Tidecks, R.: Physica B **296**, 301 (2001)
15. Maurer, D., Tsurkan, V., Horn, S., Tidecks, R.: J. Appl. Phys. **93**, 9173 (2003)

Hyperfine Interact (2011) 202:63–71
DOI 10.1007/s10751-011-0339-z

Design and construction of an electromechanical velocity modulator for Mössbauer spectroscopy

A. A. Velásquez · A. Carmona · D. Velásquez · L. Ángel

Published online: 25 August 2011

Abstract In this paper we report the design, construction and characterization of an electromechanical velocity modulator for application in Mössbauer spectroscopy. The modulator was constructed with copper coils, Neodymium magnets, steel cores and polymeric membranes. The magnetic field in the driving and velocity sensing stages was analyzed by the finite element method, which showed a linear relation between the magnetic field in the region of motion of both coils and the position of the coils within the steel cores. The results obtained by computational simulation allowed us to optimize geometries and dimensions of the elements of the system. The modulator presented its first resonance frequency at 16.7 Hz, this value was in good agreement with that predicted by a second order model, which showed a resonant frequency of 16.8 Hz. The linearity of the velocity signal of the modulator was analyzed through an optical method, based on a Michelson–Morley interferometer, in which the modulator moved one of the mirrors. Results showed a satisfactory linearity of the velocity signal obtained in the sensing coil, whose correlation with a straight line was around 0.99987 for a triangular reference waveform.

Keywords Mössbauer spectroscopy · Velocity modulator · Finite element method · Optical calibration

A. A. Velásquez (✉) · A. Carmona
Grupo de Electromagnetismo Aplicado,
Universidad EAFIT, A.A 3300, Medellín, Colombia
e-mail: avelas26@eafit.edu.co

D. Velásquez · L. Ángel
Grupo de Óptica Aplicada, Universidad EAFIT,
A.A 3300, Medellín, Colombia

1 Introduction

One of the main components of any Mössbauer spectrometer is the velocity modulator, which moves the radioactive source with a controlled velocity in order to produce the Doppler change in the energy of the gamma rays, required to achieve the resonance with the nuclear states of Mössbauer isotopes present in the absorber. Some types of modulators have been reported in the literature, among them is the Doppler modulator system developed by Kankeleit [1], which is based on applied concepts of the electromagnetic theory, namely: Lorentz force for the driving coil and the Faraday's induction law for the velocity sensing coil. A more recent modulator, built with a piezoelectric actuator was developed by Molins et al. [2], whose main advantage is its small size, close to 3 cm, which gives the possibility to introduce the Mössbauer source and the absorber inside a cryostat when a low temperature measure is necessary, allowing to obtain better counting rates than those obtained when the source and absorber are located in opposite sides of the cryostat window. There are also some commercial electromechanical modulators such as the early model A.S.A K-3 Linear Motor developed by Austin Sciences Associates [3], the models MA-260, MA-260S and MVT-1000, developed by Wissel GmbH [4], the electromechanical transducer and sensing system of the MsAa-3 RENON Mössbauer Spectrometer [5]. Other important design to mention is the velocity modulator present in the mini spectrometer MIMOS II, sent to Mars in 2003, as one of the instruments of the Spirit and Opportunity robots, which is based on the electromechanical design proposed by Kankeleit [6]. This modulator moves two ^{57}Co sources, one of which is used for calibration of the Mössbauer spectrometer and the other is used to irradiate the samples present in rocks and soils. The success of the mission of MIMOS II in the Mars exploration evidences the convenience of the electromechanical systems for exigent operation conditions, including field applications for in situ analysis. No much information additional to the above mentioned about velocity modulators for Mössbauer spectroscopy is available, however electromechanical designs seem to be most popular due to its simple and stable architecture and known mathematic model, which can be well approximate for a second order system [7]. Taking in account the importance for developing functional and low cost systems, useful to make investigation in material science, we report in this article the design and construction of an electromechanical velocity modulator for Mössbauer spectroscopy, which incorporates some advantages of the rare earth magnets (NdFeB), such as its high *BH* product compared with that of AlNiCo and CuNiFe magnets of comparable dimensions, as well as functional, reproducible and low cost elements for the oscillating structure. The behavior of the magnetic field within the steel cores was simulated by the finite element method in order to test the linearity of the driving and sensing stages. The process of the design and characterization of the velocity modulator is described through this paper.

2 Design of the prototype

2.1 Internal architecture

The starting point of our design was the model reported by Kankeleit [1], which constitutes the basis for most electromechanical modulators reported in the literature.

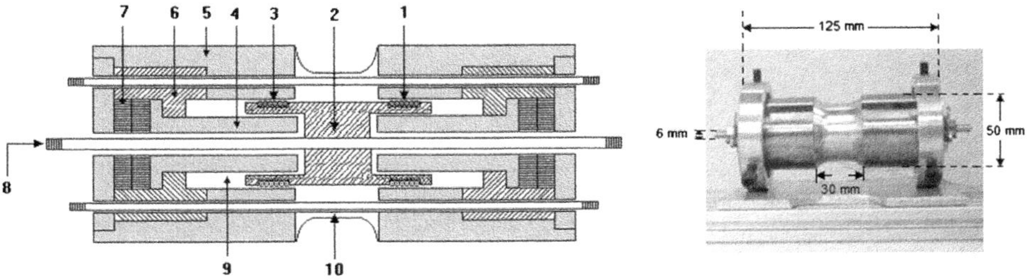

(a) (b)

Fig. 1 **a** Velocity modulator scheme: *1* sensing coil, *2* nylon support, *3* driving coil, *4* internal core, *5* external core, *6* aluminum bushing, *7* NdFeB magnets, *8* aluminum rod, *9* air cavity for the coil motion and *10* aluminum coupling housing, **b** View of the prototype

This design is based on a symmetrical structure of loudspeaker type, in which the actuator (driving stage) and the sensing stage (velocity stage) are similar and they are located in opposite ends of the structure. The main difference between driving and sensing stages is the electrical resistance of the coils, which is of the order of 1 Ω for the driving coil and 100 Ω for the sensing coil. Each coil moves through a steel core with cylindrical shape, which is magnetized by permanent magnets located concentrically with the core. The magnetic permeability of the steel core increases the intensity of the magnetic field $\boldsymbol{B}$ through the coils and gives a definite direction and magnitude to the magnetic field which crosses them. We used NdFeB magnets with disc shape, this choice obeys to the fact that the high product $\boldsymbol{BH}$ of this material gives us a magnetic field higher than that obtained with CuNiFe, AlNiCo and SmCo magnets of comparable dimensions [8], allowing to have a low current consumption for having a reasonable magnetic force on the driving coil, a convenient signal to noise ratio in the velocity sensing coil and a low size design.

2.2 Driving and sensing coils

The Fig. 1 presents the scheme and the external view of the velocity modulator. The oscillating structure is composed by driving and sensing coils as well as a cylindrical nylon support used for winding both coils. The driving coil was made with 120 turns of copper wire, reference AWG 36, which was wound onto an end of a nylon cylinder with thickness 2 mm. The sensing coil was made with 500 turns of copper wire, reference AWG 42. The electrical resistances of the driving and sensing coils were (8.6 ± 0.1) Ω and (250.4 ± 0.1) Ω respectively. Both coils oscillate inside an air cavity surrounded by an internal steel core and an external steel core, this last also serves as housing of the structure.

2.3 Driving and sensing cores

The architecture of driving and sensing cores is also sketched in Fig. 1a. Both structures are identical, they are composed of a solid axis of steel, reference ASI-SAE 1020, centered on a cylindrical cavity of the same material. An aluminum bushing was employed in order to keep concentric both steel structures, as well as to align the NdFeB magnets with the inner core. To analyze the linearity between the position

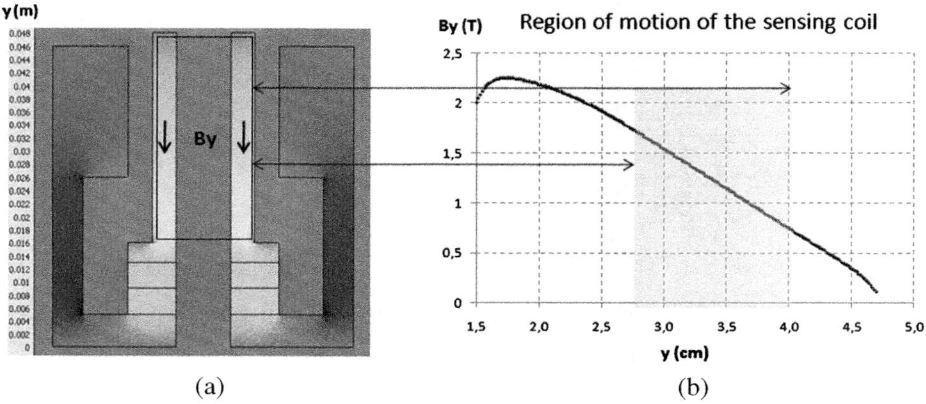

Fig. 2 **a** Region of the sensing core where the magnetic field B is calculated, **b** Relation between the magnetic field B and the position y parallel to the steel core axis

of each coil and the magnetic flux ϕ_B which crosses it, the magnetic field within the cores was simulated by the finite element method [9, 10]. This method calculates the magnetic field $\boldsymbol{B}$ in the region where the coil moves, limited by the boundary conditions imposed by us. In our case, the region of interest was a cylinder of 3 cm of length and 2 cm of diameter, which represents the portion of the internal core where the motion of the coils takes place. A view of the cross section where the magnetic field was calculated is presented in Fig. 2. The sensing coil moves in the interval [2.7, 4.0] cm, where the magnetic field $\boldsymbol{B}$ varies linearly with the y coordinate. A linear relation between $\boldsymbol{B}$ and $\boldsymbol{y}$ produces a linear relation between the voltage induced in the sensing coil, $V(t)$, and the instantaneous velocity $v(t)$ of this coil according to the Faraday's law:

$$V(t) = \alpha \frac{d\phi_B}{dt} = \alpha \frac{d(\beta y)}{dt} = \alpha\beta \frac{dy}{dt} = \delta v(t) \tag{1}$$

Where δ is the sensibility of the sensing coil, which depends on parameters such as its area, the number of turns and the magnetic properties of the core.

In the driving stage the force responsible for the oscillation of the moving structure is that due to the radial component of the magnetic field present in the air gap between the internal and the external cores. To analyze the linearity of this force with the current in the driving coil, the behavior of the radial component of the magnetic field in the air gap was also analyzed by the finite element method; the results are presented in Fig. 3. The component B_r varied less than 1% in the interval $y \in [2.7, 4.0]$ cm, where the driving coil motion takes place, this behavior gives a linear relation between the magnetic force resulting on the driving coil and the current flowing through it.

2.4 Permanent magnets

Permanent magnets of NdFeB with energy density $(BH)_{max} = 35$ MGOe and Curie temperature of 553 K were used in our design. The magnets have ring shape with

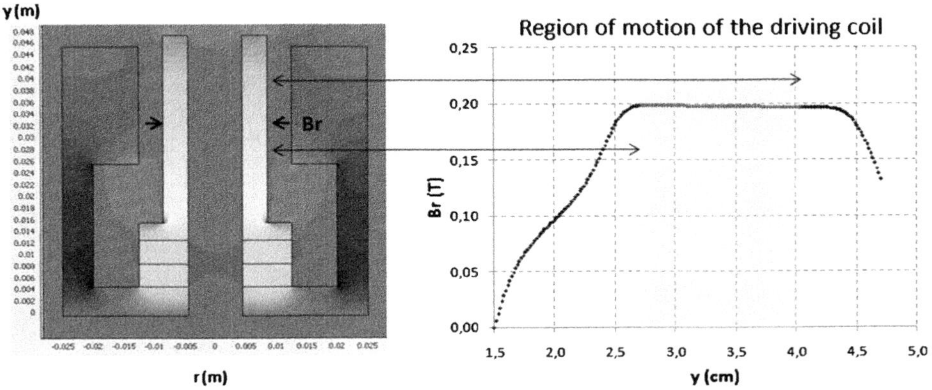

(a) (b)

Fig. 3 **a** Region of the driving core where the radial component B_r of the magnetic field is calculated, **b** relation between B_r and the position y

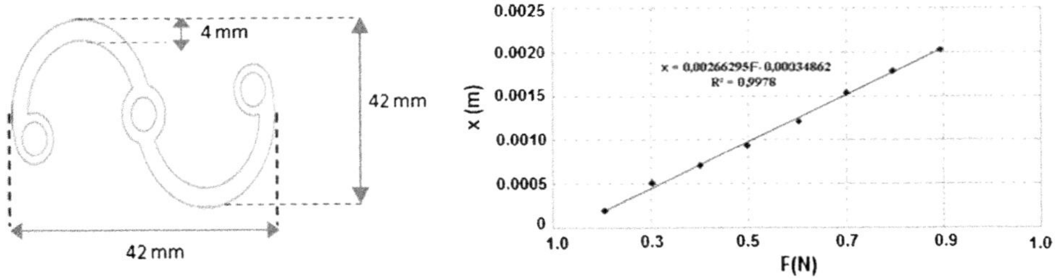

(a) (b)

Fig. 4 **a** Design of the polymeric membranes for the sustentation system of the oscillating structure, **b** elongation of the membranes vs. force applied to the aluminum rod

25 mm of outer diameter, 9 mm of inner diameter and 4 mm of thickness. The magnetic induction field had its maximal value of 0.24 T in the center of the magnet, which was measured with a teslameter PHYWE with an axial Hallsonde series 13610.01.

2.5 Polymeric membranes

The Fig. 4a presents the scheme of the elastic membranes used for the sustentation of the oscillating structure. The membranes were made of polyethylene terephthalate, with thickness 0.5 mm and density 1.425 g/cm^3.

The elastic constant of the membranes system was measured through a mechanical experiment, in which a variable mass was suspended of the lower end of a thread while the upper end of the thread was attached to one end of the aluminum rod. The small displacements of the rod were measured with a microscope. Data of rod displacement vs. applied force are presented in Fig. 4b. A linear fitting of the

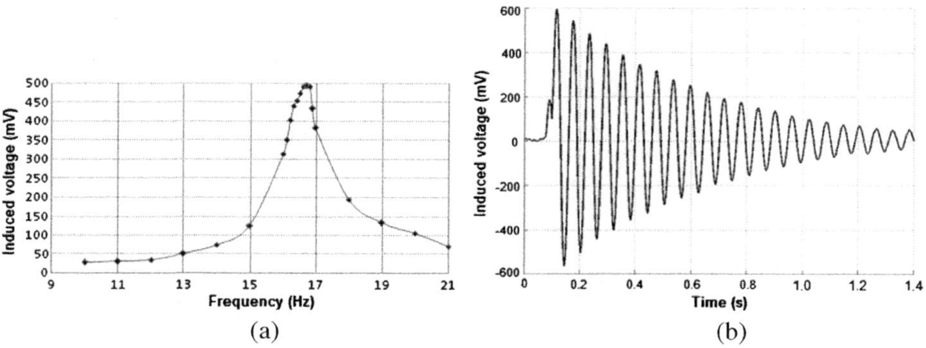

Fig. 5 **a** Frequency response of the velocity modulator, **b** impulse response

experimental data gave the value $k_1+k_2 = (376 \pm 4)$ N/m for the elastic constant of the membranes system.

3 Characterization of the velocity modulator

3.1 Resonance frequency

Resonance frequency of the modulator was measured by stimulating the driving coil with a sinusoidal signal with 50 mV of amplitude, provided by a wave generator Agilent 33220A-20 MHz. The signal induced in the sensing coil was recorded with a digital storage oscilloscope Tektronicx TDS 3052B-500 MHz. The graph of amplitude of the signal induced in the sensing coil vs. frequency of the stimulus signal is presented in Fig. 5a, from this graph a resonance frequency of 16.7 Hz was found.

3.2 Impulse response

In order to have an idea of the mathematical model of the velocity modulator in terms of its transference function, a mechanical impulse was given to the system, which consisted in a brief shove on the aluminum rod. The signal induced in the sensing coil was recorded with the storage oscilloscope and presented in Fig. 5b, the response obtained was characteristic of a second order system with a damping term. According to the impulse response function, the mathematical model of the system can be approximate by the equation:

$$m\frac{d^2x}{dt^2} + b\frac{dx}{dt} + (k_1 + k_2)\,x = f\,(t) \tag{2}$$

Where $m = (33.6 \pm 0.1)$ g is the mass of the oscillating structure, b is the damping parameter; $k_1 + k_2$ is the elastic constant of the membranes system and $f(t)$ is the impulsive force communicated to the rod. The damping parameter b was estimated from the equation of the envelope of the oscillating signal presented in Fig. 5b. The envelope was well fitted by an exponential decreasing function, from which the value

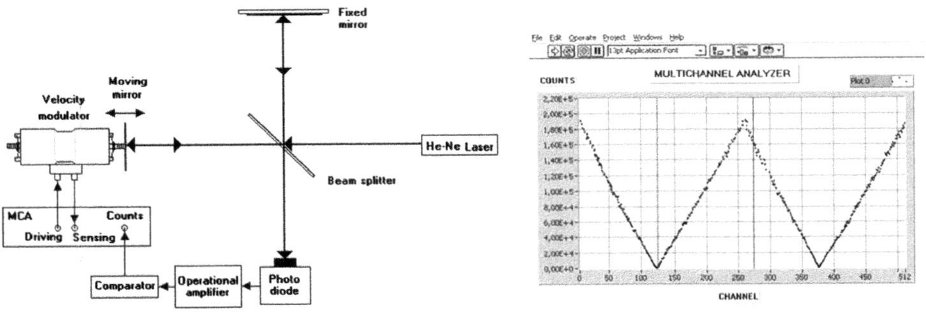

(a) (b)

Fig. 6 **a** Scheme of the Michelson–Morley interferometer used to measure the velocity signal of the modulator, **b** Number of counts vs. channel registered by the multichannel analyzer

$b = (0.296 \pm 0.004)$ kgs^{-1} was obtained. According to the mathematical model (2) the resonance frequency of the system is $f_R = (16.8 \pm 0.1)$ Hz. The discrepancy between the measured resonance frequency and the frequency predicted by the mathematical model is 0.6%, which shows that the velocity modulator can be well described by a second order linear system.

3.3 Linearity of the velocity signal

The linearity of the velocity modulator was analyzed through an absolute velocity calibration method, which has been reported in the literature [11–14]. This method is based on a Michelson–Morley Interferometer, in which the first mirror is fixed to one arm of the interferometer while the second mirror is clamped to the oscillating rod of the velocity modulator. A coherent laser beam splits in two beams when it pass through a beam splitter, the first beam is directed towards the fixed mirror and the second one is directed towards the moving mirror clamped to the modulator rod. The reflected beams are again combined in the beam splitter and its interference pattern is observed on a screen where a photodiode is located. The rings of the pattern alternate their intensity when the mirror moves a length of $\lambda/4$, being λ the wavelength of laser light used. The rod of the modulator oscillated according to a triangular reference signal of 6 Hz and 512 channels programmed in our multichannel analyzer (MCA). The basic architecture of the interferometer is schematized in Fig. 6a. For the experiment we used a He–Ne laser with 20 mW of power and $\lambda = 632.8$ nm, a beam splitter cube with 50%–50% transmitted intensity, a fixed mirror with $\lambda/4$ flatness, a photodiode ODD-660W with 1 ns of response time and spectral sensibility bandwidth 600–700 nm, and a counting fringe system designed in our laboratory.

The photodiode was located in the center of the pattern and the pulses detected were converted to logical TTL pulses through a voltage comparator circuit based on a high speed operational amplifier AD8051 with 110 MHz bandwidth and 145 V/μs slew rate. Logical pulses were sent to the counting module of the multichannel analyzer, which stored the counts produced in each channel of the reference velocity signal of the velocity modulator. The graph of the counts vs. channel in which they are registered is presented in Fig. 6b. A linear fit of the data in the interval of channels

128–256 showed a correlation coefficient $R^2 = 0.99987$. The fitting of the data in the interval of channels 256–384 showed a correlation coefficient $R^2 = 0.99985$. From the graph, the magnitude of the velocity of the oscillating structure in the velocity channel i can be obtained from the equation [11]:

$$v_i = \frac{\lambda N_i}{2nt_{dwell}} \tag{3}$$

Where λ is the wavelength of the laser light used, N_i is the average number of logical pulses per cycle registered by the MCA in the channel i, $n$is the refraction index of the media and t_{dwell} is the time interval during which the counts are stored for each channel of the reference signal. The parameters of our experiment were: $\lambda = 632.8$ nm, $n \approx 1.0$ and $t_{dwell} = (162.0 \pm 0.5)$ μs. Counts were registered for a time of 100 min, during this time 197,010 pulses were registered in the channel 256, giving an average value $N_{256} = 5.5$. With these parameters we obtained v = (10.69 ± 0.01) mm/s, which is the maximal velocity of the modulator for the conditions of the experiment.

4 Conclusions

We have designed a velocity modulator apt for use in Mössbauer spectroscopy, whose linearity deviation was better than 0.02% over a velocity scale range from −10.69 mm/s to 10.69 mm/s, for a triangular reference signal. The system implements very functional, accessible and reproducible components, such as Neodymium magnets, polymeric membranes, copper coils and steel cores. The simulation of the system by the finite element method allowed us to find convenient geometries and dimensions for steel cores, magnets and coils in order to have a linear response of the driving and sensing stages of the velocity modulator. The system was well described by a linear second order mathematical model, with a resonance frequency of (16.7 ± 0.1) Hz. The presented design is an attractive choice for the users of the Mössbauer spectroscopy technique, who are interested in the design and construction of functional and low cost instrumentation system for their laboratories.

References

1. Kankeleit, E.: Feedback in electromechanical drive system. Mössbauer Eff. Methodol. **1**, 47 (1965)
2. Casas, L.L., Molins, E., Roig, A.: Miniaturization of a Mössbauer spectrometer using a piezo-transducer and a solid state detector. Hyperfine Interact. **141/142**, 125 (2002)
3. Austin Science Associates, Inc, A.S.A Mössbauer spectrometer: K-3 linear motor, instruction manual. Austin, TX (1997)
4. Wissel. Mössbauer driving system 260: Instruction Manual. Germany (1996)
5. Górnicki, R., Blachowski, A., Ruebenbauer, K.: In: Proceedings Nukleonika, vol. 52(1), p. S7 (2007)
6. Klingelhöfer, G., Bernhardt, B., Foh, J., Bonnes, U., Rodionov, D., De Souza, P.A., Schröder, Ch., Gellert, R., Kane, S., Gütlich, P., Kankeleit, E.: The miniaturized Mössbauer spectrometer MIMOS II for extraterrestrial and outdoor terrestrial applications: a status report. Hyperfine Interact. **144/145**, 371 (2002)
7. Yadagiri Reddy, P., Narayan Reddy, P., Seshakumari, Ch., Rama Reddy, K.: On the transducer function of an electromechanical transducer an application to the Mössbauer velocity drive. Nucl. Instrum. Methods. **213**, 381–385 (1983)

8. Herbst, J.F., Croat, J.J.: Neodymium–iron–boron permanent magnets. J. Magn. Magn. Mater. **100**, 57 (1991)
9. Jianming, J.: The Finite Element Method in Electromagnetics. Wiley, New York (1993)
10. COMSOL: COMSOL Multiphysics, AC/DC Module User's Guide, Version 3.5: http://www.ewp.rpi.edu/hartford/~lemcon/Comsol_terminalvelocity.pdf. Accessed 10 Feb 2010
11. Kwater, M.: Laser calibration of the Mössbauer spectrometer velocity: improvements in design. Hyperfine Interact. **116**, 53 (1998)
12. Yoshimura, T., Syoji, Y., Wakabayashi, N.: Absolute velocity control of a Mössbauer spectrometer by utilizing a laser interferometer. J. Phys. E: Sci. Instrum. **10**, 829–833 (1977)
13. Player, M.A., Wodhams, FWD.: An improved interferometric calibrator for Mössbauer spectrometer drive systems. J. Phys. E: Sci. Instrum. **9**, 1148 (1976)
14. Vandenberghe, R.E.: Mössbauer Spectroscopy and Applications in Geology. International Training Centre for Post-Graduate Soil Scientists, p. 25. State University Gent, Belgium (1990)

Hyperfine Interact (2011) 202:73–79
DOI 10.1007/s10751-011-0340-6

Effects of particle size on the structural and hyperfine properties of tin dioxide nanoparticles

F. H. Aragón · R. Cohen · J. A. H. Coaquira · G. V. Barros · P. Hidalgo · L. C. C. M. Nagamine · D. Gouvêa

Published online: 25 August 2011

Abstract In this work, we report on the study of SnO_2 nanoparticles prepared by a polymer precursor method. X-ray diffraction (XRD) data analysis evidenced the formation of only the tetragonal rutile-type phase for the as-grown and thermally annealed samples. A mean grain size of about 11 nm for the as-prepared sample has been determined. This mean size increases after the thermal annealing and with the annealing temperature. The room temperature Mössbauer spectra (MS) were well fitted using a quadrupole splitting (QS) distribution. The isomer shift (IS) tends to increase when the grain size decreases. That increase has been associated to the extra s-electron density generated by the oxygen vacancies.

Keywords X-ray diffraction · Mössbauer spectroscopy · SnO_2 nanoparticle

F. H. Aragón · J. A. H. Coaquira (✉) · G. V. Barros
Núcleo de Física Aplicada, Instituto de Física,
Universidade de Brasilia, Brasilia, DF 70910, Brazil
e-mail: coaquira@unb.br

R. Cohen · L. C. C. M. Nagamine
Instituto de Física, Universidade de São Paulo,
São Paulo, SP 05508-090, Brazil

P. Hidalgo
Faculdade Gama- FGA, Sector Central Gama, Universidade de Brasília,
Brasília, DF 72405-610, Brazil

D. Gouvêa
Department of Metallurgical and Materials Engineering, Escola Politécnica,
Universidade de São Paulo, São Paulo, SP 05508-900, Brazil

1 Introduction

In recent years, physical properties of semiconducting oxides SnO_2, ZnO, WO_3 and In_2O_3, have been studied in connection with a search for new materials for solid-state gas sensors [1]. Nowadays, materials with an average grain size of less than 50 nm have attracted considerable scientific interest because of the improvements in a variety of properties related to the grain size. Among those compounds, nanoparticulated SnO_2 is used most extensively since the conductivity of SnO_2 is very sensitive to the surface state in the temperature range of 300–800 K, in which redox reactions can take place. On the other hand, the surface of SnO_2 nanoparticles exhibits good absorption properties and reactivity due to the presence of free electrons in the conduction band and the presence of surface and bulk oxygen vacancies and active chemisorbed oxygens [1]. A main drawback of SnO_2-based chemical sensors is their low selectivity, which does not allow one to separate the contribution made by a particular type of molecules in the gas phase to the total electric signal. In order to improve that selectivity, SnO_2 nanoparticles are doped with transition metals [2, 3], and that doping introduces a strong effect on the electronic and catalytic properties of the surface. It is known that the generation of oxygen vacancies associated with the particle size reduction or to the charge compensation introduces free electrons to the nanoparticulated system. Therefore, the investigation of properties related with the particle size and their size dependence is important from the technological point of view as well as for scientific interest. In this work, the structural and hyperfine properties of SnO_2 nanoparticles of different sizes (from 10 to 45 nm) are reported and discussed.

2 Experimental details

SnO_2 nanoparticles were produced by a polymer precursor method (this method was also known as Pechini's method), in which tin dioxide is obtaining from the precursor $SnCl_2.H_2O$. Details of the sample preparation are published elsewhere [3]. In order to obtain samples with different particle sizes, the samples were submitted to isothermal annealing at different temperatures in air environment (at 500°C, 700°C, 800°C and 900°C) for two hours. The phase formation and mean grain size have been determined by XRD using a Bruker diffractometer, model D8 Advanced with Cu-$K\alpha$ radiation. Mössbauer spectroscopy measurements were carried out with a constant acceleration spectrometer and using a $Ca^{119}SnO_3$ as the radiation source. For calibration, a natural Sn foil was used as absorber and the calibration error was estimated to be less than $\approx$0.5%. All spectra was processed at room temperature and the isomer shift (IS) is with respect to the $CaSnO_3$ source. The analysis of the spectra was obtained with a least-square fitting routine assuming a Lorentzian lineshape and considering the hyperfine splitting of I=3/2, 1/2 nuclear levels in the usual way.

3 Results and discussion

The XRD patterns of SnO_2 nanoparticles were analysed using the Rietveld refinement method. The analysis of these data indicate the formation of the cassi-

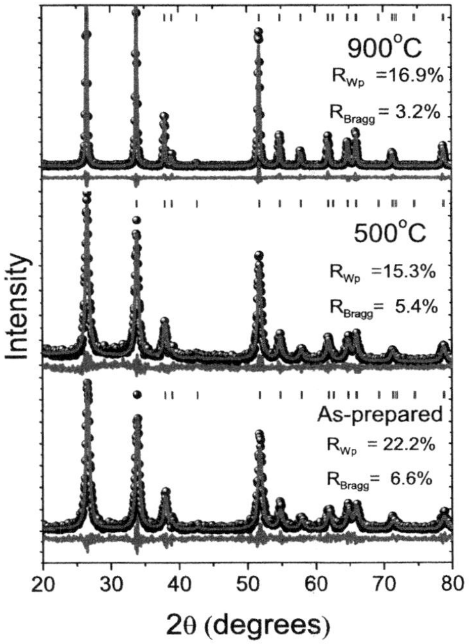

Fig. 1 Room temperature XRD patterns of SnO_2 samples submitted to different annealing temperaturs. The observed and calculated data are represented by the *points* and *solid lines*, respectively. The *green solid line* represents the difference between the observed and calculated data. The R values are included in the plots

terite phase, with no evidence of other crystalline or amorphous phase. XRD patterns of samples annealed at different temperatures are shown in Fig. 1, the linewidth (full width at half maximum) of the diffraction peaks tends to decrease as the annealing temperature increases. This effect can be attributed to the increase of the particle size and/or to changes in the residual strain.

In order to obtain additional information, the Rietveld refinement was carried out using the Thompson-Cox-Hastings pseudo-Voigt function (TCH-pV) given by [4]:

$$TCH - pV = \eta L + (1 - \eta)G \tag{1}$$

where, L and G represent the Lorentzian and Gaussian peak functions respectively, and η is a mixing parameter. The line broadening related to the instrumental contribution was corrected by adequately subtracting the linewidth of a standard sample (Si single crystal) from the linewidth of the samples. The deconvolution of the TCH-pV provides the linewidths for the Gaussian and Lorentzian components, which are given by:

$$H_G = (U\tan^2(\theta) + V\tan\theta + W + Z/\cos^2\theta)^{1/2} \tag{2}$$

$$H_L = X\tan\theta + Y/\cos\theta \tag{3}$$

where U, V, W, Z and Y are refinable parameters and can be used to determine the mean crystallite size and microstrain [4]. The mean crystallite size, residual strain and some structural parameters are shown in Table 1.

Table 1 Mean particle size, residual strain and some structural parameters obtained from the Rietveld refinement of thermally annealed SnO_2 nanoparticles. The R_{Bragg}, which determines the quality of the refinement, are also included. The number between parentheses are the uncertainties

Annealing temp. (°C)	Mean size (nm)	Residual strain (%)	a/c	$x = y$	R_{Bragg} (%)
0	11(3)	0.205	1.487	0.309	6.6
500	12(3)	0.085	1.487	0.302	5.0
700	17(3)	0.119	1.486	0.305	3.0
800	34(3)	0.080	1.487	0.304	5.0
900	45(3)	0.112	1.487	0.305	3.1

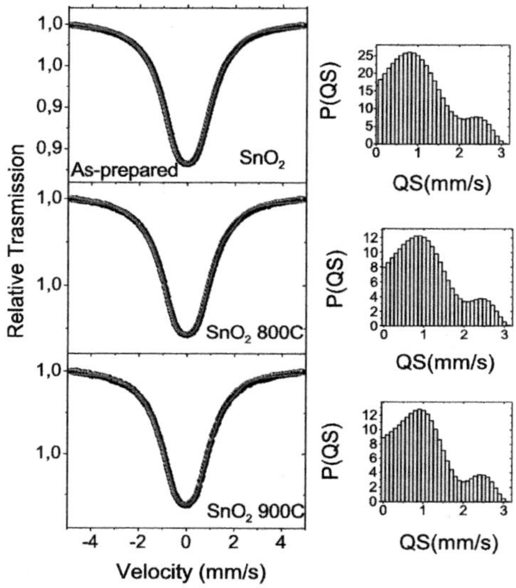

Fig. 2 Room-temperature Mössbauer spectra of SnO_2 nanoparticles. The *points* represent the experimental data and the *lines* represent the fit to a distribution of QS. The QS distributions are shown at the right-hand side

An increase in the annealing temperature drives to an increase in the particle size. More detailed analysis of the particle growth process will be published elsewhere. The value of the residual strain determined for the as-prepared sample is larger than those values obtained for the annealed ones. This result is in agreement with the larger value of the internal parameter (x) that determines the position of oxygen atoms which form the octahedron. Larger values of x means a more distorted octahedron, since an ideal octahedral coordination requires $x = 0.2929$ [5]. The R_{Bragg} values quantify the quality of the refinement (acceptable values are in the range 3–9%).

In Fig. 2 are shown the room-temperature Mössbauer spectra (MS) for the as-prepared sample (no annealed), and thermally treated samples at 800 and 900°C. Within the experimental resolution, all spectra seem to show no magnetic components and the preliminary analysis indicates that the whole set is well resolved by fitting with a single doublet. However, the linewidth of the spectrum of the as-prepared sample is ~ 1.7 mm/s. This linewidth is larger than that expected for bulk SnO_2 and suggests the presence of a distribution of quadrupole splitting (QS) in

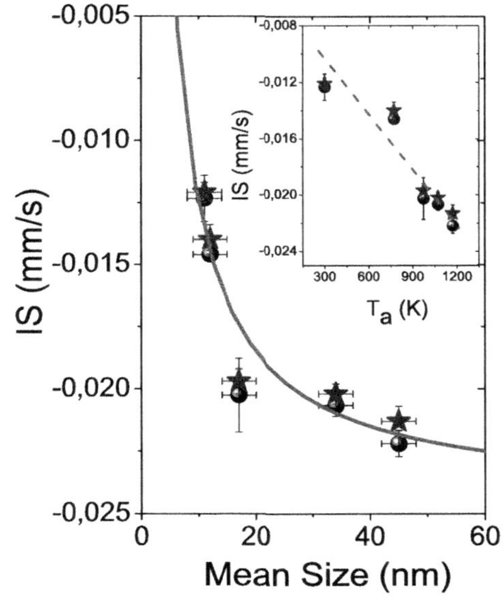

Fig. 3 Dependence of the isomer shift (IS) on the mean particle size. The *inset* shows a linear decrease of the IS as a function of the annealing temperature. The *solid spheres* are the IS values determined from the fit of the MS to a QS distribution and the *stars* are the IS values obtained from the fit to only one doublet. The *solid line* represents the fit made using the Eq. 4

our samples [6]. The fit of the Mössbauer spectra with a distribution was carried out using the histogram method of the NORMOS DIST program. The results are shown in Fig. 2. For each sample, the QS's distributions show assymetrical shapes with a broad peak centered at lower QS values. An isomer shift (IS) of $\sim$ 0.011 mm/s and an average value for the QS distribution of +0.83 mm/s are obtained for the as-prepared sample. These values are slightly larger than those reported for bulk SnO_2 [7, 8]. When the annealing temperature is increased the average value of QS shows a statistical fluctution and no clear tendency has been determined. Moreover, the IS values show a decreasing behavior when the annealing temperature is increased as shown in the inset of Fig. 3. Since higher annealing temperature produces larger particle size, as determined from XRD analysis and corroborated by transmission electron microscopy images, the IS tends to decrease as the particle size grows. This behavior is shown in the main panel of Fig. 3. Since the relative variation of the IS as a function of the particle size is small, the reproducibility has been tested by measuring at least twice each sample. The experimental points plotted in Fig. 3 are the average values of IS.

The size dependence of IS is unexpected since no size dependence of the IS was reported in nanoparticulated systems [9]. However, preliminary studies of Ni-doped SnO_2 nanoparticles [10, 11] suggest the presence of oxygen vacancies, even for the undoped system, which are likely located at the particle surface. It is known that when oxygen vacancies are generated in the SnO_2 system, electrons are transformed from p character to s character by exciting them from the valence band into the conduction band [8, 12]. Therefore, oxygen vancancies provide with extra s-electrons to the nearby Sn nucleus. As the isomer shift is a direct measure of the s-electron density at the nucleus, the increase in the oxygen vacancies induces an increase in the IS. Assuming that the measured IS is composed by two terms: (i) A contribution associated with the s-electron density as in the bulk system and (ii) a contribution coming from the s-electron density related to oxygen vacancies. The first contribution must be related to the core region of the particle, wherein no defects are assumed to exist, and the second contribution must be related to the grain surface region wherein the oxygen vacancies are located. This core-shell model has been used previously to interpret the Mossbauer data of nanoparticulated SnO_2 [13]. Accordingly, the total IS is given by:

$$IS = IS_{\text{surf}} + IS_{\text{bulk}} \tag{4}$$

As the IS_{surf} contribution is proportional to the density of oxygen vacancies and that density is proportional to the surface-to-volume ratio, we propose that $IS_{\text{surf}} = K/D$, where K is a constant and D is the diameter of the particle. Using the mean particle size determined from XRD analysis, we determined the following values: $IS_{\text{bulk}} = -0.024(1)$ mm/s and $K = 0.120(6)$ mm.nm/s. In Fig. 3 it is shown the data fit. The IS_{bulk} value is in agreement with reported values for bulk SnO_2 system [8]. It is worth mentioning that the density of oxygen vacancies strongly depends on the atmosphere the experiment is carried out. Since our experiments were carried out in air, which is considered as a strongly oxidizing atmosphere, the stoichiometric surface of SnO_2 has 5-coordinated Sn atoms which in air are sites for oxygen absorption. It means that the number of oxygen vacancies is relatively small and concentrated mainly in subsurface sites [10]. This must be the reason for the small relative variation of the IS as a function of the particle size. Moreover, additional Mössbauer measurements, carried out under different oxygen gas flow, are needed in order to elucidate this issue. The effect of oxygen vacancies is to modify the neighborhood of Sn atoms, introduce distortions, and lead to a decrease in the ionic character of the Sn^{4+} state in the $Sn-O$ bond [13]. Since the number of oxygen vacancies is expected to be small, it implies to have a very small population of tin atoms near to an oxygen vacancy. Therefore, Sn^{4+} atoms have about the same nearest surrounding, which implies to have almost the same quadrupole splitting, even when the particle size is changed. However, additional studies are required in order to confirm the effects of the oxygen vacancies on the hyperfine parameters IS and QS. Low temperature Mössbauer measurements should help us to improve our assessment since smaller linewidths are expected. Those measurements are currently underway and the results will be published elsewhere.

4 Conclusion

XRD analysis of thermally annealed SnO_2 powders indicate the formation of the tetragonal rutile-type phase. A higher annealing temperature produces a larger grain size growth. Room-temperature Mössbauer spectra are well-resolved by a distribution of QS's. The IS value for the smaller particle size is larger than that expected for bulk SnO_2 and tends to decrease as the particle size increases. It suggests the presence of an additional contribution to the IS which is associated with the oxygen vacancies, which are generated at the surface of the particle.

Acknowledgements The authors thank the Brazilian agencies CAPES, CNPq and FAPDF for the financial support.

References

1. Runyantselva, M.N., Safonova, O.V., Boulova, M.N., Ryanbova, L.I., Gaskov, A.M.: Russ. Chem. Bull. **52**, 1217 (2003)
2. Aragón, F.H., Coaquira, J.A.H., Hidalgo, P., Brito, S.L.M., Gouvêa, D., Castro, R.H.R.: J. Non-Cryst. Solids **356**, 2960 (2010)
3. Hidalgo, P., Castro, R.H.R., Coelho, A.C.V, Gouvêa, D.: Chem. Mater. **17**, 4149 (2005)
4. Paiva-Santos, C.O., Cavalheiro, A.A., Cilense, M.A., Varela, J.A., Silva, M.T., Mascarenhas, Y.P.: Adv. X-ray Anal. **44**, 38 (2000)
5. Robertson, J.: J. Phys. C: Solid State Phys. **12**, 4767 (1979)
6. Aragón, F.H., Coaquira, J.A.H., Candela, D.S., Baggio Saitovitch, E., Hidalgo, P., Gouvêa, D., Morais, P.C.: J. Phys.: Confer. Ser. **217**, 012079 (2010)
7. Grzeta, B., Tkalcec, E., Goebbert, C., Takeda, M., Takahashi, M., Momura, K., Jaksic, M.: J. Phys. Chem. Solids **63**, 765 (2002)
8. Stjerna, B., Granqvist, C.G., Seidel, A., Häggströn, L.: J. Appl. Phys. **68**, 6241 (1990)
9. Bødker, F., Mørup, S.: Europhys. Lett. **52**, 217 (2000)
10. Aragón, F.H., Coaquira, J.A.H., Hidalgo, P., da Silva, S.W., Brito, S.L.M., Gouvêa, D., Morais, P.C., Raman, J.: Spectroscopy (2010). doi:10.1002/jrs.2802
11. Aragón, F.H., Coaquira, J.A.H., Hidalgo, P., Brito, S.L.M., Gouvêa, D., Castro, R.H.R.: J. Phys., Condens. Matter **22**, 496003 (2010)
12. Cox, D.F., Fryberger, T.B., Semancik, S.: Phys. Rev. B **38**, 2072 (1988)
13. Shek, C.H., Lin, G.M., Lai, J.K.L.: Nanostruct. Mater. **11**, 831 (1999)

Hyperfine Interact (2011) 202:81–87
DOI 10.1007/s10751-011-0341-5

Mössbauer studies of Haltern 70 amphorae from Castro do Vieito, North of Portugal, and of amphora sherds from kilns in the Roman provinces Hispania Baetica and Lusitania

B. F. O. Costa · A. J. M. Silva · F. E. Wagner · U. Wagner

Published online: 30 September 2011

Abstract Haltern 70 amphora sherds from Castro do Vieito and from kilns in the Roman provinces Baetica and Lusitania were studied by Mössbauer spectroscopy, EDX and XRD. An amphora whose handle was carefully studied must have been fired reducingly at a temperature of or above about 900°C and cooled in oxidizing conditions in the end. Oxidation and hematite formation went hand in hand, and the amount of hematite formed depended sensitively of the speed of penetration of the oxygen and the cooling rate in the interior of the ceramic body.

Keywords Haltern 70 amphorae · Mössbauer spectroscopy · X-ray diffraction · X-ray fluorescence analysis

1 Introduction

Haltern 70 is the name of a particular type of amphorae that were produced in the Hispania Baetica Roman province during the second half of the first century BC and the first century AD. Recently, the excavations of Castro do Vieito (Northwest of Portugal, Rio Lima estuary), an indigene settlement from the beginning of the Roman occupation of the region during the early Imperial Period, provided the most extended collection of Haltern 70 amphorae of all the roman world. The origin of

B. F. O. Costa (✉)
CEMDRX, Physics Department,
University of Coimbra, 3004-516 Coimbra, Portugal
e-mail: benilde@ci.uc.pt

A. J. M. Silva
Centro de Estudos Arqueológicos das Universidades de Coimbra e do Porto/CAM,
Palácio de Sub-Ripas, 3004-395 Coimbra, Portugal

F. E. Wagner · U. Wagner
Physics Department, Technical University of Munich, 85747 Garching, Germany

this type of ceramics is still under discussion, although their production places are usually assigned to Baetica (several locations) or to Lusitania, both Roman provinces of Hispania. In previous papers [1, 2] we have started an archaeometric study on Haltern 70 amphora sherds from Castro do Vieito, and from the Roman provinces Hispania Baetica and Lusitania. Traditionally the amphorae were produced locally and used for the shipment of agricultural products.

Here we report mainly on results of Mössbauer spectroscopy (MS) obtained at ambient temperature and 4.2 K for amphora sherds of the Haltern 70 type from Castro do Vieito, which have previously been studied only at ambient temperature. For comparison, we engaged in Mössbauer studies of amphora sherds found at a number of kiln sites from the first century AD at various locations in the Hispania Baetica and Lusitania provinces. Some X-ray diffraction (XRD) and energy dispersive X-ray analyses (EDX) will also be presented.

2 Experimental

The sherds were ground in a mortar and Mössbauer absorbers of about 200 mg/cm^2 thickness in lucite holders were prepared. ^{57}Fe Mössbauer spectra were recorded in transmission geometry using a standard spectrometer with a sinusoidal velocity waveform. Measurements at 4.2 K were carried out in a liquid He bath cryostat. The source of about 25 mCi of ^{57}Co in a Rh matrix was always at the same temperature as the absorber. The spectra were least-squares fitted with appropriate superpositions of Lorentzian lines.

X-ray diffraction (XRD) measurements were performed at room temperature (RT) using Cu K_α radiation ($\lambda = 0.154184$ nm), and scanning electron microscopy (SEM) combined with energy dispersive X-ray analysis (EDX) was used on fracture surfaces, as described in [1, 2].

3 Results and discussion

Twelve Haltern 70 sherds obtained from the excavation of Castro do Vieito and twelve sherds found in kilns of Baetica and Lusitania were studied. References [1, 2] report previous studies. The Castro do Vieito specimens were reddish brown in color, were no more than 10 cm long and all exhibited a slight curvature. The sherds from kilns were found as large fragments of amphorae, like rims, handles and foot ends. The archeological codes for the items are used in this paper. The sherds found in excavations from Castro do Vieito are rather soft and could easily be fractured.

In all the samples, the most significant feature is the sandy texture, sometimes with big inclusions of silicates. EDX results show the presence Si content of all samples is inferior or equal to 25 wt%. This confirms that sand was added to the raw clay, probably as a temper [3]. In Vieito sherds as well as in sherds from kilns of S. Lourenço and Olhos (Lusitania), Arva and Orippo (Baetica) the Si concentration is about 20%. As observed by EDX measurements, the concentration of Ca in all samples is relatively low, being less than 5 wt%. Vieito sherds have less than 1 wt% Ca, in spite of not being tough. The analyses of Vieito sherds also reveal that almost no Na is present in the samples, and that the Al concentrations are about 9 wt%. This

Fig. 1 Sample Olhos_26, a fragment of an amphora handle from a kiln in Lusitania

is also observed in sherds from Baetica like Arva, Orippo (Medium Guadalquivir River) and Puerto Real (Bay of Cadix) as well as from Lusitania like Olhos (Mouth of the Guadiana River).

All major inclusions were identified as quartz or feldspar. In the quartz (SiO_2) no impurities were detected. The feldspar exhibits the lines of Si, Al, Fe and Ca.

The XRD measurements confirm the EDX results. Quartz (SiO_2) is the main component of all the samples. All of them show traces of other crystalline phases, usually of low concentration and difficult to identify. In Vieito samples and some samples from kilns of Baetica (e.g., Puerto Real, Arva, Pinguele) and Lusitania like Olhos, pyroxene was found. The poor crystallinity and/or low concentration of Fe do not allow the detection of this element by XRD.

About 20 samples were studied by Mössbauer spectroscopy. The recorded spectra are quite uniform. The room temperature spectra mainly exhibit a quadrupole doublet of ferric iron, a weaker doublet of ferrous iron and minor magnetic contribution, mainly of hematite. The 4.2 K spectra of all amphorae studied so far exhibit mainly a broad magnetically split pattern that is very similar to those observed recently in Celtic pottery [4] and attributed to a vitreous phase forming under reducing firing conditions.

All samples except a sample from a kiln in Lusitania, Olhos_26, exhibit only one hue of color. Olhos_26 (Fig. 1) is an oval section through a handle (large diameter about 5 cm) that exhibits differently colored rims: The outer rim is reddish brown, the core is brown, and the intermediate layer is rather vividly red. The three layers of Olhos_26 were separated, ground, and measured separately.

EDX results show that Si and Al exist in great quantities in the three layers, but there is less Al in the external layer than inside. XRD reveals quartz and pyroxene. Also pyroxene is less in the external layer than inside the handle. All samples, as observed by SEM, exhibit big inclusions of quartz. These inclusions have a very smooth surface like pebbles, as seen in Fig. 2. When a fracture occurs in the sample, the inclusions are left intact and do not fracture themselves, being stronger than their connection to the matrix (Fig. 3).

Mössbauer spectra of the three sample layers of Olhos_26 are shown in Fig. 4. The room temperature measurements show that all regions mainly contain hematite

Fig. 2 Representative scanning electron micrographs of a fresh fracture surface of a sherd from a kiln in Lusitania, Olhos_26, showing round inclusions

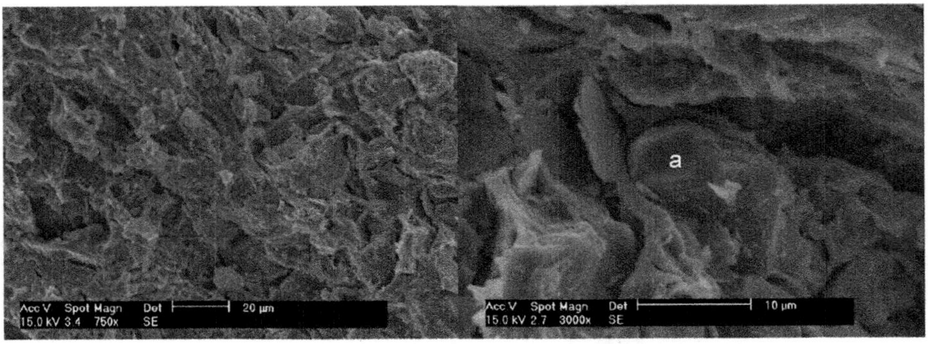

Fig. 3 Scanning electron micrographs of a fresh fracture surface of a sherd from a kiln in Lusitania, Olhos_26 (external layer) showing mostly the matrix (*right*). On the *left*, a particle of feldspar (**a**) is shown with a larger magnification

and a doublet of ferric iron with a quadrupole splitting (Δ) of about 0.85 mm/s. There is also a rather minor contribution of a broad magnetic pattern fitted by a Gaussian distribution of hyperfine fields and attributed to badly crystallized and/or nanoparticulate hematite and perhaps some maghemite. This attribution is based on the observation made in other cases that most of this broad pattern contributes to the narrow hematite pattern at 4.2 K. Some divalent iron is present in the core (about 7%); in the intermediate and outer layers it is practically absent.

The latter observation already suggests that the firing of the amphora was done in reducing conditions with an oxidation only at the end of the firing cycle, presumably during the cooling period. Hence the oxidation is complete anywhere except in the core. This already indicates that the cooling was faster than the penetration of the oxygen into the ceramic body. When the oxygen reached the core, the temperature was already too low for complete oxidation. This also explains the variations in the occurrence of hematite (14% in the outer region, 43% in the intermediate layer, and only 9% in the core).

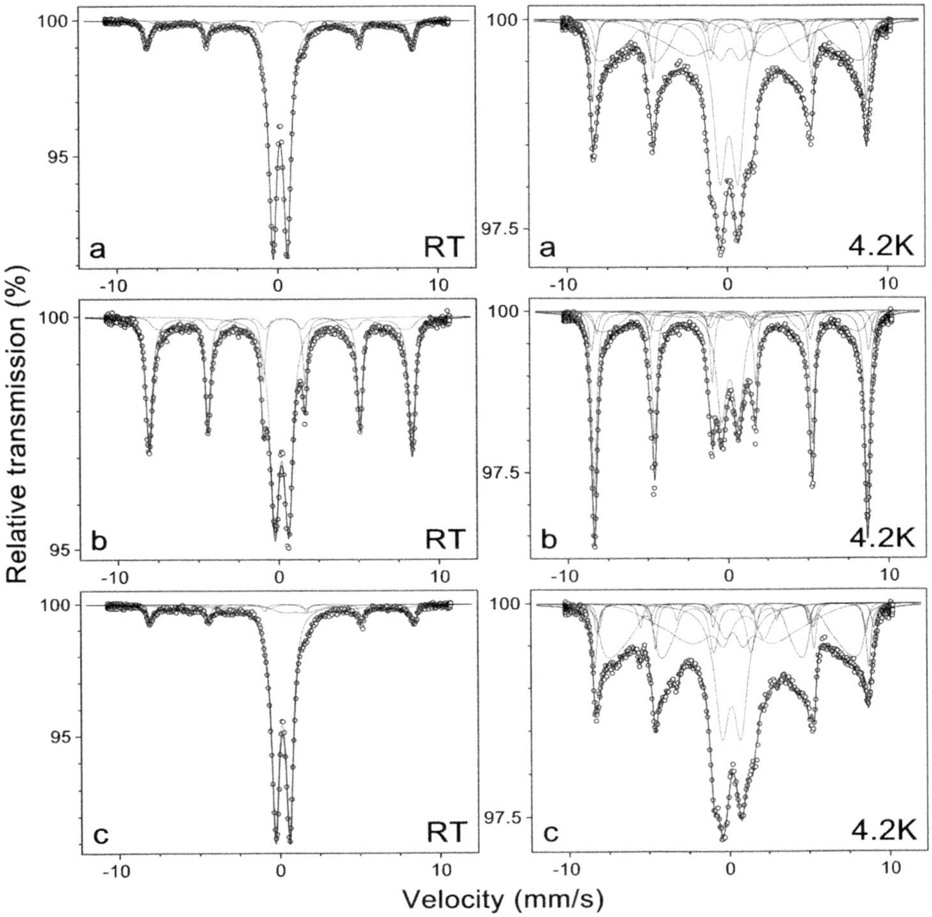

Fig. 4 ^{57}Fe Mössbauer spectra of Olhos_26. RT spectra are shown on the left, 4.2 K spectra on the right. Spectra (**a**), (**b**) and (**c**) represent the external, intermediate and core layer, respectively

According to this hypothesis, already mentioned in [4], a vitreous phase forms during the reducing firing at or above 800°C. The iron is dissolved in this amorphous phase as Fe^{2+} (or perhaps also present to some extent as hercynite). When the conditions become oxidizing at the end of the firing cycle, the ferrous iron oxidizes. The trivalent iron is less soluble in the vitreous phase and therefore tends to exsolve and form hematite. For this to occur, diffusion of the iron in the vitreous phase is required, which takes some time and occurs the faster the higher the temperature. When the cooling is too fast, oxidation may take place, but the formation of hematite may not occur because of sluggish diffusion.

The latter would have occurred in the surface of the handle: The oxygen could readily penetrate, but at the same time the cooling was too fast and only very little hematite formed. The remaining ferric iron would still be dissolved in the vitreous phase.

In the intermediate layer, the cooling was slower and hence there was more time for diffusion and the formation of hematite after the oxidation of the iron. In this way much hematite could form.

Finally, in the core the oxygen arrived later, still in time for nearly complete oxidation of the ferrous iron, but when this occurred, the temperature was already too low for effective diffusion and hence for the formation of hematite.

The 4.2 K measurements confirm the above hypothesis. As mentioned above, for hematite to exsolve from the vitreous phase, the iron must form oxidic clusters, i.e., regions where there is more iron than elsewhere. At these spots, which may form by nucleation and growth, there will be iron-rich clusters which are, however, not yet crystallized hematite. At RT they are paramagnetic or superparamagnetic, yielding the ferric doublet. When the sample is cooled, these clusters will order magnetically, probably becoming antiferromagnetic, speromagnetic, or a spinglass. They may still be superparamagnetic but on further cooling their superparamagnetism will get blocked. This occurs between about 40 and 30 K [4]. At 4.2 K the vitreous phase therefore gives a broad magnetic pattern for these clusters or a doublet for the iron that is still paramagnetic due to its low concentration.

The 4.2 K spectra were fitted with a ferric doublet, a broad magnetic pattern approximated by two static Gaussian distributions of hyperfine fields, and three narrow magnetic sextet corresponding to hematite that undergoes the Morin transition (weak), hematite that does not (quite strong), and maghemite (weak).

In the surface there is 7% of total hematite, about 1.4% of maghemite, 27% of ferric doublet and the rest in the broad magnetic pattern (about 55%).

In the intermediate region there is 44% of total hematite, 8% of maghemite, 22% of ferric doublet and the rest (ca. 22%) of broad magnetic distribution.

In the core there is 21% of ferric doublet, 7% of ferrous doublet, 11% of hematite and 4% of maghemite. The remainder (53%) is the broad magnetic distribution.

Additionally, in the core there is a very weak contribution of metallic iron (about 3%). It might arise from iron nanoparticles that are superparamagnetic at RT and therefore not showing in the RT spectrum. From studies of Celtic coin moulds [5], it is known that metallic iron may form in clays above about 900°C under reducing conditions and may escape oxidation during cooling and also during burial of the ceramics because it is well enclosed in a glassy phase. This would indicate that quite high firing temperatures have been used in the making of the amphora.

4 Conclusions

The amphora whose handle was studied must have been fired reducingly at a temperature of or above about 900°C and cooled in oxidizing conditions in the end. Oxidation and hematite formation went hand in hand, and the amount of hematite formed depended sensitively of the speed of penetration of the oxygen and the cooling rates in the interior of the ceramic body.

References

1. Costa, B.F.O., Pereira, G., Silva, A.J.M., Ramos Silva, M., do Carmo, S.J.C.: J. Phys. Conf. Ser. **217**, 2060 (4 pp.) (2010)

2. Costa, B.F.O., Silva, A.J.M., Ramalho, A., Pereira, G., Ramos Silva, M.: X-ray Spectrometry (2011, in print)
3. Gonzales, J., Gonzales, C., Gonzales, M.: Monografias del CASC **5**, 159–164 (2004)
4. Friederich, J., Gebhard, R., Krause, R., Wagner, F.E., Wagner, U.: J. Phys. Conf. Ser. **217**, 12064 (4 pp.) (2010)
5. Gebhard, R., Bott, R., Distler, N., Michálek, J., Riederer, J., Wagner, F.E., Wagner, U.: Hyperfine Interact. **154**, 199 (2004)

Hyperfine Interact (2011) 202:89–95
DOI 10.1007/s10751-011-0360-2

Characterization of the corrosion products formed on mild steel in acidic medium with *N*-octadecylpyridinium bromide as corrosion inhibitor

N. Nava · N. V. Likhanova · O. Olivares-Xometl · E. A. Flores · I. V. Lijanova

Published online: 26 October 2011

Abstract The characterization of the corrosion products formed on mild steel SAE 1018 after 2 months exposure in aqueous sulfuric acid with and without corrosion inhibitor *N*-octadecylpyridinium bromide has been carried out by means of transmission ^{57}Fe Mössbauer spectroscopy and X-ray powder diffraction (XRD). The major constituent of the rust formed in this environment without corrosion inhibitor is goethite (α-FeOOH). The samples with *N*-octadecylpyridinium bromide contain rozenite and large amounts of melanterite in the corrosion layers.

Keywords Mössbauer spectroscopy · X-ray powder diffraction · *N*-octadecylpyridinium bromide · Corrosion products

1 Introduction

Ionic liquids (ILs) have attracted considerable attention in recent 30 years. These compounds possess a large number of physicochemical properties [1–4]; namely, good electrical conductivity, solvent transport and a relatively wide electrochemical window. Ionic liquids have important electrochemical applications: as electrolytes

N. Nava (✉) · N. V. Likhanova · E. A. Flores
Instituto Mexicano del Petróleo, Dirección de Investigación y Posgrado,
Eje Central Lázaro Cárdenas No. 152, Col. San Bartolo Atepehuacan,
México D.F. 07730, México
e-mail: tnava@imp.mx

O. Olivares-Xometl
Facultad de Ingeniería Química, Benemérita Universidad Autónoma de Puebla,
Av. San Claudio y 18 Sur, Col. San Manuel, Ciudad Universitaria, Puebla, 72570, México

I. V. Lijanova
Instituto Politécnico Nacional, CIITEC,
Cerrada Cecati S/N, Col. Santa Catarina de Azcapotzalco,
México D.F. 02250, México

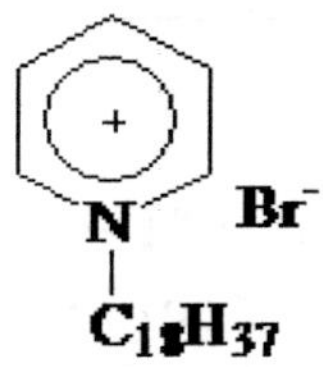

Fig. 1 Structure of *N*-octadecylpyridinium bromide

in lithium batteries, solar cells, and fuel cells; as solvents for the electrodeposition of metals, alloys and semiconductors. Ionic liquids and different types of surfactants are reported to show corrosion resistant behavior in acidic medium [5–10], for example in HCl [11–13] and H_2SO_4 solution [14, 15]. Derivatives of alkylpyridinium bromide were reported as corrosion inhibitor in aqueous media of sulfuric acid [16, 17]. However, no substantial information is available on the nature of corrosion products formed on mild steel using *N*-octadecylpyridinium bromide as corrosion inhibitor in acid medium. In a previous work [18] we reported the inhibitive capability of the mentioned ionic liquid. As we know, alkylpyridinium bromide is capable of delaying metallic corrosion of carbon steel in aqueous 1 M H_2SO_4, their inhibiting efficiency was evaluated by polarization scans and supported by surface analysis. The objective of the present work is to investigate the corrosion products formed on mild steel at room temperature in 1 M H_2SO_4 in the presence of ionic liquid (*N*-octadecylpyridinium bromide) by X-ray diffraction and Mõssbauer spectroscopy.

2 Experimental

2.1 Synthesis, characterization and evaluation of corrosion inhibitor

The tested inhibitor, namely *N*-octadecylpyridinium bromide (PyC18Br) was synthesized and characterized according to the previously described procedure [18]. The molecular structure of PyC18Br is shown in Fig. 1.

2.2 Corrosion products analysis

Carbon steel coupons of 5.08 × 2.54 × 2.54 cm of mild steel (C–0.17%, Mn–0.40%, P–0.040%, S–0.050%, Fe balance, wt.%) were exposed to 1000 mL 1 M H_2SO_4 for 80 days at room temperature with (100 ppm) and without corrosion inhibitor; the corrosion products were then brushed from steel surface to collect a few milligrams to be analyzed by XRD and Mössbauer spectroscopy.

X-ray patterns were obtained in a Siemens D500 diffractometer equipped with a copper anode X-ray tube; the Ka radiation was selected with a diffracted beam monochromator ($K\alpha = 1.5405$ Å), covering a diffraction angle range of $2\theta = 4$–70 in 0.02° increments. Crystal size was determined by the Scherrer equation [19].

Mössbauer spectra were taken at 300 and 77 K with conventional Mössbauer spectrometer operated in transmission geometry in triangular mode. The source was ^{57}Co in Rh matrix with an activity of about 925 MBq (25 mCi). The velocity scale and all the data are refer to metallic α-Fe absorber at 300 K. The absorption spectra were computer fitted by using the NORMOS-SITE and NORMOS-DISTRI program [20].

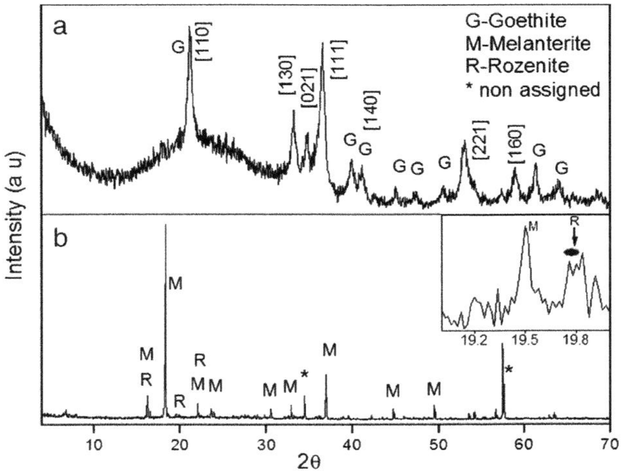

Fig. 2 X-ray powder diffraction patterns of mild steel after exposure to acidic environment for the **a**) blank and **b**) inhibited material by the addition of 100 ppm of PyC18Br

Table 1 Crystal size (Å) in some reflections of goethite

(Reflections)	(110)	(130)	(021)	(111)	(140)	(221)	(160)
Crystal size (Å)	148	164	154	130	152	76	101

3 Results and discussion

3.1 X-ray powder diffraction

In order to have basic information of phases present in the corrosion products, XRD patterns were recorded. The XRD pattern of the blank material (Fig. 2a) contains all the peaks referred to in the card No. JCPDS No 81-0464, thus indicating the presence of goethite phase (α–FeOOH), which is commonly formed in aqueous 1 M H_2SO_4; additionally the pattern shows a broad asymmetric peak in the range $2\theta = 13$–32° and it may be assigned to Na_2SO_4 phase (JCPDS No. 78–1883).

Scherrer equation was employed in order to get a rough estimate of the crystallite size of the goethite on some reflections as is shown in Table 1; as can be seen goethite phase is of nanometric size.

Figure 2b displays the pattern for the rust material studied after the addition of PyC18Br. Peaks at $2\theta = 16.1$, 18.0, 22.0, 23.8, 30.4, 32.8, 37.0, 39.5, 44.8, 49.4 and 54.3° were assigned to melanterite ($FeSO_4$ $7H_2O$) phase and peaks of low intensity at $2\theta = 16.1$ and 19.7° correspond to rozenite ($FeSO_4$ $4H_2O$) phase, (JCPDS No. 72–1106 and JCPDS No. 73–1428), respectively. Apparently, intensity of the peaks of melanterite at 18.0, 37.0 is increased and additionally, pattern shows peaks at 34.5, 57.5 which cannot be assigned to any card JCPDS. These peaks may be ascribed to the presence of sulfated form of the ionic liquid. Probably, the adsorbed species of iron sulfate (like melanterite or rozenite) interact with the cations of IL and form sulfate-anions salts which may coalesce on the metallic surface to make up an adsorptive film of electrostatic character and consequently this prevents iron dissolution to continue.

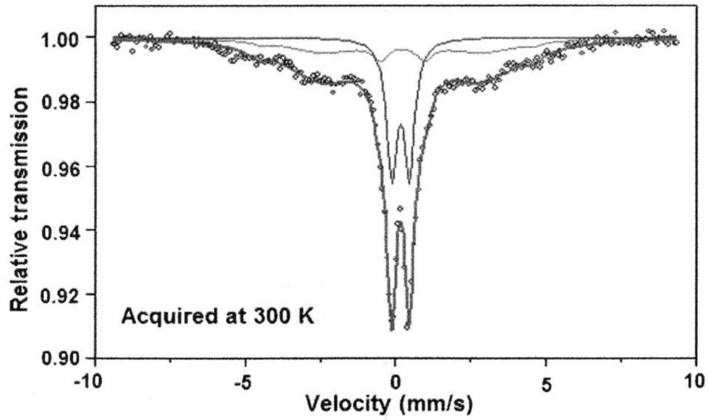

Fig. 3 Mössbauer spectrum of mild steel rust after exposure to the 1 M H_2SO_4 environment for the blank

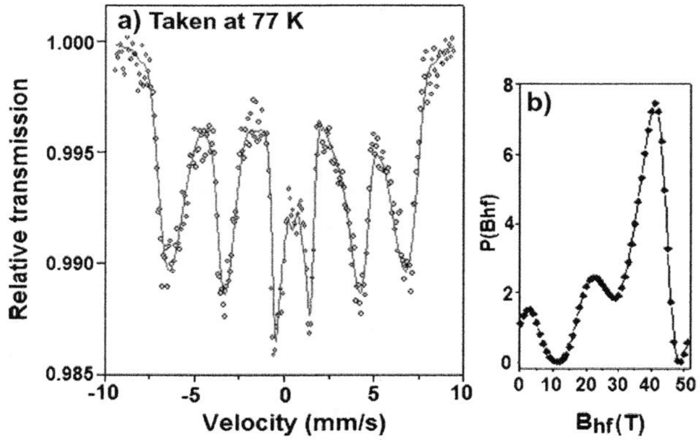

Fig. 4 Mössbauer spectrum of: **a**) blank, acquired at 77 K and **b**) the corresponding distributions of probability for the hyperfine magnetic field

3.2 Mössbauer spectroscopy

Figure 3 shows the Mössbauer spectrum of blank, taken at 300 K; the spectrum is characterized by one broadened sextet which is typical for weakly interacting magnetic particles and one intense doublet which might be associated with nanoparticles (Table 2) [21–24]; the collapsing sextet (broadened signal) is fitted with a low mean hyperfine magnetic field of Bhf $= 28.2$ T, having an $\delta = 0.20$ mm/s, $\Delta = 0.19$ mm/s and $\Gamma = 0.64$ mm/s. The quadrupole interaction of 0.19 mm/s is typical of the magnetically ordered goethite with nanocrystalline structure [25]. The reduced hyperfine magnetic field compared with well-crystallized goethite (around 35 T), can be generally assigned to small particles with a crystal size just above 100 Å and/or varying crystallinity of goethite [26].

The doublet with an $\delta = 0.25$ mm/s and $\Delta = 0.59$ mm/s, can arise from any paramagnetic and/or superparamagnetic iron oxide phase, including the superparamagnetic particles of goethite with crystal size below 100 Å.

In order to get additional information about this doublet, the Mössbauer spectrum of the sample was taken at 77 K (Fig. 4a). The resulting Mössbauer spectrum consists of a single magnetic phase, showing very broad resonant lines which deviate

Table 2 Mössbauer parameters

Sample	Phase	Temperature (K)	δ (mm/s)	Δ (mm/s)	Γ (mm/s)	Bhf (Tesla)	Relative area (%)
Blank	Goethite	300	0.20	0.19	0.64	28.2	67
			0.25	0.59	0.42	–	33
		77	0.33	0.30	0.38	41[b]	
PyC18Br	Rozenite[a]	300	1.17	3.24	0.30	–	–
	Melanterite[a]	300	1.17	3.24	0.30	–	–

Error: δ = ±0.01 mm/s; Δ = ±0.01 mm/s; Γ = Bhf = ±0.2 T

δ isomer shift given relative to *α*-Fe at RT; *Δ* Quadrupole splitting; *Γ* line width; *Bhf* hyperfine magnetic field

[a]Only one doublet was used to fit both structures

[b]Maximum value of the HFD: rozenite δ = 1.16 mm/s, Δ = 3.13 mm/s and melanterite δ = 1.17 mm/s, Δ = 3.30 mm/s

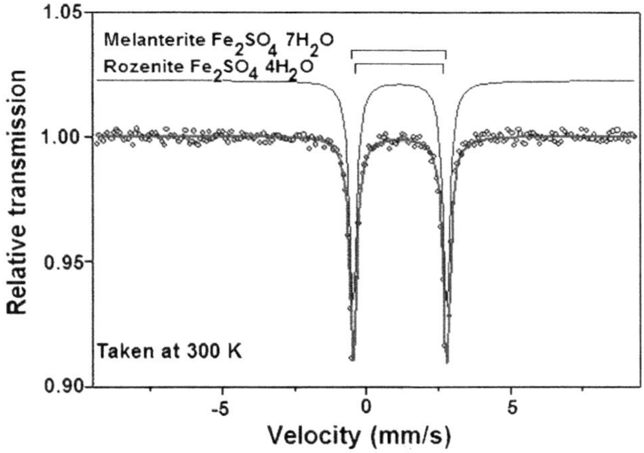

Fig. 5 Mössbauer spectrum of mild steel rust after exposure to the 1 M H_2SO_4 environment for inhibited material at 100 ppm of PyC18Br

from Lorentzian line shapes, suggesting magnetic order inside the iron compound particles, as it is usually observed in fine particles. In this sense, the doublet measured at room temperature can be associated with superparamagnetic particles of goethite with crystal size below 100 Å. As known, this broadened sextet (Fig. 4a) is fitted by using a hyperfine field distribution (HFD) as shown in Fig. 4b. The maximum probability of the HFD is at Bhf = 41 T. The first and intermediate peaks of the HFD appearing at 3 and 22 T, are surely due to mathematical oscillations and cannot be associated with any magnetic nanoparticle.

Thus as a result of the HDF analysis, the Mössbauer spectrum recorded at 77 K confirms the origin of the doublet detected at room temperature arising from nanoparticles of goethite which contribute 33% of the total goethite content with crystal size below 100 Å and 67% of the same compound with crystal sizes above 100 Å, (Table 2).

The Mössbauer spectrum of the PyC18Br sample is shown in Fig. 5, the spectrum was intend to be fitted by using two doublets, representing rozenite ($FeSO_4$ $4H_2O$) and melanterite ($FeSO_4$ $7H_2O$) phases, but both structures have almost similar

Mössbauer parameters ($\delta = 1.16$ mm/s, $\Delta = 3.13$ mm/s and $\delta = 1.17$ mm/s, $\Delta = 3.30$ mm/s, respectively) [27] and it was not possible to separate each other so only one doublet was used to fit the spectrum, representing both structures.

In the PyC18Br sample, results of XRD revealed that hydrated iron sulfates were formed, confirming this ionic liquid inhibits the Fe-oxyhydroxide formation. Hydrated iron sulfates initially formed on the metallic surface probably remain bonded to IL cations, and they might eventually form sulfates of ionic liquids in solution.

4 Conclusions

The corrosion products of mild steel in aqueous 1 M H_2SO_4 solution with/without *N*-octadecylpyridinium bromide were studied by X-ray diffraction and Mössbauer spectroscopy. The results revealed that:

1. According to the XRD patterns melanterite and rozenite were formed in presence of the PyC18Br inhibitor, whereas goethite with different crystal size was obtained without the inhibitor.
2. According to the Mõssbauer analysis, it was confirmed the formation of iron sulfates in presence of PyC18Br; whereas the corrosion products such as iron oxyhydroxides were detected in the absence of the IL.
3. It was confirmed from the Mössbauer analysis at 300 and 77 K that goethite had two crystal sizes: below and above 100 Å.

References

1. Zhao, H.: Innovative applications of ionic liquids as "green" engineering liquids. Chem. Eng. Commun. **193**, 1660–1677 (2006)
2. Zhang, S., Sun, N., He, X., Lu, X., Zhang, X.: Physical properties of ionic liquids: Database and evaluation. J. Phys. Chem. Ref. Data **35**(4), 1475–1517 (2006)
3. Fernicola, A., Scrosati, B., Ohno, H.: Potentialities of ionic liquids as new electrolyte media in advanced electrochemical devices. Ionics **12**(2), 95–102 (2006)
4. Tsuda, T., Hussey, C.L.: Electrochemical applications of room-temperature ionic liquids. Interface **16**, 42–49 (2007)
5. Yue, G.K, Lu, X.M., Zhu, Y.L., Zhang, X.P., Zhang. S.J.: Surface morphology, crystal structure and orientation of aluminium coatings electrodeposited on mild steel in ionic liquid. Chem. Eng. J. **147**, 79–86 (2009)
6. Caporali, S., Fossati, A., Lavacchi, A., Perissi, I., Tolstogouzov, A., Bardi, U.: Aluminium electroplated from ionic liquids as protective coating against steel corrosion. Corros. Sci. **50**(2), 534–539 (2008)
7. Birbilis, N., Howlett, P.C., MacFarlane, D.R., Forsyth, M.: Exploring corrosion protection of Mg via ionic liquid pretreatment. Surf. Coat. Technol. **201**(8), 4496–4504 (2007)
8. Zhang, Q., Hua, Y.: Corrosion inhibition of mild steel by alkylimidazolium ionic liquids in hydrochloric acid. Electrochim. Acta **54**(6), 1881–1887 (2009)
9. Ashassi-Sorkhabi, H., Es'haghi, M.: Corrosion inhibition of mild steel in acidic media by [BMIm]Br Ionic liquid. Mater. Chem. Phys. **114**(1), 267–271 (2009)
10. Hegazy, M.A., Abdallah, M., Ahmed, H.: Novel cationic gemini surfactants as corrosion inhibitors for carbon steel pipelines. Corr. Sci. **52**(9), 2897–2904 (2010)
11. Noor, E.A., Al-Moubaraki, A.H.: Thermodynamic study of metal corrosion and inhibitor adsorption processes in mild steel/1-methyl-4[4′(-X)-styryl pyridinium iodides/hydrochloric acid systems. Mater. Chem. Phys. **110**(1), 145–154 (2008)

12. Negm, N.A., Sabagh, A.M.A., Migahed, M.A., Bary, H.M.A., Din, H.M.E.: Effectiveness of some diquaternary ammonium surfactants as corrosion inhibitors for carbon steel in 0.5 M HCl solution. Corr. Sci. **52**(6), 2122–2132 (2010)
13. Al-Sabagh, A.M., Tantawy, N.S., Nasser, N.M., Mishrif, M.R.: Corrosion inhibition efficiency in relation to micellarinteraction parameters of cationic/nonionic surfactant mixtures for carbon steel pipelines in 1M HCl solution. J. Dispers. Sci. Technol. **30**(10), 1411–1423 (2009)
14. Morad, M.S., Hermas, A.A., Obaid, A.Y., Qusti, A.H.: Evaluation of some bipyridinium dihalides as inhibitors for low carbon steel corrosion in sulfuric acid solution. J. Appl. Electrochem. **38**, 1301–1311 (2008)
15. Bhrara, K., Kim, H., Singh, G.: Inhibiting effects of butyl triphenyl phosphonium bromide on corrosion of mild steel in 0.5 M sulphuric acid solution and its adsorption characteristics. Corros. Sci. **50**(6), 2747–2754 (2008)
16. Saleh, M.M., Atia, A.A.: Effects of structure of the ionic head of cationic surfactant on its inhibition of acid corrosion of mild steel. J. Appl. Electrochem. **36**(8), 899–905 (2006)
17. Migahed, M.A.: Electrochemical investigation of the corrosion behaviour of mild steel in 2 M HCl solution in presence of 1-dodecyl-4-methoxy pyridinium bromide. Mater. Chem. Phys. **93**(1), 48–53 (2005)
18. Likhanova, N.V., Domínguez-Aguilar, M.A., Olivares-Xometl, O., Nava-Entzana, N., Arce, E., Dorantes, H.: The effect of ionic liquids with imidazolium and pyridinium cations on the corrosion inhibition of mild steel in acidic environment. Corros. Sci. **52**(6), 2088–2097 (2010)
19. Klug, H.P., Alexander, L.E.: X-ray Diffraction Procedures. Wiley, New York (1974)
20. Brand, R.A.: Improving the validity of hyperfine field distributions from magnetic alloys. Part I: Unpolarized source. Nucl. Instrum. Methods Phys. Res. B **28**(3), 398–416 (1987)
21. Murad, E., Schwertmann, U.: Influence of Al substitution and crystal size on the room-temperature Mössbauer spectrum of hematite. Clays Clay Miner. **34**(1), 1–6 (1986)
22. Murad, E.: The characterization of goethite by Mõssbauer spectroscopy. Am. Mineral. **67**, 1007–1011 (1982)
23. Schwertmann, U., Murad, E.: Effect of pH on the formation of goethite and hematite from ferrihydrite. Clays Clay Miner. **31**(4), 277–284 (1983)
24. Schwertmann, U., Cambier, P., Murad, E.: Properties of goethites of varying crystallinity. Clays Clay Miner. **33**(5), 369–378 (1985)
25. Madsen, D.E., Hansen, M.F., Koch, C.B., Mørup, S.: Interactions between goethite particles subjected to heat treatment. J. Phys. Condens. Matter **20**(13), art. No. 135215 (2008)
26. Cabral-Prieto, A., Reyes-Felipe, A.A., Siles-Dotor, M.G.: Synthesis and characterization of nanophasic goethite. Nanostruct. Mater. **10**(2), 311–326 (1998)
27. Rothstein, Y., Dyar, M.D., Schaefer, M.W., Lane, M.D., Bishop, J.L.: In: 36th Annual Lunar and Planetary Science Conference, League City, Texas, abstract No. 2108, 14–18 March (2005)

Hyperfine Interact (2011) 202:97–106
DOI 10.1007/s10751-011-0361-1

Effect of titanium ion substitution in the barium hexaferrite studied by Mössbauer spectroscopy and X-ray diffraction

Pamela Quiroz · Bernd Halbedel · Angel Bustamante · Juan C. González

Published online: 25 August 2011

Abstract A series of M-type barium hexaferrite has been synthesized in a glass melt by partially substituting the Fe_2O_3 with TiO_2 for investigation of their structure. The glass melt has the basic composition (mol%): 40 BaO + 33 B_2O_3 + (27-x) Fe_2O_3 + x TiO_2 with x = 0, 3.6, 5.4 and 7.2 mol% TiO_2. The substituted ferrites were studied by means of X-ray diffraction, Mössbauer spectroscopy and vibration sample magnetometer. X-ray diffraction studies revealed that not all samples have a single ferritic phase, a small second phase corresponding to $BaTi_6O_{13}$ was also observed to form. The Mössbauer spectra changed from magnetically ordered (x = 0) to magnetically ordered with strong line broadening. Moreover, the broadening increases with TiO_2 content. The Mössbauer parameters suggested that Ti^{4+} occupies the 2a and 12k crystal sites, and the Ti^{4+} substitution on the 2b and $4f_2$ site also occurs at high melt dopings. Therefore, coercivity and saturation magnetization decreased.

Keywords Barium hexaferrite · Mössbauer spectroscopy · Substitution · X-ray diffraction

P. Quiroz · B. Halbedel
Department of Inorganic-Nonmetallic Materials,
Institute of Materials Engineering, Ilmenau University of Technology,
P.O. Box 100565, 98684 Ilmenau, Germany

P. Quiroz
e-mail: pamela.quiroz-penaranda@tu-ilmenau.de

A. Bustamante (✉)
Laboratory of Ceramics and Nanomaterials, Faculty of Physical Sciences,
San Marcos National University, P.O. Box 14-149, 14 Lima, Peru
e-mail: angelbd1@gmail.com

J. C. González
Surfaces Research Group-Interfaces and Thin Films,
Materials Science Institute of Sevilla - CSIC, 41092 Sevilla, Spain

1 Introduction

Due to the development of new technologies for radar and microwave communication, the research on microwave absorbing materials has increased in recent years. The main applications of these materials intend to reduce the human exposure to microwaves by means of absorbing coatings [1–3]. Ti-doped barium hexaferrite (Ti-doped BHF) powder is an efficient absorber of electromagnetic waves in the microwave spectrum. To optimize the Ti-doped BHF microwave absorption properties an increase in the saturation magnetization (M_S) and a decrease in the coercivity ($_JH_C$) are required [2]. These properties depend on the localization of the Ti^{4+} in the barium hexaferrite (BHF) structure. According to the literature [4, 5], the Ti^{4+} ions preferentially occupy the octahedral $4f_2$ sites of the BHF structure when the Ti-doped BHF is synthesized using the ceramic route, especially at low doping rates.

Since the magnetic moment of the Ti^{4+} ions is zero and the spin direction of $4f_2$ site in the BHF structure is down, M_S must increase with the doping concentration of titanium, while $_JH_C$ must decrease, which corresponds our expectations of lower magnetocrystalline anisotropy. However with higher doping rates of Ti^{4+}, M_S decreases [6]. A possible explanation for this effect is that the Ti^{4+} ions could be located also in other lattice sites like 12k, 2b or 2a with spin-up directions. Other possible explanations could be: a disordered collinear magnetic structure, due to the change of the exchange interactions attributed to the immediate neighbours, and the formation of other phases. In this work, the Ti ions doping the BHF structure is studied by X-ray diffraction and Mössbauer spectroscopy.

2 Experimental

The Ti-doped BHF production comprises three main stages [2, 6, 7]. In the first stage, doped flakes are synthesized using the glass crystallization technique (GCT) with the basic composition (mol%): 40 BaO + 33 B_2O_3 + (27-x) Fe_2O_3 + x TiO_2, in which the chosen melt dopings were x = 0, 3.6, 5.4 and 7.2 mol% TiO_2 [6]. In the second stage, the flakes are annealed at a temperature determinated by Differential Scanning Calorimetry (DSC), taking into account the exothermic peaks that appear in the resulting diagrams [6]. Finally in the third stage, the material is powdered to obtain particles of a suitable size for the acid treatment, used to completely dissolve the borate glass matrix separating the solid unsolved ferrite particles.

The cation locations in the BHF structure were analyzed with X-ray diffraction (XRD) and Mössbauer spectroscopy. The ^{57}Fe Mössbauer spectra were obtained at room temperature using a constant acceleration spectrometer with ^{57}Co in an Rh matrix and 1024 channels. The isomer shifts values are reported relative to metallic iron (α-Fe) taken at room temperature (RT).

The variation of the melt doping concentrations and of the crystallization parameters has a strong influence on the magnetic properties and on the formation of new phases. These effects are investigated with X-ray diffraction (XRD) and vibration sample magnetometer (VSM). The XRD patterns were obtained using a Bruker diffractometer with Co (K_α) radiation ($\lambda = 0.17889$ nm). The diffraction data for the

Fig. 1 Crystal structure of Barium hexaferrite ($2^*BaFe_{12}O_{19}$) showing a unit cell and positions of ionic sites in the four spinel blocks (SRS*R*) and relative orientation of magnetic moments (↑↓) of Fe^{3+} ions

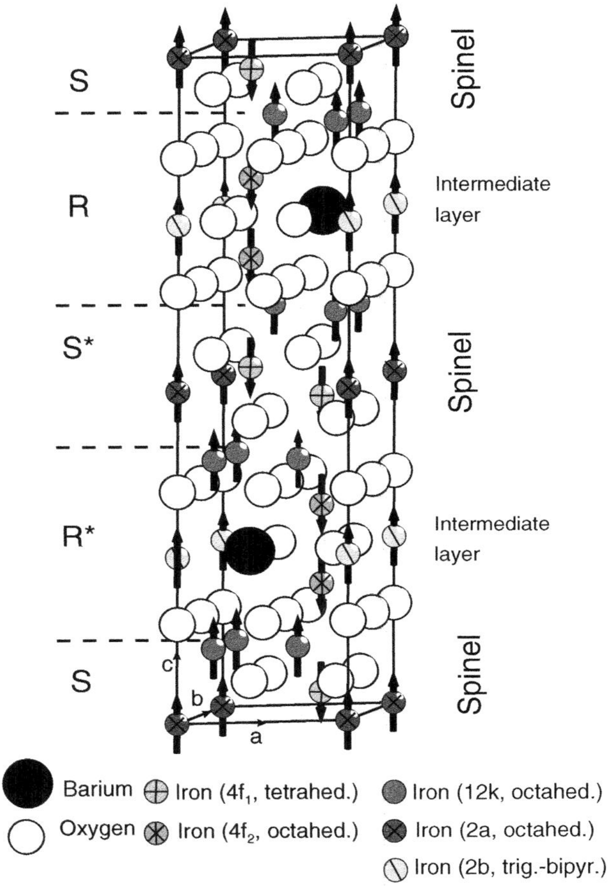

Rietveld refinement were collected at RT in Bragg-Brentano geometry. The data were collected over the diffraction range $2\theta = 15°...90°$, with step $\Delta 2\theta = 0.02°$. We use Co–Kα_1 and Co–Kα_2 radiations with wavelengths 0.1788965 and 0.1792850 nm, respectively, and an intensity ratio of 0.497. The background was refined using a fifth-order polynomial equation. A pseudo-Voigt function (linear combination of Gaussian and Lorentz functions) was used to describe the reflection profiles. During refinement absorption and microabsorption were neglected. The R_s fitting values (R_p, R_{wp} and R_{exp}) and the χ^2 were used as the numerical criteria for a good fitting [8]. To ensure the convergence, the Rietveld refinements were performed in the following order: (1) scale factor, sample displacement and background, (2) cell parameters, (3) peak shape, half width and asymmetry parameter, (4) atomic position parameters, (5) site occupancies, and (6) isotropic thermal parameters.

The magnetic properties were determined at RT and up to a maximum applied field of $1.28 \cdot 10^6$ A/m by using a 7312 LakeShore Vibrating Sample Magnetometer (VSM) [2].

Table 1 Number of ions per unit formula, coordination and spin direction for the Fe^{3+} cation sublattices of M-type BHF

Lattice site	Coordination	Number of Fe^{3+} ions per unit formula	Block	Spin direction
12k	Octahedral	6	R-S	Up ↑
2b	Trigonal–bipyramidal	1	R	Up ↑
$4f_2$	Octahedral	2	R	Down ↓
$4f_1$	Tetrahedral	2	S	Down ↓
2a	Octahedral	1	S	Up ↑

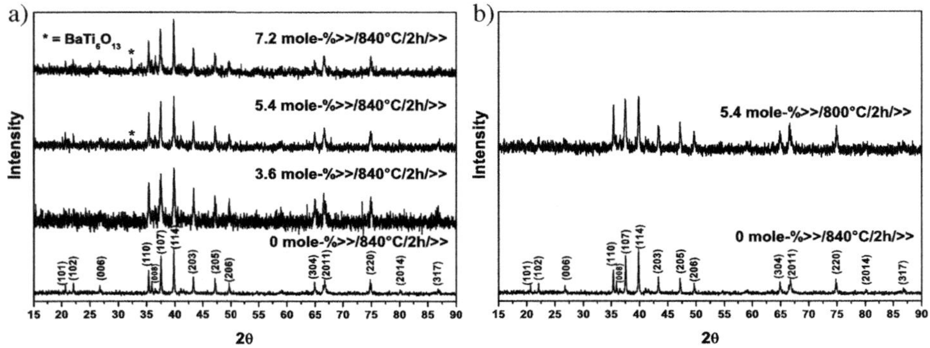

Fig. 2 **a** XRD of undoped and Ti-doped BHF powders tempered at 840°C. **b** XRD of the undoped and 5.4 mol% TiO_2 sample tempered at 800°C. *Lines* represent the standard BHF PDF card no. 39–1433. *Principal peak for $BaTi_6O_{13}$ (PDF 01-070-1567)

3 Results and discussions

Figure 1 shows the crystal structure of BHF with a unit cell greatly elongated in the c direction [9]. It is composed by the superposition of four spinel blocks SRS*R*. The spinel block R or the inverse block R*contain the heavy Ba^{2+} ions. The closed-packed Ba^{2+} and O^{2-} ions are large and nonmagnetic, and the Fe^{3+} ions are located in the interstices. The Fe^{3+} ions are distributed among five different lattice sites as listed in Table 1 [10]. The 12k sites are located at the interface between the S-R, R-S*, S*-R* and R*-S blocks, whereas the $4f_2$ and 2b sites are located into the R and R* blocks, and the $4f_1$ and 2a sites are located into the S and S* blocks, as illustrated in the figure.

Each Fe^{3+} ion in the structure contributes differently to the magnetic properties depending on their magnetic interactions with the O^{2-} ions [2, 6, 11]. When the Ti^{4+} enters in the structure and substitutes some Fe^{3+} ions, the magnetocrystalline anisotropy field changes.

The X-ray diffractograms of the Ti-doped BHF powders tempered at 840°C are shown in Fig. 2a. The positions of all the peaks in the diffractograms correspond to the BHF structure (PDF card No. 39-1433). The presence of a new phase in the 5.4 and 7.2 mol% samples, related to $BaTi_6O_{13}$ is also revealed. Analyzing the effect of the tempering temperature (Fig. 2b) for the 5.4 mol% TiO_2 sample tempered at

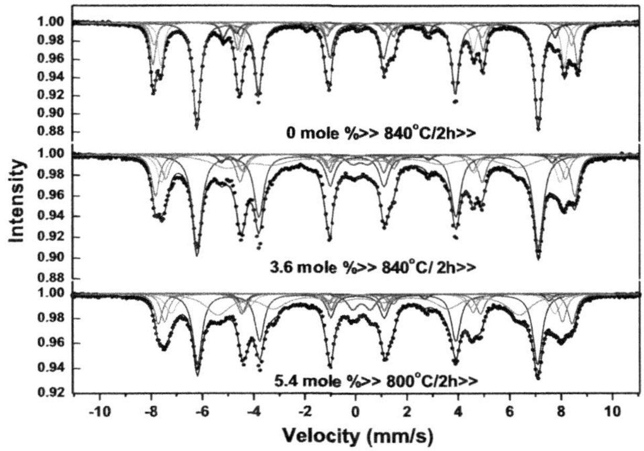

Fig. 3 Mössbauer spectra of the undoped and Ti-doped BHF powders at RT. (the *different colors* indicate different lattice sites)

800°C, the peak at $2\theta = 32.4°$ is not observed. The diffraction peaks for this sample correspond only to the BHF pattern.

In order to study the effect of the Ti^{4+} ion substitution in the BHF, the samples with only the ferritic phase (0 mol% TiO_2 and 3.6 mol% TiO_2 tempered at 840°C and 5.4 mol% TiO_2 tempered at 800°C) were analyzed by Mössbauer spectroscopy and VSM. The other samples are not discussed in this paper, because part of the Ti^{4+} ions is also present in the second non-ferritic phase ($BaTi_6O_{13}$) leading to not reliable analysis.

Figure 3 shows RT Mössbauer spectra of the undoped and Ti-doped BHF. The spectra were fitted with five sextets denoted to $4f_1$, $4f_2$, 12k, 2a and 2b sites belonging to the BHF structure. The hyperfine parameters for each site, deduced from the computer fitting as a function of mol% TiO_2, are listed in Table 2.

The undoped BHF hyperfine parameters are in good agreement with previous published data [10, 12–15]. The substitution of Fe^{3+} to Ti^{4+} mainly affects the 12k site. This is also detected by sextet with broad width, designed as 12k′ which has also been observed by Meaz and Bender Koch [16]. The samples containing Ti^{4+} ions show two doublets D1 and D2. Moreover, an equivalent amount of Fe^{3+} ions is expected to be converted into Fe^{2+} ions, however a small D1 is observed for the undoped BHF.

The substitution mechanism of Ti^{4+} and the presence of Fe^{2+} in the BHF structure is still under debate. The Ti^{4+} substitution requires the presence of Fe^{2+}, which has been demonstrated by cerimetry measurements (see Table 3). From theoretical calculations, it is assumed that a similar amount of Ti^{4+} and Fe^{2+} ions exists to maintain electrical neutrality in the BHF structure. As shown in the experimental data in the Table 3, an increase in the melting doping concentration leads to an increase of Fe^{2+} ions. This is consistent with the theory and with the statement that Ti^{4+} ions are part of the ferritic phase. Furthermore, a small amount of Fe^{2+} was identified in the undoped BHF sample, which is denoted with D1 in the Mössbauer spectra. For the melt doping 5.4 mol%, the Fe^{2+} content obtained experimentally is much higher than theoretical calculation. In addition, the absorption area in the Mössbauer spectra indicates a ratio Fe^{2+}/Fe^{3+} which is lower than the ratio obtained by cerimetry and ICP. The high Fe^{2+} content in this sample may come from the

Table 2 Mössbauer hyperfine parameters obtained at RT of the undoped and Ti-doped BHF powders

Melt doping (mol%)	Tempering temperature (°C)	Site	Isomer shift δ (mms^{-1}) ±0.03	Quadropole shift ε (mms^{-1}) ±0.03	Hyperfine field B_{hf} (Tesla) ±0.02	Width of line Γ (mms^{-1}) ±0.02	Relative area (%) ±0.2
0	840	12k	0.37	0.21	41.3	0.33	50.3
		$4f_1$	0.28	0.11	48.7	0.26	16.6
		$4f_2$	0.40	0.09	51.4	0.26	15.8
		2a	0.37	0.02	50.5	0.30	9.8
		2b	0.31	1.10	40.1	0.27	5.9
		D1	0.18	2.09		0.27	1.5
3.6	840	12k	0.38	0.20	41.3	0.44	38.9
		12k′	0.47	0.30	35.2	1.60	19.0
		$4f_1$	0.27	0.03	48.2	0.41	10.8
		$4f_2$	0.40	0.08	50.7	0.37	14.6
		2a	0.33	0.17	48.3	0.38	9.0
		2b	0.28	1.05	40.0	0.32	2.5
		D1	0.18	2.09	–	0.32	2.8
		D2	0.31	0.60	–	0.58	2.3
5.4	800	12k	0.38	0.19	41.1	0.45	31.5
		12k′	0.46	0.15	36.5	1.13	24.9
		$4f_1$	0.31	0.09	48.1	0.49	21.7
		$4f_2$	0.40	0.08	50.4	0.40	10.9
		2a	0.47	0.12	36.1	0.83	12.3
		2b	0.49	1.22	37.9	0.36	2.5
		D1	0.19	2.13	–	0.40	2.9
		D2	0.34	0.56	–	0.40	3.8

presence of glass phase which is not identified by XRD. The origin of the high Fe content measured in the undoped BHF powders is currently not well understood.

The hyperfine field splitting (B_{hf}) are shown in Fig. 4a. When increasing Ti^{4+} substitution, the internal magnetic field decreases. This decrease could be attributed to the smaller magnetic moment (almost zero) of Ti^{4+} which replaces the Fe^{3+} ($5\mu_B$) ions. This also causes the reduction of super-exchange interaction between neighboring cations (Ti^{4+}, Fe^{2+} and Fe^{3+}) and oxygen ions. In contrast, the B_{hf} corresponding to the site 12k remains almost constant (around 41.2 T). Figure 4b shows that the experimental occupation area of Fe^{3+} decreases. Despite the Ti^{4+} concentration increases, a decrease on B_{hf} is not observed. Therefore, we assumed that substitution occurs at this site. As observed from the melting-doping 0 mol% corresponds at the 12k site the 50.3% for the total area of the iron in the octahedral sites and with a B_{hf} of 41.3 T. Moreover, for the melt doping 3.6 mol%, a decrease on the total area to 38.9% is observed, but it always occurs with the same B_{hf} within the error margin. The total area difference between 0 mol% and 3.6 mol% should represent a B_{hf}-change in a unique crystallographic site with octahedral coordination, because there is no change in the crystal structure, as it was demonstrated by XRD. Thus, the new site, called 12k′ [16], represents the presence of Ti^{4+} in octahedral coordination with lower B_{hf} (35.2 T). This explains the progressive weakening of the magnetic moments of iron at these sites, since it allows a non-magnetic ion. For

Table 3 Comparison of the Fe^{2+} and Fe^{3+} in the Ti-doped BHF powders

Melt doping (mol%)	Temper temperature (°C)	Theoretical			Experimental					
		Fe (mol%)	Fe^{2+} (mol%)	Fe^{3+} (mol%)	Fe (mol%) ICP analysis	Ti (mol%) ICP analysis	Fe^{2+} (mol%) cerimetrie	Fe^{3+} (mol%)	Fe^{2+}/Fe^{3+} (%) cerimetrie	Fe^{2+}/Fe^{3+} (%) Mössbauer
0.00	840	37.5	–	37.5	47.59	0	0.72	46.87	1.53	1.52
3.60	840	34.82	2.68	32.14	34.99	2.52	2.15	32.84	6.55	5.38
5.40	800	33.30	4.16	29.14	35.41	4.45	8.73	26.68	32.72	7.18

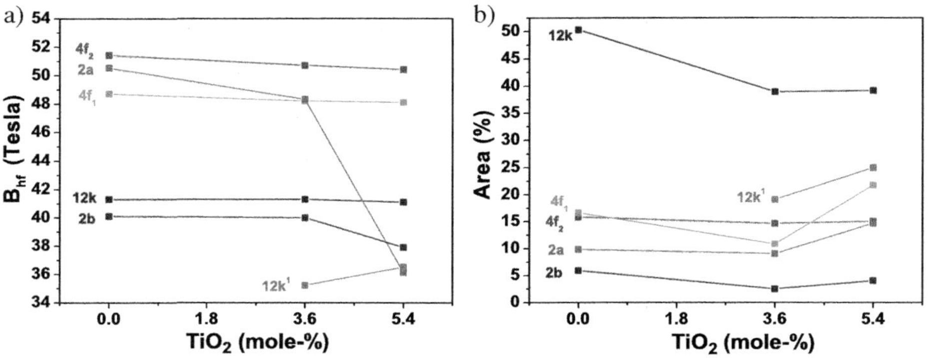

Fig. 4 **a** B_{hf} on the five Fe^{3+} lattice sites and **b** Fe^{3+} ion occupation area ratio on the five sites of the undoped and Ti-doped BHF powders

Table 4 Lattice constants (a and c) of the undoped and Ti-doped BHF powders as a function of the melt doping obtained by Rietveld refinement

Melt doping (mol%)	a	c	a/c
0	5.891 ±0.00019	23.2061 ± 0.005	0.2539
3.6	5.8863 ± 0.00021	23.2695 ± 0.005	0.2530
5.4	5.8865 ± 0.00025	23.2901 ± 0.005	0.2527

5.4 mol%, the B_{hf} value at 12k′ site is approximately 36.5 T which agrees with that obtained by other authors [17]. The split in the 12k site may be attributed to different neighborhoods of the iron in this crystallographic site.

Analyzing the 2a site, a decrease of B_{hf} (from 50.5 T for the melt doping 0 mol% to 48.3 T for the melt doping 3.6 mol%) is observed, indicating the presence of Ti^{4+}. Note that, at 5.4% mole doping; both, the hyperfine field and the area, decrease confirming the substitution of Ti^{4+} by Fe^{3+}. Probably, there is a certain margin of error in the calculation of the area because more intense doublets appear, especially D2. This represents a significant amount of iron, which is not added to any sextets.

The Ti^{4+} doping into the BHF structure changes the lattice parameters, mainly the c parameter. The incorporation of the Ti^{4+} ion occurs at the 2a and 12k Wyckoff sites. The 2a site of Fe^{3+} splits into two depending on the change of the c-coordinate [18]. Furthermore, the value of the occupancy of Fe^{3+} at the 2a site is not exactly the unity (it is 0.95). Rietveld refinement suggests that Ti^{4+} occupies mainly the 2a site rather than the 12k site. The Wyckoff fit better with the Ti^{4+} ion forming an octahedron than double trigonal pyramids, neither tetrahedral sites like 2b or $4f_1$. We also explored the possibility of doping the $4f_2$ site, but no improvement in the χ^2 was detected.

Table 4 shows the lattice parameters, a and c, as a function of the Ti^{4+} content obtained by Rietveld analysis. The increase in the lattice parameter c means that the unit cell stretches perpendicularly to the basal plane. The decrease on the ratio a/c is attributed to the small increase of the anisotropy field. In addition to the reduction of the magnetocrystalline anisotropy due to the substitution of iron by titanium, a decrease in the $_JH_C$ has resulted. This is in agreement with the magnetic results (Fig. 5). The magnetic hysteresis curves shows that the original hard magnetic

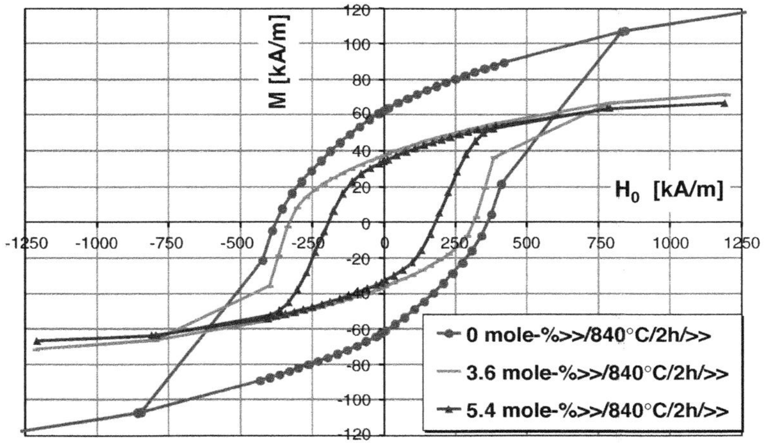

Fig. 5 Magnetic hysteresis curves for the undoped and Ti-doped BHF powders

behaviour of the undoped BHF powders is shifted to soft magnetic by means of high doping content. In effect, it is clearly observed that Ti-doped BHF samples exhibit a decrease of ${}_JH_C$ and M_S.

4 Conclusion

Ti^{4+} ions enter in the BHF lattice without causing great distortion in the crystalline structure. In the case of 5.4 and 7.2 mol% TiO_2 samples tempered at 840°C, a fraction of Ti^{4+} forms an additional phase corresponding to $BaTi_6O_{13}$. However, if the 5.4 mol% TiO_2 sample is tempered at 800°C, a ferritic phase is formed; the Ti^{4+} ions enter in the BHF structure, and ${}_JH_C$ decreases. There is no variation in the 3.6 mol% TiO_2 sample with respect to the undoped BHF. Nevertheless, the corresponding decrease of ${}_JH_C$ may be attributed to the increase in the hexagonal c axis. Because Fe^{3+}/ Fe^{2+}/ Ti^{4+}–O–Fe^{3+}/ Ti^{4+}/ Fe^{2+} bonds are longer than Fe^{3+}–O–Fe^{3+} bonds, then the superexchange interaction between neighbouring cations, and consequently ${}_JH_C$, minimizes. As described in [6], the observed decrease in ${}_JH_C$ may be originated but the reduction in the grain size. Mössbauer spectroscopy reveals that for the 3.6 mol% TiO_2 sample, the Ti^{4+} ions are placed on the octahedral 2a and 12k site. Whereas, for high melt doping substitution is detected on the octahedral-$4f_2$ and trigonal-bipyramid-2b sites. Nevertheless, the spectrum of this sample exhibits strong asymmetry and the fit is only a first approximation [17]. The high Fe^{2+} content in this sample may come from a BaO-FeO-B_2O_3 phase which is not identified by XRD. Future work will be improving the preparation procedure to avoid the residual glass phase. It is also important to perform the annealing in a reducing atmosphere in order to increase the Fe^{2+} in the structure and thus enhance the Ti^{4+} doping in the BHF structure.

References

1. Ruan, S.-P., Xu, B.-K., Suo, H., Wu, F.-Q., Xiang, S.-Q.: J. Magn. Magn. Mater. **212**, 175–177 (2000)
2. Halbedel, B., Hülsenberg, D., Belau, St., Schadewald, U.: cfi/Ber. DKG **82**(13), 182–188 (2005)
3. Rohde & Schwarz, News 199/09, pp. 70–72 (2009)
4. Mariño-Castellanos, P.A., Anglada-Rivera, J., Cruz-Fuentes, A., Lora-Serrano, R.: J. Magn. Magn. Mater. **280**, 214–220 (2004)
5. Brabers, V.A.M., Stevens, A.A.E., Dalderop, J.H.J., Simga, Z.: J. Magn. Magn. Mater. **196/197**, 312–314 (1999)
6. Quiroz, P., Halbedel, B.: In: CD Proceedings 54th International Scientific Colloquium, TU Ilmenau, isle Verlag Ilmenau, paper_id_048 (2009)
7. Schadewald, U., Halbedel, B., Romanus, H., Hülsenberg, D.: Mat.-wiss. u. Werkstofftechnik **37**(11), 941–944 (2006)
8. Young, R.A.: The Rietveld Method. Oxford University Press, Oxford (1993)
9. Townes, W.D., Fang, J.H., Perrota, A.J.: Zeitschrift für Kristallographie **125**, 437–449 (1967)
10. Albanese, G., Watts, B.E., Leccabue, F., Diaz Castañon, S.: J. Magn. Magn. Mater. **184**, 337–343 (1998)
11. Rösler, S., Wartewig, P., Langbein, H.: Cryst. Res. Technol. **38**(11), 927–934 (2003)
12. Morish, A.H., Zhou, X.Z., Yang, Z., Zeng, H.H.-X.: Hyperfine Interact. **90**, 365 (1994)
13. Zhou, X.Z., Morish, A.H., Yang, Z., Zeng, H.H.-X.: J. Appl. Phys. **75**, 5556 (1994)
14. Lanje, N.Y., Kulkarni, D.K., Rewatkar, K.G.: Mater. Lett. **47**, 125 (2001)
15. Pereda, J.A., Isalgué, A., Tejada, J., Litterst, F.J., Obradors, X.: Hyperfine Interact. **28**, 569 (1986)
16. Meaz, T.M., Bender Koch, C.: Hyperfine Interact. **156/157**, 341–346 (2004)
17. Meaz, T.M., Bender Koch, C.: Hyperfine Interact. **166**, 455–463 (2005)
18. Obradors, X., Collomb, A., Pernet, M., Joubert, J.C., Isalgué, A.: J. Magn. Magn. Mater. **44**, 118–128 (1984)

Hyperfine Interact (2011) 202:107–115
DOI 10.1007/s10751-011-0342-4

Smooth driving of Mössbauer electromechanical transducers

A. Veiga · M. A. Mayosky · N. Martínez · P. Mendoza Zélis · G. A. Pasquevich · F. H. Sánchez

Published online: 18 October 2011

Abstract Quality of Mössbauer spectra is strongly related to the performance of source velocity modulator. Traditional electromechanical driving techniques demand hard-edged square or triangular velocity waveforms that introduce long settling times and demand careful driver tuning. For this work, the behavior of commercial velocity transducers and drive units was studied under different working conditions. Different velocity reference waveforms in constant-acceleration, constant-velocity and programmable-velocity techniques were tested. Significant improvement in spectrometer efficiency and accuracy was achieved by replacing triangular and square hard edges with continuous smooth-shaped transitions. A criterion for best waveform selection and synchronization is presented and attainable enhancements are evaluated. In order to fully exploit this driving technique, a compact microprocessor-based architecture is proposed and a suitable data acquisition system implementation is presented. System linearity and efficiency characterization are also shown.

Keywords Mössbauer · Instrumentation · Spectrometer · Transducers · Constant-velocity

A. Veiga (✉) · M. A. Mayosky
Facultad de Ingeniería, Universidad Nacional de La Plata, La Plata, Argentina
e-mail: veiga@fisica.unlp.edu.ar

A. Veiga · N. Martínez · P. Mendoza Zélis · G. A. Pasquevich · F. H. Sánchez
Instituto de Física La Plata, CONICET, La Plata, Argentina

M. A. Mayosky · N. Martínez
Comisión de Investigaciones Científicas de la Provincia de Buenos Aires, La Plata, Argentina

N. Martínez · P. Mendoza Zélis · G. A. Pasquevich · F. H. Sánchez
Facultad de Cs. Exactas, Universidad Nacional de La Plata, La Plata, Argentina

1 Introduction

Double loudspeaker-type electromechanical transducers have become a standard for Doppler energy modulation of the radioactive source in Mössbauer experiments [1–6]. These devices have proved to be reliable and relatively inexpensive. Required velocity resolution for hyperfine observations (better than 10 μm/s) can be obtained with a careful adjustment of the corresponding drive. Transducers success relies mainly on their excellent performance in traditional constant-acceleration (CA) experiments, in which case the velocity reference is a zero DC triangular voltage waveform that must be tuned to the natural oscillating frequency of the transducer. In such case the operation of the oscillating device is favored by the low harmonic content of the required velocity.

However, in constant-velocity (CV) and programmable-velocity (PV) experiments [7–9], the reference has a much higher harmonic content. This causes the performance to drop drastically, due to transient responses that deteriorate the whole spectrometer efficiency.

In the following sections we present an analysis of the velocity control problem, we propose a technique based on the manipulation of the harmonic content of the reference waveform, and finally we present the design, implementation and characterization of a programmable data acquisition system that enables the application of the proposed technique.

2 Operation fundamentals and characterization

Regular electromechanical transducers are damped oscillating systems that exhibit longitudinal displacements proportional to the current applied to a driving coil enclosed in a uniform magnetic field. The voltage induced on a second coil is used to sense the resulting velocity to provide a control feedback loop. Due to this derivative behavior, their transfer function presents a cusp-like frequency response [1] centered at the mechanical resonance (10–20 Hz), where minimum attenuation and phase shift occur. An electrical resonance coincident with phase inversion (near 10 KHz) limits the gain margin to 40 dB. In order to flatten the frequency response around the natural frequency, driving methods apply pole-zero compensation to provide additional 20 dB in gain margin. Feed-forward control techniques are used to improve low frequency behavior. However, low gain at high frequencies and null DC sensitivity are inherent characteristics of these devices that can not be easily overcome. Moreover, several electrical and mechanical nonlinearities further deteriorate far-from-resonance operation.

Given an abrupt change in reference, as required during CV returns and slope changes in CA mode, the low gain at high frequencies is the cause of long error stabilization periods. PV operation is more affected as it requires CV waveforms to be periodically modified during the experiment to achieve channel selection. In this case it is important to select fixed frequency operation, to avoid that a change in velocity should alter the frequency requirements imposed to the feedback system. Otherwise compensation will require adjustments, turning automatic operation unfeasible. In order to avoid that, the harmonic content of the reference waveform should remain

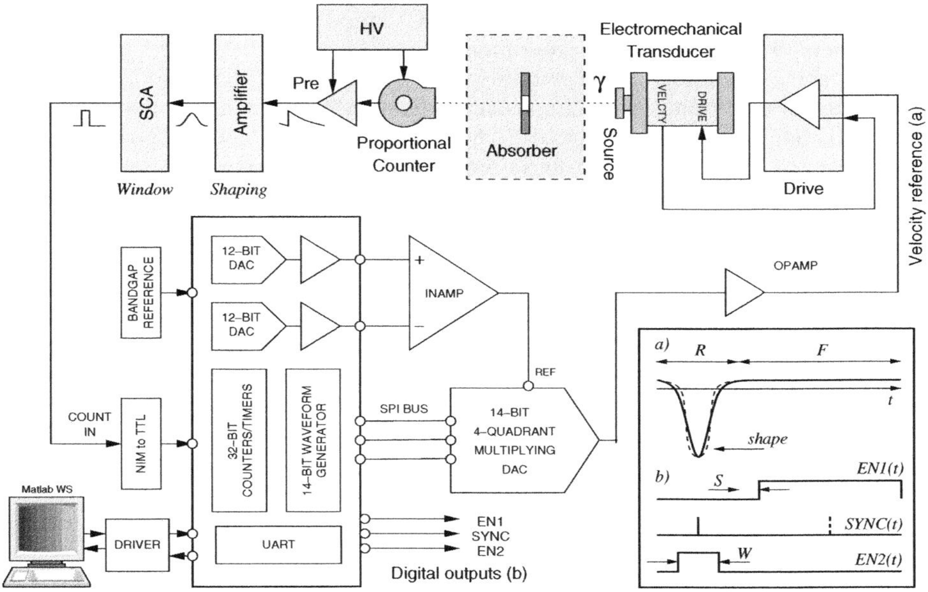

Fig. 1 Schematic diagram of the proposed data acquisition system. Traditional driving stage and detection of resonant photons (*shaded*) are included in order to complete a programmable-velocity spectrometer. *Inset* outputs of the digital waveform generator for one period, operating in PV mode: **a** analog velocity reference and **b** digital synchronization signals required for automatic operation. All parameters (R, F, S and W) are programmable in order to optimize transducer behavior. Return trajectory shape is also configurable so its frequency content can be tuned with control system frequency response. *Dotted line* in (**a**) represents a *softer* trajectory, with the same effective displacement than solid line, that can better suit a massive system

unaltered for all forward velocities, being restricted by the frequency characteristic of the transducer.

The solution proposed is to provide a smooth periodic flat-topped null-media waveform with the appropriate frequency content, which is linearly amplified to get the required forward velocity. Parameters of the proposed reference are shown in Fig. 1 together with the whole experimental configuration. As presented in the inset, duty-cycle $F/(F+R)$, efficiency $(F-S)/(F+R)$ and *shape* must remain constant, providing identical behavior for all channels, thus enabling programmable operation. A drawback of fixed frequency operation is the degradation of potential efficiency improvement in low velocity channels. This technique equals efficiency for all the velocity range.

Frequency content considerations can also be applied to CA operation, where velocity reference is a null media triangular voltage that introduces a transient response after the change of slope. A smooth reference can improve efficiency reducing the settling time.

In order to achieve smoothness, the velocity reference should observe as many continuous derivatives as possible. But narrowing the reference frequency content around transfer cusp implies widening the return trajectory in the time domain. Fourier transform localization property [10] shows that it is not possible to arbitrarily concentrate both a function and its frequency spectrum. An efficient velocity reference (a large flat section in CV mode) will demand a fast return movement, which will degrade smoothness.

The driving coil has a power limit given by its limited dimensions, as it must be designed to actuate inside a constant magnetic field gap, a fact that also limits the maximum allowed displacement. Then, the driving stage output power must be limited in order not to exceed these limits. In a constant magnetic field, the force (i.e. acceleration) is proportional to the driving current. Being the driving coil highly inductive, voltage is proportional to current first derivative (i.e. velocity reference second derivative). In this context, it is advisable not only to select waveforms that require a continuous voltage and derivatives. It is also important to require a limited voltage magnitude. A third derivative must also be considered in order to address output amplifier maximum slew rate. If the driving unit fails to deliver required output voltage in the feedback loop (because of coil protection or slew rate limitations) the recovery time is degraded.

In consequence, for each electromechanical transducer with a given feedback layout, there is a triple compromise between reference smoothness (minimum bandwidth), duty-cycle (minimum reversing trajectory) and third-order derivative magnitude (amplifier slew rate). As these three variables can not be minimized simultaneously, a balance must be found in order to get a maximum efficiency. It was found that the optimal waveform depends on transducer and driving unit characteristics, but some general rules can be established.

As a first practical approximation, it must be considered that transducer characteristics impose frequency and power limits to the optimal reference waveform. These limits can be roughly calculated and must be implemented within a reasonable security margin. Fundamental frequency of the velocity reference must be selected slightly below transducer resonance (that can be easily measured) in order to adjust the center of the reference frequency content with the transducer response. Additionally, transducer power capabilities limit the maximum amplitude for reversing trajectory in the highest velocity channel. Given this limited amount of deliverable power, duty-cycle and waveform are hardly related.

Once power and frequency limits are secured, CA or CV waveform must be selected in order to maximize efficiency. As shown, smoother waveforms demand the smaller driving power and can maximize useful spectrum region without saturation of the driving system. The optimal balance between softness and duty-cycle will be different for each individual transducer and driver pair.

Several functions were considered as candidates for optimal return trajectory in CV/PV operation or slope change in CA mode. A first useful group of trajectories can be built using n-order splines. In this case rising the order for a fixed duty-cycle increases softness but also increases maximum required velocity; the same behavior present the even powers of a sine half cycle. With a different approach, portions of ascending frequency sines or cosines can be combined with increasing order polynomials to form an adequate returning (i.e. two stretches of quarter period cosine superimposed to a parabola with offset can provide a continuous third derivative with moderate amplitudes). In terms of the localization property, self-dual functions [10] show a special behavior: Gaussian shaping presents the minimum product between time and frequency standard deviations; being sharp and asymptotic in the extremes, net trajectory is reduced and requires relatively high velocities at the center. Other exponential or asymptotic functions (like $e^{-1/t}$, $e^{-\exp(-t)}$, $e^{-1/(1-t2)}$, *erf* or *tanh*) can be carefully spliced together in order to compose a soft return trajectory. A more systematic method to produce trajectories is to propose an n-order polynomial with $n + 1$ constrains (trajectory limits, null media and m-order continuous derivatives).

Derivatives and FFT were evaluated for these functions considering a varying duty cycle. Quantization and sampling frequency effects were also studied. A set of best performing waveforms was selected for experimental testing in commercial velocity transducers and drivers. The study was conducted using an ad-hoc designed waveform generator that was later incorporated to the spectrometer implementation that will be presented in the last section. It implements memory stored tables with different high resolution base shapes, programmable duty-cycle and efficiency, null-media, and full dynamic range resolution for all velocities. Its output can be externally amplified preserving resolution (14-bit) for channel selection. Sampling frequency (200 KHz) was selected well above sensitivity of the electromechanical transducers. Response to selected waveforms was tested for two different configurations: (a) Austin Science Associates K-3 linear motor with drive unit, and (b) WissEl MA250 velocity transducer with MR350 drive unit. Both arrangements exhibited comparable behavior. Imposing up to a continuous third-order derivative gives the best performance, as increasing this order, even though improving smoothness, unnecessarily forces the slew rate of the driving amplifier. Shapes of a fourth power of a sine semi cycle present also a good behavior, presenting a short underdamped transient response. Efficiencies up to 75% for the first arrangement and 85% for the second were achieved in both horizontal and vertical layouts.

3 Synchronization and stability considerations in PV mode

Being reversing time not negligible in CV/PV operation, inhibition for the counting system must be provided by the waveform generator during $R + S$ reversing interval (see Fig. 1 inset). Both counter and live-time timer must be paused while the transducer returns and stabilizes the recording velocity. Digital *EN1* signal can be used to implement this functionality. Duty-cycle of *EN1* is equivalent to counting efficiency, as usual in CV operation. But in order to achieve efficient programmable operation, two additional timing aspects must be considered.

Channel selection is accomplished by proportionally amplifying the base reference waveform. In order not to alter the mean value of the pickup displacement, channel changes must be performed either gradually, what is time consuming, or in precise synchronization with the trajectory. Given the limited low frequency sensitivity, a DC perturbation would only be addressed by the pickup springs, introducing a long term stabilization interval of several periods. Changing amplitude exactly in the trajectory center is the only way to avoid this problem, as a change in waveform amplitude does not require a change in position. When a new velocity is requested to the generator, it must be held until the center of the trajectory is reached. With that purpose a second synchronization signal must be provided by the generator, indicating it is ready to accept changes, as shown in Fig. 1 inset (*SYNC*). The reversing trajectory center is preferred, with the drawback of being the maximum velocity point. If transient response can not be stabilized, forward trajectory center can be used (dotted *SYNC*), with the drawback that channel commutation must be implemented while recording.

Under certain circumstances the CV/PV return trajectory can also be used to record resonant absorption (reverse-counting technique). This can be used to improve efficiency, but some remarks must be done regarding the usefulness of the results. As velocity precision can not be guaranteed in this region, it must be taken into account that reference is only an approximation of actual velocity. Despite this uncertainty, the technique can be useful for background recording. For example, if absorption has been recorded in the last absorption line of a sextet, experimentalist can be confident that negative-half background rate is being recorded during the fastest return path (when duty cycle is 80%, maximum return velocity is about four times the programmed velocity). To exploit this characteristic, a second ratemeter and a digital vetoing signal can be added to count events that occur only during the fastest return zone. This inhibition signal must be symmetrical with the trajectory center during the return path (maximum velocity), with a single programmable parameter W (see *EN2* in Fig. 1). If reverse-counting provides useful information, instrument efficiency can be redefined as $(F + W - S)/(F + R)$ and it can be considered to provide a direct measurement of relative resonant absorption. In a real application (e.g. sixth line temperature tracking) a 10% W can be used to rise from 80% to 90% the spectrometer efficiency. In such a case, every eight cycles at programmed velocity (forward-counting), one equivalent background channel will be obtained with reverse-counting.

This technique can also be used to mitigate stability issues that are present during PV operation. This mode enables the recording of a single channel of the spectrum for an arbitrarily long period of time before moving to the following point of interest. If background rate is not being recorded, eventual changes in the emission rate (i.e. due to thermal instability) can not be differentiated from resonant absorption changes [12]. Moreover, radioactive source fading can not be neglected for long term experiments dealing with small Mössbauer effects, being mean life of ^{57}Co of about one year (i.e. 0.3%/day). These inconveniences are not present when the whole spectrum is swept several times per second (as in CA mode), as the effect of a slow counting rate drift is spread uniformly in all the channels, keeping the spectrum shape unaltered. An additional background channel can be recorded under PV operation in order to monitor these phenomena, but efficiency will be degraded if only a few channels of interest are being considered. Background reverse counting can be a good alternative in such cases. For the same reason, if several channels are being recorded in PV mode, alternating between them at relatively small intervals is advisable.

4 Implementation of a programmable data acquisition system

An implementation in the form of a programmable-velocity scaler (PVS) is presented. It integrates the programmable digital waveform generator, a synchronized multichannel counter and a serial interface. Nuclear Instrumentation Module (NIM) standard was used for final implementation.

PVS is designed as a direct replacement of the usual Multichannel Scaler (MCS) in Mössbauer experiments. Figure 1 shows a schematic diagram of the scaler. Its main components are a 14-bit 4-quadrant precision multiplying DAC and a 32-bit RISC microcontroller that includes two embedded 12-bit DAC and several 32-bit timer/counters. Programmable base waveforms are stored in microcontroller memory and sent to DAC via SPI bus. Chanel selection for PV mode (or amplitude for CA mode) is accomplished by driving the external DAC reference with both embedded DAC through an instrumentation amplifier (INAMP), preserving shape. Additional circuitry is provided for NIM pulse conditioning and output stage.

The module supports CA, CV and PV operation modes. PV operation is based in velocity/live-time pairs. While counting photons from the SCA (forward and reverse, using enable signals), the reference is driven for the required number of cycles in order to get the live counting time. Once live-time is reached, both forward and reverse counting are returned through the serial link. The module is ready to accept a new velocity/live-time request, while keeps driving the transducer in the last velocity with halted ratemeters. When the new request is accepted, reference is synchronously updated and counting restarts.

User interface was designed to interact with high level scientific programming languages that provide tools for experiment setup, on-line control and dynamic observations. Since no hard real-time restrictions exist (a delay in serial communications degrades overall efficiency but not quality of the results) any high level language that can handle serial links can be used for building a custom application program. An I/O library was developed including procedures for the basic functionality. Additional functions were also developed to aid in measurement automation, like region of interest recording, channel swept, etc. Combining these tools with high level programming language capabilities, advanced real-time processing can be achieved in a medium-sized workstation without considerable efficiency degradation. For example, a least squares spectrum fit can be used every minute for measurement monitoring, introducing a negligible efficiency loss. A lower-level language implementation of the library is also available for embedded applications development.

5 Performance and applications

In order to characterize efficiency, linearity and energy resolution of the instrument in PV mode, a contrast procedure was conducted. A well known spectrum was recorded with traditional equipment and compared with the same spectrum recorded with the new scaler. Both calibrations were performed recording a complete α-Fe 12 μm iron foil spectrum for one hour. A range from -8 to 8 mm/s was programmed in both cases. Common instrumental to both experiments was proportional counter, ^{57}Co source mounted on MA250 transducer with MR350 driving unit, 142PC preamplifier, 572A amplifier and 420A single channel analyzer. The detection system was set in order to output 10^4 s^{-1} background rate.

The CA spectrum was recorded with a Nucleus MCS-II multichannel scaler board operating in 1,024 channels (512 after folding). The PV spectrum was taken replacing MCS for PVS (with 85% duty-cycle and 80% efficiency) where 512 channels were swept at one second per channel. Figure 2 shows the normalized superposition of both results. All six absorption minimums resulted coincident in less than one channel, evidencing excellent linearity and resolution. Regarding efficiency, programmable-velocity spectrum accumulates 80% of the statistics, a fact that is not evident in the figure. Repeatability resulted satisfactory providing detector thermal stability.

In a closer look (see fifth line detail in Fig. 2) the new spectrum presents sharper lines. After fitting considering the sample thickness, PV line width reduces from

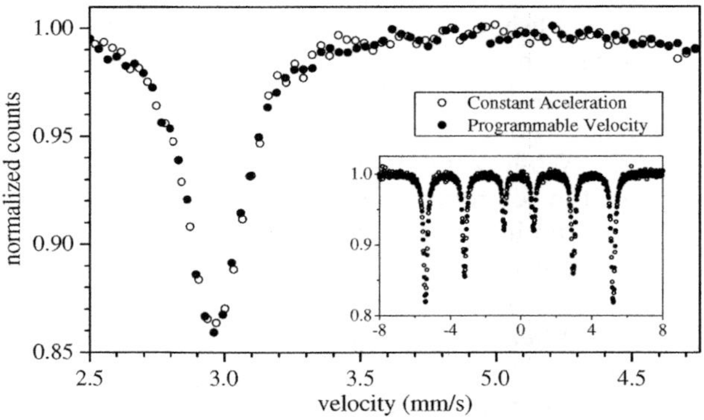

Fig. 2 Superimposed plot of CA and PV calibration runs on α-Fe 12 μm iron foil for one hour at room temperature. The first one (*open dots*) was recorded with a Nucleus MCS-II board and the second one (*solid dots*) with the new scaler operating in PV mode, using common detection and driving systems. In the *inset*, full range evidences equivalent linearity and energy resolution. The main figure presents a detail of the fifth absorption line, evidencing a small improvement in line resolution that can be explained based on the different operational mode

0.214 ± 0.006 to 0.208 ± 0.006 mm/s and sixth line relative effect increases a 2%. This small improvement originates in the CV operational nature. For each CA interval the velocity reference is a ramp between two consecutive channels, not a constant, so counting occurs in an energy interval rather than in a single energy. Every CA channel records the mean absorption within this small velocity range. Alternately, CV technique holds a single energy value for the entire counting interval, providing enhanced resolution. This is evidenced by deeper lines that improve signal to noise spectrum ratio, compensating in part the efficiency loss due to CV operation.

The programmable capabilities of this instrument have been successfully applied to the determination of magnetic hyperfine field temperature dependencies [11, 12], crystallization kinetics [13] and Fe magnetic moments dynamics [14]. The ability to interact with a high-level algorithm between absorption records (channel/live-time pairs) expands the application scope to dynamic experiments. For example, a group of selected channels can be displaced in energy while the spectrum region of interest shifts in energy as a consequence of some external parameter manipulation (e.g. temperature). Mössbauer Line Tracking (MLT) is a recently developed technique [15–17] designed to continuously locate a resonant absorption line during the experiment. In such application the counting efficiency is critical since it imposes dynamic limits. Reducing the time per channel or the number of recorded channels can help to improve responsiveness at expenses of increasing statistical uncertainty. Programmable capabilities of the presented equipment open a door for discussions on optimal region of interest design [18].

References

1. Cranshaw, T.E.: Mössbauer spectroscopy. J. Phys. E: Sci. Instrum. **7**, 497–507 (1974)
2. Kankeleit, E.: Velocity spectrometer for Mössbauer experiments. Rev. Sci. Instrum. **35**, 194–197 (1964)

3. Gaitanis, N., Kostikas, A., Simopoulos, A.: A constant acceleration Mössbauer spectrometer with velocity range selectivity. Nucl. Instrum. Methods **75**, 274–276 (1969)
4. Lipkin, J., Schechter, B., Shtrikman, S., Treves, D.: Inexpensive automatic recording Mössbauer spectrometer. Rev. Sci. Instrum. **35**, 1336–1339 (1964)
5. Sarma, P.R., Sharma, A.K., Tripathi, K.C.: A constant velocity Mössbauer spectrometer free of long-term instrumental and radioactive decay drifts in the count rate. Nucl. Instrum. Methods **164**, 591–593 (1979)
6. Rubin, D.: Constant acceleration transducer employing negative feedback for use in Mössbauer experiments. Rev. Sci. Instrum. **33**, 1358–1360 (1962)
7. Seberini, M.: A constant velocity Mössbauer drive. J. Phys. E: Sci. Instrum. **21**, 641–647 (1988)
8. Veiga, A., Martínez, N., Mayosky, M., Spinelli, E., Mendoza Zélis, P., Pasquevich, G.A., Sánchez, F.H.: A constant-velocity Mössbauer spectrometer with controlled temperature swep. Rev. Sci. Instrum. **73**, 3579–3583 (2002)
9. Veiga, A., Martínez, N., Mendoza Zélis, P., Pasquevich, G.A., Sánchez, F.H.: Advances in constant-velocity Mössbauer instrumentation. Hyperfine Interact. **167**, 905–909 (2006)
10. Papoulis, A.: The Fourier Integral and its Applications. McGraw-Hill, New York (1962)
11. Mendoza Zélis, P., Pasquevich, G.A., Sánchez, F.H., Veiga, A., Martínez, N.: A new application of Mössbauer effect thermal scans: determination of the magnetic hyperfine field temperature dependence. Phys. Lett. A **298**, 55–59 (2002)
12. Pasquevich, G.A., Mendoza Zélis, P., Fernández van Raap, M.B., Sánchez, F.H.: Hyperfine field temperature dependence of Fe3Si from Mössbauer thermal scans. Physica B **354**, 369–372 (2004)
13. Saccone, F.D., Rodríguez Torres, C.E., Pasquevich, G.A., Fernández van Raap, M.B., Sánchez, F.H.: Crystallisation kinetics of B-rich mischmetal-Fe-B nanocomposite ribbons. Physica B **354**, 237–240 (2004)
14. Pasquevich, G.A., Mendoza Zélis, P., Sánchez, F.H., Fernández van Raap, M.B., Veiga, A., Martínez, N.: Determination of the iron magnetic moments dynamics in the nanocrystalline ribbons $Fe_{90}Zr_7B_3$ by Mössbauer magnetic scans. Physica B: Condensed Matter. **384**, 348–350 (2006)
15. Veiga, A., Pasquevich, G.A., Mendoza Zélis, P., Sánchez, F.H., Fernández van Raap, M.B., Martínez, N.: Experimental design and methodology for a new Mössbauer scan experiment: absorption line tracking. Hyperfine Interact. **188**, 137–142 (2009)
16. Mendoza Zélis, P., Pasquevich, G.A., Veiga, A., Fernández van Raap, M.B., Sánchez, F.H.: A quasi-continuous observation of the α-transition of Fe1+xS by Mössbauer line tracking. Hyperfine Interact. **195**, 161–165 (2010)
17. Mendoza Zélis, P., Pasquevich, G.A., Sánchez, F.H., Veiga, A., Ceolin, M., Cabrera, A.F., Coronado-Miralles, E., Monrabal-Capilla, M., Galan-Mascaros, J.R.: Mössbauer thermal scan study of a spin crossover system. J. Phys.: Conf. Series **217**, 012017 (2010)
18. Pasquevich, G.A., Veiga, A., Mendoza Zélis, P., Sánchez, F.H.: Optimal configuration for programmable Mössbauer experiments. J. Phys.: Conf. Series **217**, 012139 (2010)

Hyperfine Interact (2011) 202:117–121
DOI 10.1007/s10751-011-0366-9

Calibration of the isomer shift of the 14.4 keV transition of ^{57}Fe

A. M. Mudarra Navarro · C. E. Rodríguez Torres · L. A. Errico

Published online: 25 August 2011

Abstract With the aim of calibrating Mössbauer spectroscopic measurements, the electron contact density was calculated on the Fe nuclear position in various iron compounds. The full-potential augmented-plane waves plus local orbitals (APW+lo) method within the local-density approximation was used, considering different approximations for the exchange and correlation potential. By comparison with experimental ^{57}Fe Mössbauer data, the calibration constant α relating measured isomer shifts to the electron contact density was derived. An isomer shift calibration constant $\alpha = -0.25_1$ a.u.^{3} mm s^{-1} has been obtained.

Keywords Fe · Ab initio · Calibration constant · Mössbauer spectroscopy

1 Introduction

Hyperfine interaction techniques such as Mössbauer spectroscopy (MS), nuclear magnetic and nuclear quadrupole resonance (NMR and NQR) or perturbed angular correlations (PAC), among others [1, 2] are based on the observation of the coupling between nuclear moments and extra-nuclear fields. These techniques are often used as a powerful tool to characterize precisely atomic sites in a given sample and to

A. M. Mudarra Navarro · C. E. Rodríguez Torres (✉) · L. A. Errico
Departamento de Física-IFLP (CONICET-UNLP), Facultad de Ciencias Exactas, Universidad Nacional de La Plata, CC 67, 1900 La Plata, Argentina
e-mail: torres@fisica.unlp.edu.ar

A. M. Mudarra Navarro
e-mail: mudarra@fisica.unlp.edu.ar

L. A. Errico
Universidad Nacional del Noroeste Bonaerense (UNNOBA), Monteagudo 2772, Pergamino, Bs. As., Argentina

obtain information about local symmetry, coordination, and valence of defect or native centers in solids, among other electronic, structural, and magnetic properties on a nanoscopic scale [3]. In particular, MS of ^{57}Fe plays an important role in the structural determination of iron containing solid state compounds.

Two magnitudes that can be determined by MS are: the isomer shift (IS), which provides information on the local chemical bond, and the quadrupole splitting (QS), which is a "fingerprint" of the local symmetry around the probe nucleus. Both quantities depend on the solid state environment. The IS is given by [1, 2]:

$$\mathrm{IS} = \alpha(\rho_A - \rho_S) \tag{1}$$

were ρ_A and ρ_S stand for the electron charge densities at the nuclear positions (the contact densities) in two different solid state environments, here denoted the absorber (A) and the source (S) materials, respectively. The QS is originated by the interaction of the nuclear quadrupole moment with the crystalline electrostatic potential. For the case of the 14.4 keV transition of ^{57}Fe, the QS is given by [1, 2]:

$$\Delta_{EQ} = \frac{eQ}{2} V_{ZZ} \left(1 + \frac{\eta^2}{3}\right)^{1/2} \tag{2}$$

where e is the elementary charge, Q is the nuclear quadrupole moment of the excited state of the Mössbauer nucleus, and V_{ZZ} and η are the main component of the electric field gradient (EFG) tensor and asymmetry parameter, respectively [1, 2]. All the information that the IS and the QS can provide about the system under study could be obtained by confrontation with an accurate prediction of the contact densities and the EFG, such as those obtained with ab initio electronic-structure calculations [4–6]. Of course, accurate values of the nuclear-quadrupole moments Q and α are essential for a correct comparison between the experimentally "determined" contact densities and EFGs. For this reason, the calibration of MS has been an issue for a long time (see for example, [7–9] and references therein). Nowadays, the more accurate scheme for the calculation of Q and α is the determination of these parameters from the comparison of calculated electron contact densities or EFGs with corresponding experimental IS and QS.

The Q value of the first excited state of ^{57}Fe has been estimated in the early days of the Mössbauer spectroscopy as lying in between +0.15 b and +0.28 b. Calculations performed later suggested $Q = +0.082$ b [10, 11]. More sophisticated models lead to $Q = +0.16$ b [12, 13]. In the case of the calibration constant α, values vary from -0.11 to -0.62 a.u.^{3} mm s^{-1} [14, 15] for the transition considered here. Ericksson and Svane [16] obtained $\alpha = -0.22$ a.u.^{3} mm s^{-1}, in agreement with previous results. More recently, U. D. Wdowik and K. Ruebenbauer [8] and R. Kurian and M. Filatov [9] reported $\alpha = -0.29$ and -0.31 a.u.^{3} mm s^{-1}, respectively for metallic iron compounds. In view of these discrepancies, in this work we present an ab initio determination of the calibration constant α of the 14.4 keV transition of ^{57}Fe including also insulators compounds. This determination was obtained comparing experimentally determined IS to calculated contact densities at Fe sites in Fe compounds using the full-potential augmented-plane waves plus local orbitals (APW+lo) method, which became the most accurate and reliable method to include solid-state effects.

Table 1 Differences in electron densities $\rho_A - \rho$ (BCC-Fe), and relative experimental IS [25] for the compounds taken into account in the present work

	$\rho_A - \rho$ (BCC-Fe), LDA (a.u^{-3})	$\rho_A - \rho$ (BCC-Fe), CW (a.u^{-3})	$\rho_A - \rho$ (BCC-Fe), PBE-sol (a.u^{-3})	IS (mm/s)
Γ−Fe	+0.38	+0.29	+0.30	−0.1
α−Fe	0	0	0	0.0
Fe_2O_3	−1.64	−1.73	−1.71	0.37
FeO	−3.28	−3.49	−3.49	0.81
$FeTiO_3$	−4.15	−4.32	−4.35	1.07

Electron densities and shifts are relative to BCC α-Fe

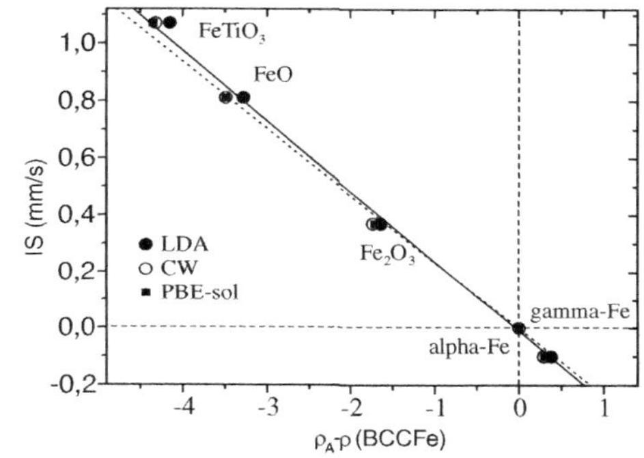

Fig. 1 Experimental isomer shift (relative to BCC-Fe) versus difference in electron density $\rho_A - \rho$ (BCC-Fe) for the studied Fe-compounds. Solid (*dashed*) line corresponds to a linear fit to the data obtained with LDA (CW/PBE-sol)

2 Method of calculation

Theoretical calculations based on the density functional theory (DFT, [17, 18]) were performed with the WIEN2K implementation [19] of the APW+lo method [20, 21] in a scalar relativistic version. Exchange and correlation effects were treated within DFT using the local-density (LDA, [22]), the generalized-gradient approximation recently proposed by Wu and Cohen (WC, [23]) and the PBE-sol [24] approximations. In this method, the wave functions are expanded in spherical harmonics inside non-overlapping atomic spheres of radius R_{MT} and in plane waves in the remaining space of the unit cell (the interstitial region). The parameter $R_{MT}K_{max}$, which controls the size of the basis set, was set to nine. Integration in reciprocal space was performed using the tetrahedron method taking up to 500 k-points in the case of oxides (20,000 k-points in the case of metallic compounds) in the full Brillouin zone. In order to evaluate the convergence of our results, we performed calculations for different muffin-tin radii, k-point sampling, and basis-set size. In all the cases, we used the experimental lattice parameters but we optimized the internal position parameters of all atoms of the structures by force minimization (for details see [5, 6]). Atomic positions were relaxed until the forces were below 0.1 mRy/au.

3 Results and conclusions

The differences in electron densities $\rho_A - \rho$ (BCC-Fe), and relative (to BCC-Fe) experimental IS for all the studied systems are listed in Table 1. The calculated values are converged to the digits shown. Of course, the calculated values are subject (besides the converge error) to systematic errors, that include, for example, the validity of the parameterizations employed for the exchange and correlation potential, the scalar relativistic approximation, and other approximations of the computational scheme. An estimate of 10% uncertainty (due to the "systematic errors" mentioned above) in the calculated contact densities seems to be conservatively safe.

The agreement between experiment and theory may be judged from plot of the experimental IS as a function the calculated electron contact densities difference $\rho_A - \rho$ (BCC-Fe), which is shown in Fig. 1. The straight line through the points in Fig. 1 confirms relationship (1). From a best linear fit we derive the calibration constant α of (1) we obtained $\alpha = -0.25_1$ a.u.^{3} mm s^{-1} in the case of LDA and $\alpha = -0.23_1$ a.u.^{3} mm s^{-1}. This error is solely due to the errors of linear regression. The result obtained for α is in good agreement with the value reported in Ref. reported in [16], but is smaller that the results reported in [8, 9], were metallic compounds were considered. Calculations including more Fe-compounds, with different ionic character and other approximations for the exchange and correlation potential are now in progress.

Acknowledgements This work was partially supported by ANPCyT, Argentina under Grant No. PICT98 03-03727, CONICET, Argentina (PIP 0002), and the Third World Academy of Sciences (TWAS, Italy, RGA 97-057). This research made use of the HP-Parallel-Computing Bose Cluster, and the computational facilities of the Physics of Impurities (PhI) group at IFLP and Departamento de Física (UNLP). L.E. and C.R.T. are members of CONICET.

References

1. Schatz, G., Weidinger, A. (ed.): Nuclear Condensed Matter Physics—Nuclear Methods and Applications, p. 63. Wiley, Chichester, England (1996)
2. Frauenfelder, H., Steffen, R.M.: In: Siegbahn, K. (ed.) Alpha-, Beta-, And Gamma-Ray Spectroscopy, vol. 2. North-Holland, Amsterdam (1966)
3. Errico, L., Rentería, M., Petrilli, H.M.: Phys. Rev. B **75**, 155209 (2007)
4. Akai, H., Akai, M., Blügel, S., Drittler, B., Ebert, H., Terakura, K., Zeller, R., Dederichs, P.H.: Prog. Theor. Phys. Suppl. **101**, 11 (1990)
5. Errico, L., Fabricius, G., Rentería, M., de la Presa, P., Forker, M.: Phys. Rev. Lett. **89**, 055503 (2002)
6. Errico, L., Fabricius, G., Rentería, M.: Phys. Rev. B **67**, 144104 (2003)
7. Svane, A., Christensen, N.E., Rodríguez, C.O., Methfessel, M.: Phys. Rev. B **55**, 12572 (1997)
8. Wdowik, U.D., Ruebenbauer, K.: Phys. Rev. B **76**, 155118 (2007)
9. Kurian, R., Filatov, M.: Phys. Chem. Chem. Phys. **12**, 2758 (2010)
10. Duff, J., Mishra, K.C., Das, T.P.: Phys. Rev. Lett. **46**, 1611 (1981)
11. Vajda, S., Sprouse, G.D., Rafailovich, M.H., Noe, J.W.: Phys. Rev. Lett. **47**, 1230 (1981)
12. Dufek, P., Blaha, P., Schwarz, K.: Phys. Rev. Lett. **75**, 3545 (1995)
13. Martínez-Pinedo, G., Schwerdtfeger, P., Caurier, E., Langanke, K., Nazarewicz, W., Söhnel, T.: Phys. Rev. Lett. **87**, 062701 (2001)
14. Simanek, E., Sroubek, Z.: Phys. Rev. **163**, 275 (1967)
15. Uher, R.A., Sorensen, R.A.: Nucl. Phys. **86** (996)
16. Eriksson, O., Svane, A.: J. Phys., Condens. Matter **1**, 1589 (1989)
17. Hohenberg, P., Kohn, W.: Phys. Rev. **136**, B864 (1964)
18. Kohn, W., Sham, L.J.: Phys. Rev. **140**, A1133 (1965)
19. Blaha, P., Schwarz, K., Madsen, G., Kvasnicka, D., Luitz, J.: WIEN2k. In: Schwarz, K. (ed.) An Augmented Plane Wave Plus Local Orbitals Program for Calculating Crystal Properties. Technical Universität Wien, Austria (1999)
20. Sjöstedt, E., Nordström, L., Singh, D.J.: Solid State Commun. **114** (2000)

21. Madsen, G.K.H., Blaha, P., Schwarz, K., Sjöstedt, E., Nordström, L.: Phys. Rev. B **64**, 195134 (2001)
22. Perdew, J.P., Burke, K., Ernzerhof, M.: Phys. Rev. Lett. **77**, 3865 (1996)
23. Wu, Z., Cohen, R.E.: Phys. Rev. B **73**, 235116 (2006)
24. Perdew, J.P., Ruzsinszky, A., Csonka, G.I., Vydrov, O.A., Scuseria, G.E., Constantin, L.A., Zhou, X., Burke, K.: Phys. Rev. Lett. **100**, 136406 (2008)
25. Dyar, M.D., Agresti, D.G., Schaefer, M.W., Grant, C.A., Sklute, E.C.: Ann. Rev. Earth Planet. Sci. **34**, 83 (2006)

Hyperfine Interact (2011) 202:123–129
DOI 10.1007/s10751-011-0374-9

Mössbauer study of carbon coated iron magnetic nanoparticles produced by simultaneous reduction/pyrolysis

Fernanda G. Mendonça · José D. Ardisson · Marcelo G. Rosmaninho · Rochel M. Lago · Juliana C. Tristão

Published online: 30 September 2011

Abstract Magnetic iron nanoparticles immersed in a carbon matrix were produced by a combined process of controlled dispersion of Fe^{3+} ions in sucrose, thermal decomposition with simultaneous reduction of iron cores and the formation of the porous carbonaceous matrix. The materials were prepared with iron contents of 1, 4 and 8 in %wt in sucrose and heated at 400, 600 and 800°. The samples were analyzed by XRD, Mössbauer spectroscopy, magnetization measurements, TG, SEM and TEM. The materials prepared at 400° are composed essentially of Fe_3O_4 particles and carbon, while treatments at higher temperatures, e.g. 600 and 800° produced as main phases Fe^0 and Fe_3C. The Mössbauer spectra of samples heated at 400° showed two sextets characteristic of a magnetite phase and other contributions compatible with Fe^{3+} and Fe^{2+} phases in a carbonaceous matrix. Samples treated at temperatures above 600° showed the presence of metallic iron with concentrations between 16–43%. The samples heated at 800° produced higher amounts of Fe_3C (between 20% and 58%). SEM showed for the iron 8% sample treated at 600–800°C particle sizes smaller than 50 nm. Due to the presence of Fe^0 particles in the carbonaceous porous matrix the materials have great potential for application as magnetic adsorbents.

Keywords Iron nanoparticles · Sucrose · Carbon coated

F. G. Mendonça · M. G. Rosmaninho · R. M. Lago · J. C. Tristão
Departamento de Química – ICEx, Universidade Federal de Minas Gerais,
Belo Horizonte, MG, 31270-901, Brazil

J. C. Tristão (✉)
Universidade Federal de Viçosa – Campus Florestal, Florestal, MG, 35.690-000, Brazil
e-mail: juliana@ufv.br

J. D. Ardisson
Laboratório de Física Aplicada, Centro de Desenvolvimento de Tecnologia Nuclear,
CDTN, Belo Horizonte, MG, 30123-970, Brazil

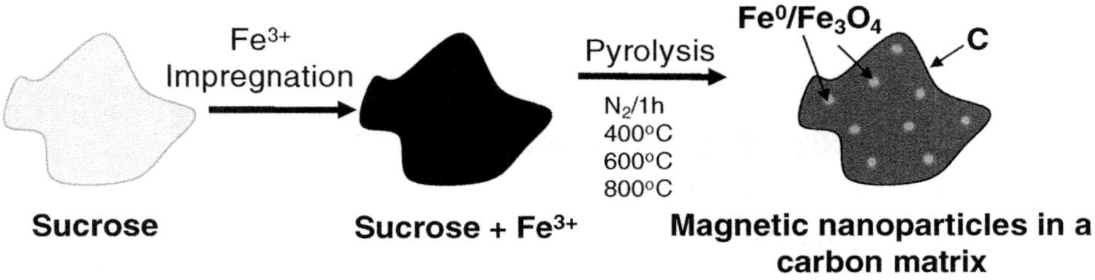

Fig. 1 Magnetic nanoparticles formation by simultaneous reduction/pyrolysis process

1 Introduction

The preparation of magnetic nanoparticles encapsulated by carbon has been actively studied [1–4]. The outside carbon coating can protect the metal nanoparticles against degradation and oxidization and provides new applications for the materials as magnetic adsorbents or catalyst supports. Carbon coated magnetic nanoparticles can be produced by different methods such as arch discharge (e.g. Fe [1], SiO_2/Fe [5], Al_2O_3/Fe [6]), plasma discharge (e.g. Fe_3C [7]), pyrolysis of metallic complexes (e.g. CNT/FeRu [8]) and explosive reaction (e.g. Fe_7C_3 [9]). All processes described above are relatively complex, demand special precursors and produce relatively low yields.

The heating and decomposition of an aqueous solution of glucose is a process to synthesize carbon-encapsulated magnetic nanoparticles, e.g. carbon coating from glucose and nanomaterials containing Ni or Fe covered by Au [10] and Fe_xO_y that were used as catalysts in the Fischer-Tropsch synthesis [11].

In this work, magnetic nanoparticles formed by Fe^0 and Fe_3O_4 immersed in a carbon matrix were produced by simultaneous processes of reduction/pyrolysis of Fe^{3+} dispersed in sucrose. In this process, Fe^{3+} is impregnated in sucrose followed by material pyrolysis at different temperatures (Fig. 1).

1.1 Experimental

The $Fe(NO_3)_3.9H_2O$ solutions containing 1, 4 and 8 wt% in iron were prepared at pH~1. In the next step, the solutions were heated and commercial sugar was added slowly. Heating was sustained until a dark paste was formed. The same procedure (sugar heating) was also performed without the iron salt. The reduction and pyrolysis at different temperature conditions, 400, 600 and 800°, were carried out with 7 g of each sample in N_2 (130 mLmin^{-1}), in a quartz tube into a horizontal furnace (BLUE M. Lindberg) at a heating rate of 10° min^{-1} and kept at the final temperature for 1 h. The samples prepared in this work were named as AXFeY where X = 1, 4 or 8% of iron and Y = 400, 600 or 800°C.

The samples were characterized by X-ray diffraction (XRD, Rigaku Geigerflex using Cu Kα radiation scanning 2θ from 10 to 80°), transmission Mössbauer spectroscopy (spectrometer CMTE MA250 with a ^{57}Co/Rh source at room temperature using α-Fe as a reference), magnetization measurements (portable magnetometer

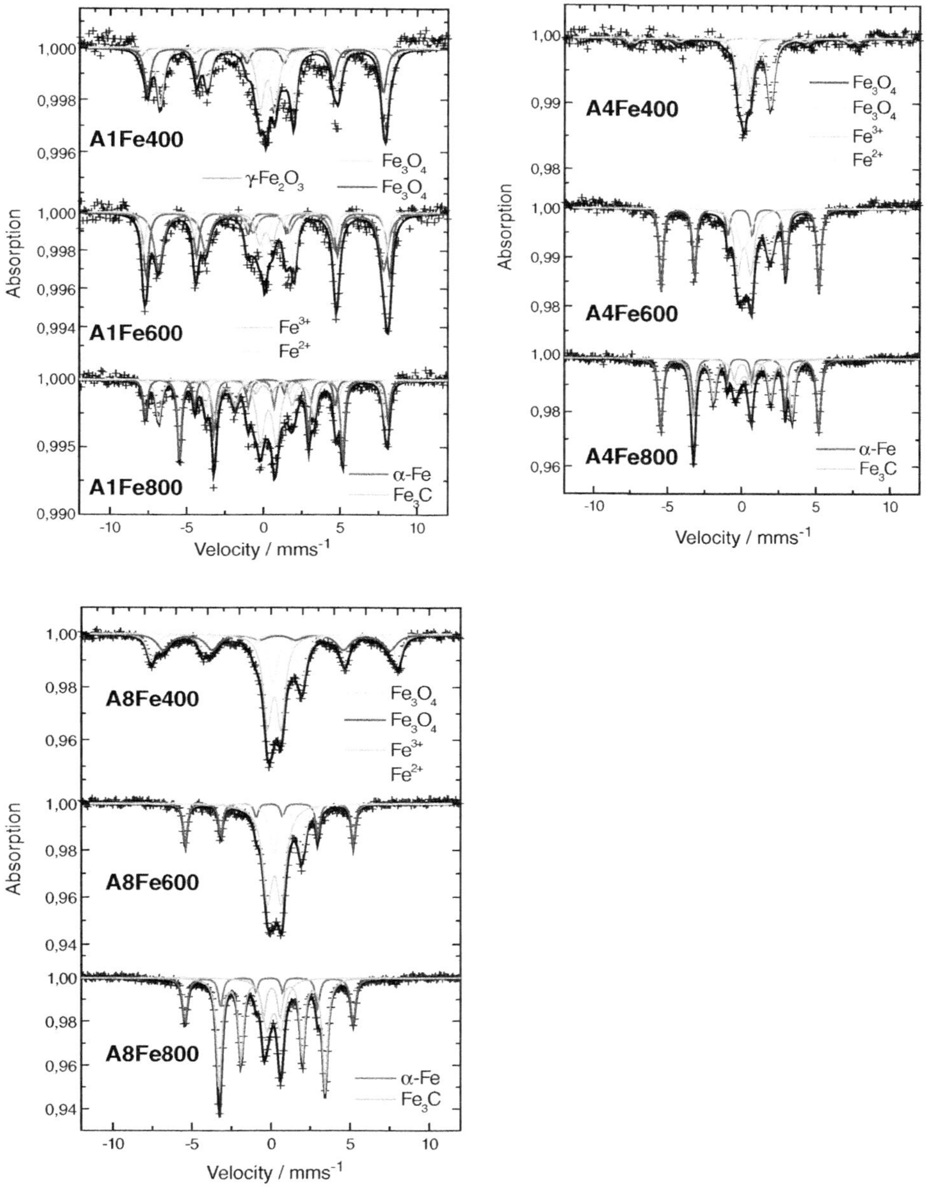

Fig. 2 Mössbauer spectra of materials containing 1 (A1Fe), 4 (A4Fe) and 8 (A8Fe) % iron obtained after sucrose pyrolysis at 400, 600 and 800°C, respectively

with a fixed magnetic field of 0.3 T), thermal analysis (Shimadzu TGA-60, constant heating rate of 10° min^{-1} under air flow of 100 mL min^{-1}), scanning electron microscopy (SEM, Quanta 200 FEG) and transmission electron microscopy (TEM, Tecnai G2–20Twin).

Table 1 Mössbauer hyperfine parameters for the materials containing 1, 4 and 8% iron obtained after sucrose pyrolysis at 400, 600 and 800°

Sample	Phase	δ/ mm s^{-1}	ε, Δ / mm s^{-1}	B_{hf} / T	RA / %
A1Fe400	Fe_2O_3	0.35(5)	−0.04(6)	51.0(5)	4(1)
	[Fe_3O_4]	0.32(4)	0.04(6)	48.3(5)	18(1)
	{Fe_3O_4}	0.56(4)	−0.01(4)	45.8(4)	28(1)
	Fe_3C	0.14(5)	0.03(5)	20.7(5)	22(1)
	Fe(3+)	0.34(6)	0.95(5)	–	16(1)
	Fe(2+)	1.14(5)	1.78(4)	–	12(1)
A1Fe600	Fe_2O_3	0.35(5)	0.04(6)	49.0(5)	22(1)
	[Fe_3O_4]	0.26(5)	0.04(5)	48.3(6)	11(1)
	{Fe_3O_4}	0.56(5)	−0.01(5)	45.8(5)	32(1)
	Fe_3C	0.14(6)	0.03(4)	20.7(4)	17(1)
	Fe(3+)	0.33(4)	0.83(4)	–	8(1)
	Fe(2+)	1.14(5)	1.82(5)	–	10(1)
A1Fe800	[Fe_3O_4]	0.30(4)	0.04(5)	48.9(5)	14(1)
	{Fe_3O_4}	0.66(5)	−0.01(5)	45.8(4)	20(1)
	Fe	0.00(4)	0.00(5)	33.0(5)	18(1)
	Fe_3C	0.14(5)	0.03(6)	20.7(5)	22(1)
	Fe(3+)	0.43(5)	0.99(5)	–	21(1)
	Fe(2+)	1.15(6)	1.82(5)	–	5(1)
A4Fe400	[Fe_3O_4]	0.28(5)	0.04(5)	48.9(5)	11(1)
	{Fe_3O_4}	0.50(5)	−0.01(5)	45.0(5)	8(1)
	Fe(3+)	0.27(5)	0.52(5)	–	33(1)
	Fe(2+)	1.15(5)	1.77(5)	–	48(1)
A4Fe600	Fe	0.00(5)	0.00(4)	33.0(5)	43(1)
	Fe(3+)	0.29(4)	0.87(5)	–	29(1)
	Fe(2+)	1.12(6)	1.80(5)	–	28(1)
A4Fe800	Fe	0.00(5)	0.00(6)	33.0(5)	41(1)
	Fe_3C	0.16(5)	0.03(5)	20.7(5)	43(1)
	Fe(3+)	0.29(5)	0.90(5)	–	16(1)
A8Fe400	[Fe_3O_4]	0.30(5)	0.04(6)	48.5(5)	22(1)
	{Fe_3O_4}	0.50(5)	−0.01(6)	44.5(4)	24(1)
	Fe(3+)	0.30(4)	0.80(5)	–	35(1)
	Fe(2+)	1.15(5)	1.77(5)	–	19(1)
A8Fe600	Fe	0.00(5)	0.00(4)	33.0(5)	22(1)
	Fe(3+)	0.30(6)	0.60(5)	–	53(1)
	Fe(2+)	1.15(5)	1.77(5)	–	25(1)
A8Fe800	Fe	0.00(4)	0.00(5)	33.0(5)	25(1)
	Fe_3C	0.16(5)	0.03(5)	20.8(5)	54(1)
	Fe(3+)	0.27(5)	0.88(6)	–	21(1)

δ isomer shift with respect to α-Fe, ε quadrupole shift, Δ quadrupole splitting, B_{hf} hyperfine Field, RA relative sub-spectral area, [ ] tetrahedral site, { } octahedral site. The numbers enclosed in brackets represent the error in the determination of the Mössbauer parameters.

2 Results and discussion

The materials were characterized after pyrolysis at different temperatures, i.e. 400°, 600° and 800°. XRD analyses showed the presence of the phase magnetite (Fe_3O_4, PDF 1-1111), besides carbon (PDF 26-1077) for the samples with different iron contents (1, 4 and 8%) treated at 400°. At higher temperatures, 600 and 800°, new

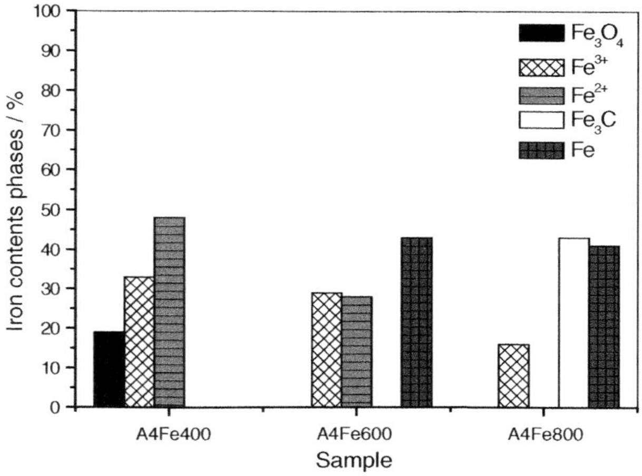

Fig. 3 Iron contents phases for the sample containing 4% iron obtained after pyrolysis at 400, 600 and 800°C

phases such as metallic iron (PDF 1-1267) and iron carbide (Fe_3C, PDF 23-1113) were formed.

The Mössbauer spectra obtained for the materials A1Fe, A4Fe and A8Fe after pyrolysis at 400, 600 and 800°C are shown in Fig. 2. The hyperfine parameters are presented in Table 1. The results show that the spectra recorded from the samples heated at 400° show the two sextets characteristic of Fe_3O_4, and that correspond as it is well-know, to the Fe^{3+} and $Fe^{2.5+}$ ions which occupy the tetrahedral and octahedral sites of its spinel related structure, plus two paramagnetic doublets which parameters characteristic of Fe^{3+} and Fe^{2+}. Phases as γ-Fe_2O_3 were not identified, except for the sample A1Fe400. For the samples treated at 600° it can be observed the presence of metallic iron with relative areas of 22% and 43% for the sample with 8 and 4% of iron, respectively. The samples heated at 800° showed larger relatives areas from the phase Fe_3C, formed from the reaction of metallic iron with carbon produced in the reaction. The relative spectral areas for Fe_3C were 20, 43 and 58% for the samples A1Fe800, A4Fe800 and A8Fe800, respectively.

Contents of iron phases shown in Fig. 3 were obtained by Mössbauer spectroscopy for the samples containing 4% of iron after pyrolysis at 400, 600 and 800°. The reaction at 400° led to the formation of magnetite, which at higher temperatures was converted to Fe^0 and Fe_3C. The dispersed Fe^{3+} and Fe^{2+} represent 33 and 48%, respectively of the iron phases identified in the materials heated at 400°. Treatment at 600 and 800° led to decrease on the Fe^{3+} and Fe^{2+} dispersed phase. Metallic iron Fe^0 is formed at temperature as low as 600° and further converted to Fe_3C at higher temperatures.

Saturation magnetization measurements showed that all materials produced were magnetic. The magnetization increased with iron content in the sample with the highest values of 14 and 22 JT^{-1} kg^{-1}obtained for the samples A4Fe800 and A8Fe800, respectively. These high magnetization values are related to the high Fe^0 concentration as revealed by Mössbauer spectroscopy. TG weight losses suggested carbon contents of 85–95 wt% for the A1Fe material treated at different temperatures,

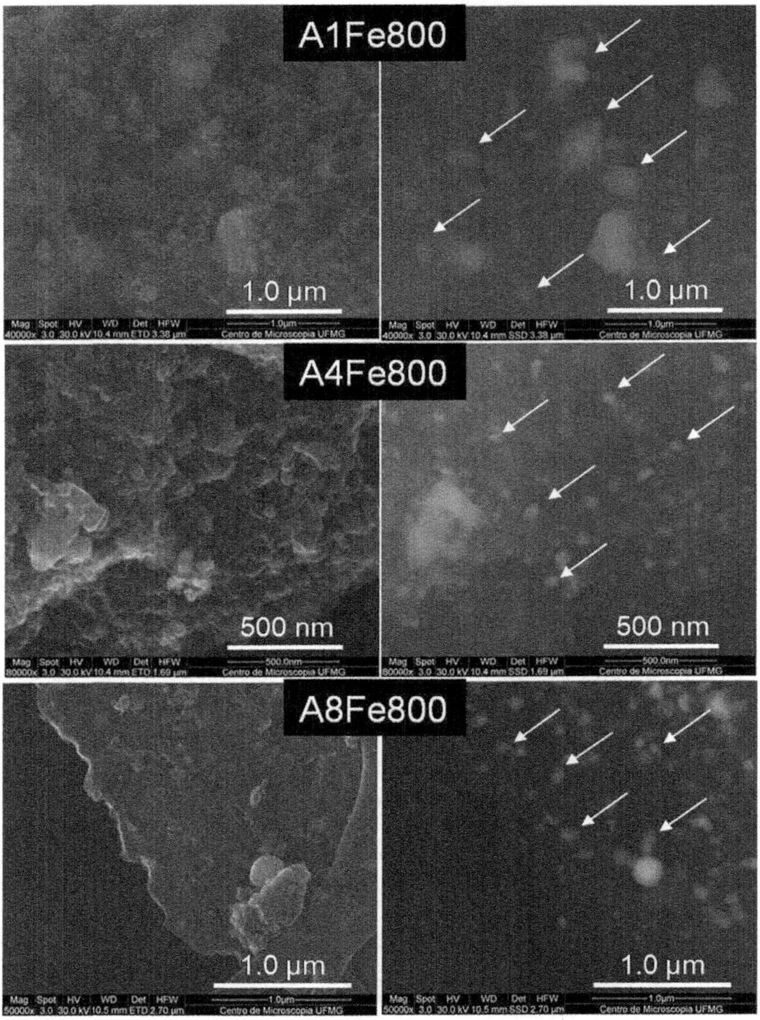

Fig. 4 SEM images of A1Fe, A4Fe and A8Fe after pyrolysis at 800°C

whereas A4Fe and A8Fe showed carbon contents between 74–90 and 61–79 wt% respectively.

SEM images recorded with backscattered electrons (Fig. 4) for the samples treated at 800°C, showed that the iron is present as spherical particles with nanometric dimension of 100 nm for the A8Fe800 and A4Fe400 samples, and some particles reaching 500 nm for the sample A1Fe800, all dispersed in the carbon matrix. Analyses of these particles by TEM (Fig. 5) showed a Fe^0 core with diameter of ca. 50 nm, encapsulated by a more organized carbon, such as graphite carbon layer structures.

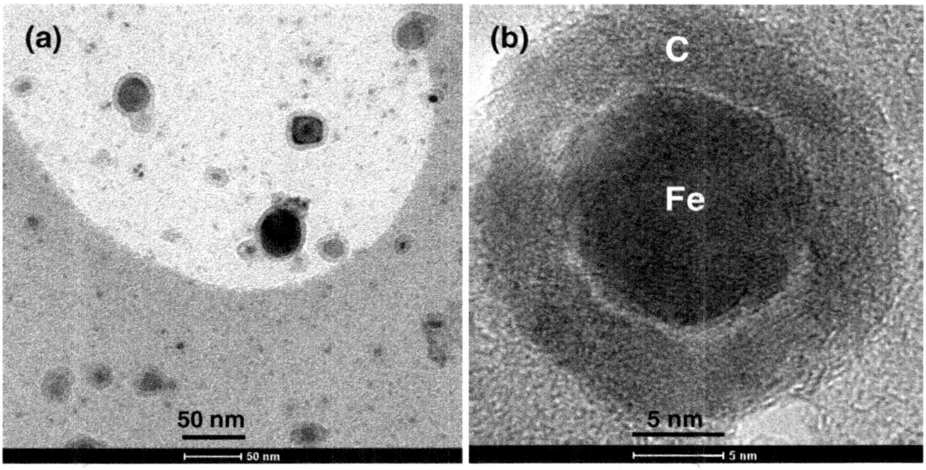

Fig. 5 TEM images of the sample A8Fe800

3 Conclusions

Magnetic nanoparticles dispersed in a carbon matrix can be produced from the pyrolysis of sucrose containing Fe^{3+}. XRD and Mössbauer spectroscopy showed that the treatment at 400° produces essentially the Fe_3O_4 as the main phase, while higher temperatures (600 and 800°C) lead to the predominant formation of the Fe^0 and Fe_3C phases for the samples with iron content of 4 and 8%. For the sample with 1% of iron, the Fe_3O_4 phase is present at all temperatures (400, 600 and 800°C) and the Fe^0 phase exists only at 800°C.

Acknowledgements The authors would like to thank the support from PRPq/UFMG, FAPEMIG, CNPq.

References

1. Qiu, J.S., Li Y.F., Wang, Y.P., An, Y.L., Zhao, Z.B., Zhou, Y., Li, W.: Fuel **86**, 267–274 (2004)
2. Chen, M., Bin, L., Huai-he, S., Lin-jie, Z.: New Carbon Materials **25**, 199–204 (2010)
3. Narkiewicz, U., Podsiadly, M.: Appl. Surf. Sci. **256**, 5249–5253 (2010)
4. El-Gendy, A.A., Ibrahim, E.M.M., Khavrus, V.O., Krupskaya, Y., Hampel, S., Leonhardt, A., Büchner, B., Klingeler, R.: Carbon **47**, 2821–2828 (2009)
5. Fernandez-Pacheco, R., Arruebo, M., Marquina, C., Ibarra, R., Arbiol, J., Santamaria, J.: Nanotechnology **17**, 1188–1192 (2006)
6. Geng, D.Y., Zhang, Z.D., Zhang, W.S., Si, P.Z., Zhao, X.G., Liu, W., Hu, K.Y., Jin, Z.X., Song, X.P.: Scr. Mater. **48**, 593–598 (2003)
7. Sergiienko, R., Shibata, E., Akase, Z., Suwa, H., Nakamura, T., Shindo, D.: Mater. Chem. Phys. **98**, 34–38 (2006)
8. Panchakarla, L.S., Govindaraj, A.: Bull. Mater. Sci. **30**, 23–29 (2007)
9. Wu, W.Z., Zhu, Z.P., Liu, Z.Y., Xie, Y.I., Zhang, J., Hu, T.D.: Carbon **41**, 317–321 (2003)
10. Wang, Z.F., Xiao, P.F., He, N.Y.: Carbon **44**, 3277–3284 (2006)
11. Yu, G., Sun, B., Pei, Y., Xie, S., Yan, S., Qiao, M., Fan, K., Zhang, X., Zong, B.J.: Am. Chem. Soc. **132**, 935–937 (2010)

Hyperfine Interact (2011) 202:131–137
DOI 10.1007/s10751-011-0353-1

Synthesis and characterization of hollow α-Fe_2O_3 sub-micron spheres prepared by sol–gel

Lizbet León · Angel Bustamante · Ana Osorio · G. S. Olarte · Luis De Los Santos Valladares · Crispin H. W. Barnes · Yutaka Majima

Published online: 30 August 2011

Abstract In this work we report the preparation of magnetic hematite hollow sub-micron spheres (α-Fe_2O_3) by colloidal suspensions of ferric nitrate nine-hydrate ($Fe(NO_3)_3{\cdot}9H_2O$) particles in citric acid solution by following the sol–gel method. After the gel formation, the samples were annealed at different temperatures in an oxidizing atmosphere. Annealing at 180°C resulted in an amorphous phase, without iron oxide formation. Annealing at 250°C resulted in coexisting phases of hematite, maghemite and magnetite, whereas at 400°C, only hematite and maghemite were found. Pure hematite hollow sub-micron spheres with porous shells were formed after annealing at 600°C. The characterization was performed by X-ray diffraction (XRD), Mössbauer spectroscopy (MS) and scanning electron microscopy (SEM).

Keywords Sol–gel method · Mössbauer spectroscopy · Hematite · Maghemite

L. León (✉) · A. Bustamante · A. Osorio · G. S. Olarte
Universidad Nacional Mayor de San Marcos, Lima, Perú
e-mail: lizbetlf@gmail.com

L. De Los Santos Valladares · C. H. W. Barnes
Cavendish Laboratory, University of Cambridge,
J.J Thomson Av., Cambridge CB3 0HE, UK

L. De Los Santos Valladares
e-mail: ld301@cam.ac.uk

L. De Los Santos Valladares · Y. Majima
Materials and Structures Laboratory, Tokyo Institute of Technology,
4259 Nagatsuta-cho, Midori-ku, Yokohama 226-8503, Japan

Y. Majima
CREST, Japan Science and Technology Agency JST,
4259 Nagatsuta-cho, Midori-ku, Yokohama 226-8503, Japan

1 Introduction

The production of magnetic nanostructures is of current interest due to their promising applications in biomedicine [1, 2], bioengineering [3], nanodevices [4] and magnetorheology [5]. In the case of iron oxides nanostructures, the formation of antiferromagnetic hematite (α-Fe_2O_3) [6], ferrimagnetic maghemite (γ-Fe_2O_3) [7] and ferromagnetic magnetite (Fe_3O_4) [8] has been identified during their production by thermal oxidation [9]. Recently, hollow iron-oxide-based microstructures have attracted high interest because of their potential applications in photonic crystals, drug delivery, catalysis and encapsulation [10]. Various techniques such as polystyrene-templates [11], sono-chemical [12], solid-state reaction [13], hydrothermal [14] and the surfactant-assisted sorvo-thermal [15, 16] method have been reported for the preparation of iron oxides hollow structures. However, faster, cleaner and template-free methods are desired to prepare hollow iron-based structures. Sol–gel is a versatile technique used for the preparation of iron oxides nanostructures [17–20]. In this technique, different nanostructures can be obtained after thermal oxidizing the gel precursors previously obtained by hydrolysis and condensation of metal-oxides in alcohol or aqueous medium [21]. In this paper, we report the preparation of hollow hematite submicron spheres by the sol–gel method. Hematite was chosen because of its abundance in nature, non-toxicity, cheap-cost and stability in ambient conditions. We studied the best annealing temperature of the sol–gel precursor to obtain uniform and high yield hollow α-Fe_2O_3 sub-microspheres.

2 Experimental

The preparation started with the suspension of colloidal ferric nitrate nine-hydrate ($Fe(NO_3)_3.9H_2O$) particles (200 ml) and mono hydrated citric acid ($C_6H_8O_7.H_2O$) (0.1 and 0.2 M concentrations) in 800 ml of distilled water following the same procedure reported elsewhere [22]. Citric acid is commonly used because of its effects in the formation, growth and shape of iron oxides nanostructures [23]. The solutions were agitated in a magnetic stirrer at 350 rpm and 70°C for a period of 48 h to evaporate the acid and water residuals. The gels were then stored for two days at 40°C to completely eliminate any residuals and possible impurities formed by hydrolysis. For thermal oxidation, the precursors were introduced in a tubular furnace (LENTON LTF-PTF Model 16/610) to be annealed in oxygen flow at different temperatures between 180°C to 600°C. The furnace was programmed to reach the required temperatures after 1 h 30 min, to remain constant for 12 h, and finally cool down at a rate of 1.4°C /min. The crystallizations of the samples were studied by X-ray diffraction (XRD). The data was collected from 20° to 65° (0.02° steps) using a powder universal diffractometer *Bruker D8 Focus* with Cu Kα radiation (1.5406 Å). The Mössbauer spectra were recorded at room temperature in the transmission geometry, using a ^{57}Co in Rh source with average activity 25 mCi and the isomer shifts were measured relative to the metal α-Fe. The morphological analysis was performed using a scanning electron microscope (*SEM–XL30 SFEG*).

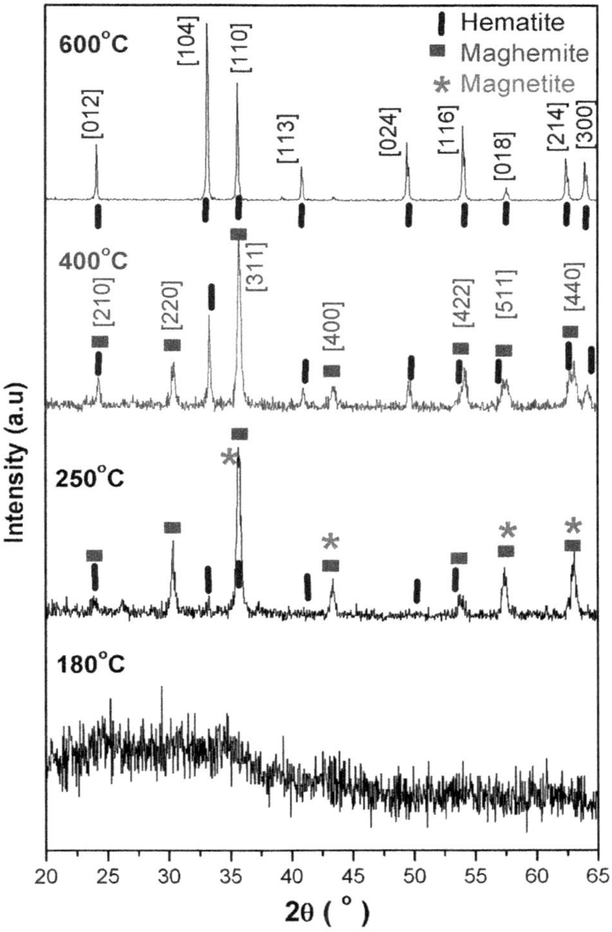

Fig. 1 XRD patterns of the samples after annealing at different temperatures. Note the hematite starts forming at 250°C

3 Results and discussion

Figure 1 shows the XRD plots of the iron oxide samples after annealing at 180, 250, 400 and 600°C in oxygen flow. Initially, when annealing at 180°C, the sample is an amorphous solid and no preferred reflections appear in the XRD. After annealing at 250°C maghemite (PDF2-No. 04-0755) and magnetite (PDF2-No. 85-1436) are formed as principal phases, whereas a small amount of hematite (PDF2-No. 86-550) was detected from its main reflection [311] at $2\theta = 33.16°$. Temperature variation from 250°C to 400°C increases the presence of hematite and its reflections with [140], [113], [024] and [300] increasing. Note that at this annealing temperature the main peak [311] belonging to maghemite slightly shifts to a higher angle, this effect has also been observed by other authors when annealing magnetite to obtain maghemite or hematite [24, 25], and it was recently proposed by Tian et al. [26] that this is a characteristic of core-shell oxidation formation. If that is the case, the maghemite

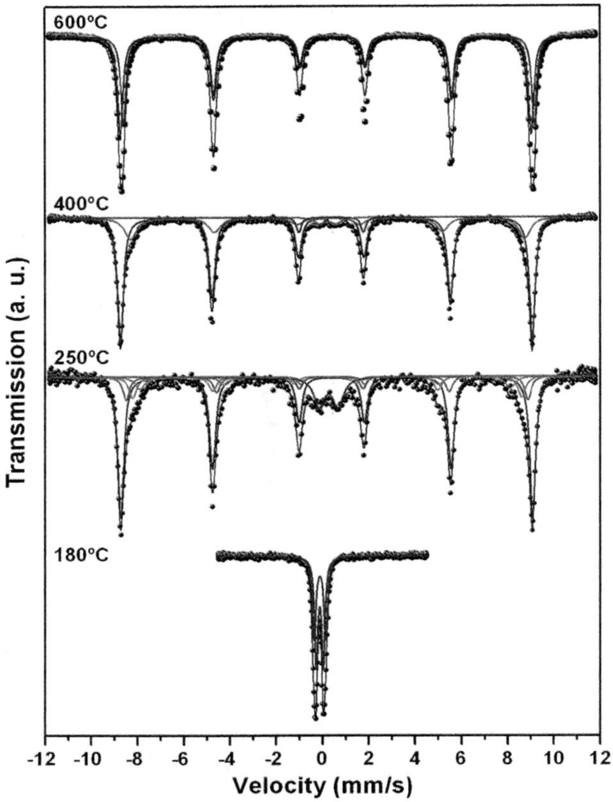

Fig. 2 Room temperature Mössbauer spectra of the sample annealed different temperatures

annealed at 400°C is shelled by the hematite oxide and it would be the previous stages where the hollow nanosphere formation occurs, as we will discuss below. The XRD in the figure verifies the transformation of maghemite to hematite under thermal oxidation, similar to that reported by other authors [27]. Eventually, after annealing the sample at 600°C, the oxidation stabilizes to hematite in which only the Miller indexes of this phase are observed in the XRD. This result agrees well with that reported by Zhang et al. [22] in which 2 h heat treatments in air around 800 K are necessary to obtain a single α-Fe_2O_3 phase. The average sizes of crystallites estimated from the line-width of the main XRD reflections by using the Scherrer's equation [28] are 48 nm, 35 nm and 24 nm when annealing at 600, 400 and 250°C respectively.

Figure 2 shows the Mössbauer spectra obtained at room temperature. The hyperfine parameters are given in the Table 1. The spectra of the sample annealed at 180°C fits well with two non-magnetic doublets, both of which correspond to the Fe^{3+} sites, one of them with $\delta = 0.47 \pm 0.01$ mm/s, QS $= 0.90$ mm/s, and the other with $\delta = 0.24 \pm 0.01$ mm/s, QS $= 0.93$ mm/s as listed in Table 1. Fitting the spectra at 250°C ($\chi^2 = 1.21$) reveals four sextets, two of them corresponding to the magnetite and the remaining two for hematite and maghemite respectively.

Table 1 Mössbauer hyperfine parameters: hyperfine field B_{hf}, isomer shift δ and quadrupole shift 2ε for samples annealed at different temperatures

Temperature annealing	Concentration of citric acid (M)	Site	δ (mm/s)	QS/2ε (mm/s)	B_{hf} (T)	Γ (mm/s)	A (%)
600°C	0.2	1	0.44	−0.07	51.20	0.26	46.4
		2	0.38	−0.14	51.24	0.27	53.6
400°C	0.1	1	0.36	−0.10	51.51	0.28	69.2
		2	0.33	−0.02	49.25	0.58	24.4
		3	0.32	0.67		0.50	3.0
		4	0.46	2.59		0.21	3.4
250°C	0.1	1	0.37	−0.11	51.53	0.28	57.6
		2	0.40	−0.12	50.24	0.30	10.4
		3	0.25	0.04	48.61	0.35	10.8
		4	0.34	−0.14	46.29	0.45	7.9
		5	0.35	0.76		0.50	10.5
		6	0.45	2.56		0.30	2.8
180°C	0.1	1	0.47	0.90		0.42	54.6
		2	0.24	0.93		0.40	45.4

$\Delta\delta = \pm 0.001$, $\Delta QS = \pm 0.01$, $\Delta B_{hf} = \pm 5T$, $\Delta\Gamma = \pm 0.01$ and $\Delta A = \pm 1\%$

Moreover, at this temperature, two additional doublets are found, one doublet corresponding to the Fe^{3+} site ($\delta = 0.35$ mm/s, QS $= 0.76$) which are associated with the superparamagnetic particles in the oxide and a doublet, corresponding to the Fe^{2+} sites ($\delta = 0.46$ mm/s, QS $= 2.56$) associated with spurious phases. The sample annealed at 400°C reveals only the presence of hematite and maghemite. The sample annealed at 600°C presents two sextets corresponding only to hematite. One of these sextets belongs to the bulk ($\delta = 0.38$ mm/s) and the other to the surface ($\delta = 0.44$ mm/s). Similar results have been reported by other authors [29]. In this work, the thermal oxidation of maghemite to hematite occurs progressively from 250°C (area 57.6%) to 400°C (area 69.2%). This transition could generate core–shell intermediates as suggested by Tian et al. [26].

Figure 3 shows different magnification micrographs of the hematite sample obtained after annealing at 600°C. Spheres of around 700 nm diameter with high yield are clearly distinguishable (Fig. 3a and b). In the figure, the arrows point to some broken spheres which enable the direct observation of the shell structure. The cavities on the broken spheres can be distinguished better in the magnifications in Fig. 3c and d, porosity on the surface is also detected. The exact mechanism of the hollow spheres formation without the use of templates is still under investigation. The growth process which leads to the hollow spheres could be described as follows. Initially, Fe^{2+} ions in the precursors react progressively with the decomposition products of the citric acid to form the iron oxides; this reaction is mediated by increasing the temperature. In our case, thermal oxidation from maghemite to hematite occurred progressively at temperatures 250° and 400°C. This transformation probably generates core–shell intermediate products as proposed by Tian et al. [26]. Eventually, when the annealing temperature scales to 600°C, the dissolution of the remaining maghemite occurs leaving the porous hematite shell. Similar results in which dissolution of different types of iron oxide cores such as goethite occurs to leave hematite hollow spheres have been recently reported by Li et al. [30].

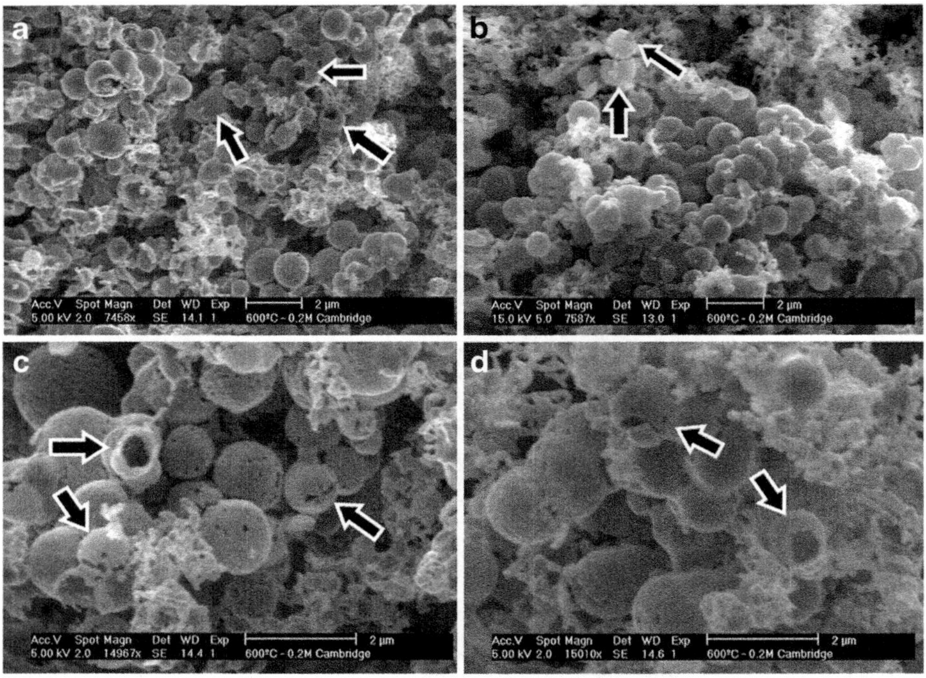

Fig. 3 SEM micrographs at different magnifications of hematite hollow spheres obtained after annealing sol gel precursors at 600°C. Some broken hollow spheres are identified by *arrows*

4 Conclusions

We prepared hollow α-Fe_2O_3 sub-micron spheres by using the sol–gel technique. During annealing the precursors obtained by sol–gel at different temperatures, magnetite, maghemite and hematite were detected by XRD and Mössbauer spectroscopy. Annealing at 250°C and 400°C resulted in a mixture of the phases. Hollow spheres of 700 nm average diameter, which are composed of hematite crystallites of around 48 nm size are obtained in this work at 600°C. Therefore, the temperature of annealing has notorious effects in controlling the type and shape of iron oxide when following the sol–gel method.

Acknowledgements L. León and A. Bustamante thank the financial support from the Scientific Initiation Programme of the University of San Marcos (Grant No. PIC—RR 06030-07). The work in Cambridge was supported by the European Union Programme ALBAN (Grant No E06D101257PE) and the Cambridge Overseas Trust. The work in Japan was supported by the Japan Society for the Promotion of Science (JSPS grant No PE10027). The work in Japan was also partially supported by Grants-in-Aid for Scientific Research on Innovative Areas (No. 20108011, π-Space) from the Ministry of Education, Culture, Sports, Science and Technology (MEXT), a Collaborative Research Project of the Materials and Structures Laboratory (Tokyo Institute of Technology) and the World Class University Program, at Sunchon National University, Korea.

References

1. Petri-Fink, A., Chatellain, M., Juilleret-Jeanneret, L., Ferrari, A., Hofmann, H.: Biomaterials **26**, 2685–2694 (2005)
2. Lee, S.-J., Jeong, J.-R., Shin, S.-C., Kim, J.-C., Chang, Y.-H., Chang, Y.-M., Kim, J.-D.: J. Magn. Magn. Mater. **272–276**, 2432–2433 (2004)
3. Lubbe, A.S., Alexiou, C., Bergemann, C.: J. Surg. Res. **95**, 200 (2001)
4. Tadić, M., Marković, D., Spasojević, V., Kusigerski, V., Remškar, M., Pirnat, J., Jagličić, Z.: J. Alloys and Comp. **441**, 291–296 (2007)
5. De Los Santos Valladares, L., Llandro, J., Lee, D. W., Mitrelias, T., Palfreyman, J., Hayward, T.J., Cooper, J., Bland, J.A.C., Barnes, C.H.W., Arroyo, J.L., Lees, M.: J. Mag. Magn. Mater. **321**, 2129–2134 (2009)
6. Amin, N., Arajs, S.: Phys. Rev. B. **35**, 10 (1987)
7. Dutta, P., Manivannan, A., Seehra, M.S.: Phys. Rev. B. **70**, 174428 (2004)
8. Liu, Q., Asare, K.O.: J. Colloid Interface Sci. **231**, 401–403 (2000)
9. Zboril, R., Mashlan, M., Petridis, D.: Chem. Mater. **14**, 969–982 (2002)
10. Lou, X.W., Archer, L.A., Yang, Z.C.: Adv. Mater. **20**, 3987 (2008)
11. Shiho, H., Kawahashi, N.: J. Colloid Interf. Sci. **226**, 91 (2000)
12. Zhu, J.J., Xu, S., Wang, H., Zhu, J.M., Chen, H.Y.: Adv. Mater. **15**, 156 (2003)
13. Wang, J.S., Yang, J.K., Bao, Y.: Powder Technol. **145**, 172 (2004)
14. Liu, X.-M., Yin, W.-D., Miao, S.-B., Ji, B.-M.: Mat. Chem. Phys. **113**, 518–522 (2009)
15. Shen, D.H., Chen, D.R., Jiao, X.L., Zhao, Y.T.: J. Mater. Chem. **13**, 2266 (2003)
16. Lian, S.Y., Wang, E.B., Gao, L., Wu, D., Song, Y.L., Xu, L.: Mater. Res. Bull. **41**, 1192 (2006)
17. Sugimoto, T., Wang, Y., Itoh, H., Muramatsu, A.: Colloids Surf. A **134**, 265–279 (1998)
18. Sugimoto, T., Sakata, K.: J. Colloid Interface Sci. **152**, 587 (1992)
19. Sugimoto, T., Khan, M.M., Muramatsu, A.: Colloids Surf. A **70**, 167 (1993)
20. Sugimoto, T., Khan, M.M., Muramatsu, A., Itoh, H.: Colloids Surf. A. **79**, 233 (1993)
21. Hench, L.L., West, J.K.: Chem. Rev. **90**, 33 (1990)
22. Zhang, J., Rong, L.X., Liu, Y., Dong, B.Z.: Mat. Sci. Eng. **A351**, 224–227 (2003)
23. Kandori, K., Kawashima, Y., Ishikawa, T.: J. Colloid Interf. Sci. **152**, 284 (1992)
24. Sun, Y.-K., Ma, M., Zhang, Y., Gu, N.: Colloids Surf A: Physicochem. Eng. Aspects **245**, 15–19 (2004)
25. Xu, J., Yang, H., Fu, W., Du, K., Sui, Y., Chen, J., Zeng, Y., Li, M., Zou, G.: J. Mag. Magn. Mat. **309**, 307–311 (2007)
26. Tian, Y., Wu, D., Jia, X., Yu, B., Zhan, S.: J. Nanomat. 837123 (2011)
27. Haneda, K., Morrish, A.H.: J. Physique, Sup. **4**, 38 (1977) C1-321–323
28. Cullity, B.D.: Elements of X-ray Diffraction, 3rd edn., Prentice-Hall, Prentice-Hall International, Upper Saddle River (2000)
29. Bercoff, P.G., Bertorello, H.R.: Appl. Phys. A. **100**, 1019–1027 (2010)
30. Li, W., Guan, J.-G., Wang, W., Tong, G.-X., Fan, X.-A.: Mat. Chem. Phys. **118**, 496–500 (2009)

Hyperfine Interact (2011) 202:139–144
DOI 10.1007/s10751-011-0354-0

Magnetic and structural properties of $Cu_{0.85}Fe_{0.15}O$ system synthesized by co-precipitation

H. D. Colorado · G. A. Pérez Alcázar

Published online: 25 August 2011

Abstract $Cu_{0.94}Fe_{0.06}O$ and $Cu_{0.85}Fe_{0.15}O$ samples were synthesized by using the co-precipitation chemical method. Starting from aqueous solutions of copper nitrate, CuO $(NO_3)_2$ $3H_2O$, iron nitrate, Fe $(NO_3)_3$ $9H_2O$ and sodium hydroxide as precipitating agent, NaOH. The precipitate of three samples for $Cu_{0.94}Fe_{0.06}O$ and five for $Cu_{0.85}Fe_{0.15}O$ of fine powder were calcined for 5 h at different temperatures. The obtained X rays diffraction patterns refined by the Rietveld method show the CuO characteristic pattern, showing that the Fe atoms enter to replace Cu atoms. Furthermore, it was obtained that the crystallite size decreases with calcination temperatures for $Cu_{0.94}Fe_{0.06}O$. The transmission Mössbauer spectroscopy showed that the samples present a disordered paramagnetic behavior due to the big value of the half-width of line of the quadrupolar splitting. Vibrating sample magnetometry confirms the paramagnetic character. The XRD results indicate that the material is nanostructured, due that the crystallite sizes are of the order of 10 nm for $Cu_{0.94}Fe_{0.06}O$ and 40 nm for $Cu_{0.85}Fe_{0.15}O$.

Keywords Mössbauer spectroscopy · X rays diffraction · Vibrating sample magnetometry · Copper oxide · Co-precipitation · Nanostructure material

1 Introduction

The thorough investigation of the properties of CuO was stimulated by the discovery of high temperature superconductors (HTSC). This is because CuO is the basis of the HTSC cuprates and exhibits properties similar to the semiconducting phase of the HTSC. Both are nearly insulating and present antiferromagnetic order at

H. D. Colorado (✉) · G. A. Pérez Alcázar
Departamento de Física, Universidad del Valle, A.A. 25360, Cali, Colombia
e-mail: herdacom@gmail.com

relatively high temperatures [1, 2]. Copper oxide awake increasingly interesting from the standpoint of practical and fundamental aspects. It has proved to be an important industrial material that can be widely used in applications such as gas sensors, magnetic storage media in solar energy transformation, semiconductors and catalysis [3–6].

By reducing the crystal size of copper oxide to a nanometric scale, unique properties that may occur tend to be significantly different from those of a bulk sample; examples of such properties are the optical, electrical, magnetic and catalytic properties. It is therefore necessary to know the physical and chemical characteristics for different grain sizes of CuO particles in the development of practical applications [7].

On the other hand, it also highlights the interest in methods of synthesis of nanoparticles directly to control the physical and chemical properties. Consequently, we have studied several methods to obtain nanostructured materials, being among the more traditional sol–gel [8], co-precipitation [9], Pechini [10, 11], and other chemical methods [12]. And because it is well established synthetic method in this work are prepared CuO samples doped with 6% and 15% Fe using co-precipitation, which has the advantage of being an easy and low budget. Although the CuO system had been widely studied, there is very little information available of this system doped with iron.

This paper presents the magnetic behavior of samples at room temperature obtained by transmission Mössbauer spectroscopy (MS) and vibrating sample magnetometry. The structural properties were studied by X-ray diffraction (XRD) and scanning electron microscopy (SEM).

2 Experimental

The production process of copper monoxide doped with Fe, was arisen by co-precipitation method. We used stoichiometric amounts of copper nitrate, $Cu(NO_3)_2.3H_2O$, and iron nitrate, Fe $(NO_3)_3.9H_2O$, starting salts for an aqueous solution under magnetic stirring at 600 rpm for 40 min at 40°C. Then, with the addition of sodium hydroxide (NaOH), co-precipitation occur releasing heat to stabilize and eventually kill a temperature of 40°C. NaOH is also used to control the pH which remained at 12.53 in the aqueous solution on magnetic stirring for 1 h. The precipitate resulting from the chemical reaction is slowly cooled to room temperature, and then transferred to a centrifuge at 2,000 rpm for 10 min in order to decant the $Cu(FeOH)_2$ and remove the $NaNO_3$ formed. This process is repeated 4 times until complete removal of $NaNO_3$. After the precipitate is dried for 192 h at 50°C and calcined for 5 h at 150°C, 200°C and 250°C for $Cu_{0.94}Fe_{0.06}O$ and at 100°C, 150°C, 200°C, 250°C and 300°C for $Cu_{0.85}Fe_{0.15}O$.

Mössbauer measurements were performed in a conventional spectrometer 57Co/Rh source of 25 mCi. All the spectra were fitted with the MOSFIT program. The $Cu_{0.94}Fe_{0.06}O$ XRD pattern was performed on a Bruker D8 Advance diffractometer with a Cu anode tube and graphite monochromator input $Cu_{0.85}Fe_{0.15}O$ detector and a pro PANalytical X'Pert diffractometer with tube Cu anode. The diffractometer was calibrated with a sample of alumina and the radiation used was Cu K-α. The 2θ angular scan was performed discretely in steps of 0.02°

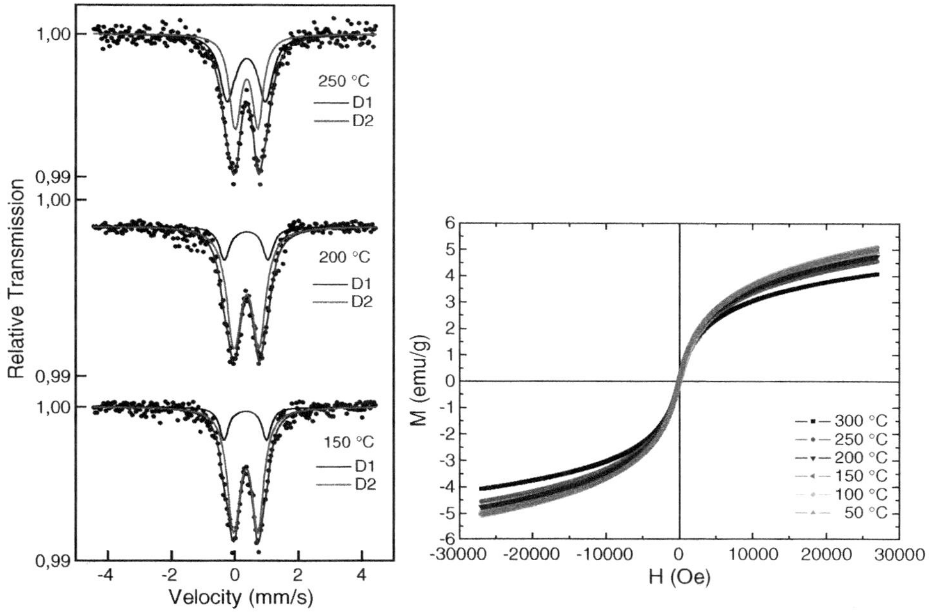

Fig. 1 Transmission Mössbauer spectra (*left*) and hysteresis cycles (*right*) of the $Cu_{0.94}Fe_{0.06}O$ and $Cu_{0.85}Fe_{0.15}O$ systems, respectively, at different calcination temperatures

every 30 s. The diffractograms were refined with the GSAS program which uses the Rietveld method.

3 Results and discussion

Figure 1 shows the Mössbauer transmission spectra and vibrating sample magnetometry for different calcination temperatures of the $Cu_{0.94}Fe_{0.06}O$ sample. Each spectrum was fitted with two doublets indicating the paramagnetic nature of the samples. At least one doublet of each spectrum has a line half width greater than 0.50 mm/s, which confirms a disordered paramagnetic behavior of the system for the temperatures considered. Similar fit was reported by Young et al. [13]. They report two doubles corresponding to sites of Fe^{+3} and Fe^{+2}, respectively. In our case the two doublets present basically the isomer shift of Fe^{+3} ($\delta \sim 0.35$ mm/s), but different quadrupole splitting ($\Delta Q1 \sim 1.25$ and $\Delta Q2 \sim 0.72$ mm/s). Then it is proposed that the second site corresponds to the Fe^{+3} sites and the first one to the Fe^{+3} sites but with a more asymmetric charge distribution. The obtained hysteresis cycles of the samples of $Cu_{0.85}Fe_{0.15}O$ present near zero coercive fields and a magnetization values which tend to the saturation. These results, and the Mössbauer ones, suggest the presence of a soft magnetic behavior of the samples. More measurements at different temperatures are necessary to determine what type of phase is the present.

Figure 2a presents the behavior of the percentage of the spectral area of each doublet with the calcination temperature. It is noted how the area of each doublet, D1 and D2, is constant between 150°C and 200°C, then the D1 area tends to increase

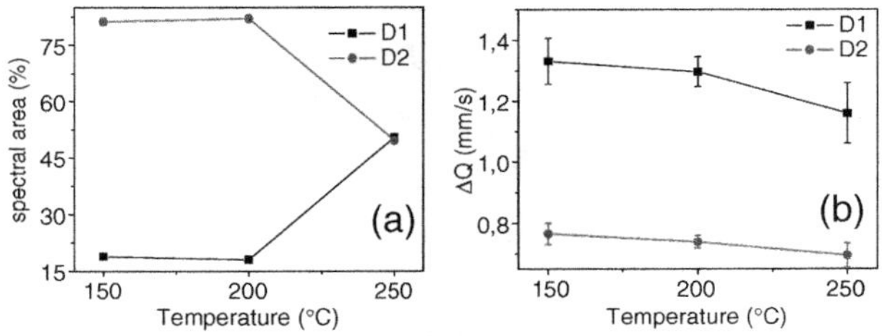

Fig. 2 Variation with the different calcination temperatures of the spectral area (**a**) and of the quadrupole splitting (**b**) of the $Cu_{0.94}Fe_{0.06}O$ system

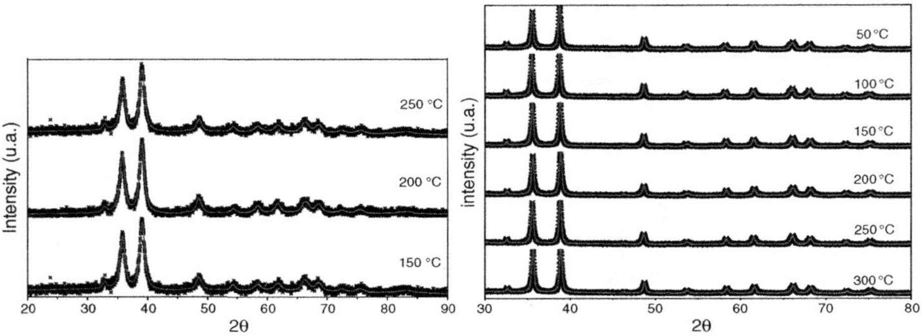

Fig. 3 XRD patterns of the $Cu_{0.94}Fe_{0.06}O$ system (*left*) and of the $Cu_{0.85}Fe_{0.15}O$ system (*right*) for different calcination temperatures

with temperature while the D2 decreases. It seems that between 200°C and 250°C there is a rearrangement of charges around the decreasing Fe^{+3} sites with greater symmetry around him. In Fig. 2b, as expected, the quadrupole splitting for different temperatures remains essentially constant.

Figure 3 shows the XRD patterns for different calcination temperatures.

As can be seen, only one phase, corresponding to the monoclinic structure of CuO, was obtained for the different calcination temperatures, and then this confirms that indeed Fe atoms replace some of Cu atoms. Figure 4 shows the crystallite sizes and lattice parameters of the $Cu_{0.94}Fe_{0.06}O$ system obtained by XRD for different temperatures.

Clearly seen in Fig. 4a how the crystallite size increases with calcination temperature taking the values 9.41 nm for 150°C, 9.53 nm for 200°C and 9.76 nm for 250°C. The lattice parameters *a* and *c* decrease with temperature but these values are higher than those of the CuO reference values, whereas *b* values are lower compared to the reference one. The lattice parameter *c* does not have a defined behavior. For the $Cu_{0.85}Fe_{0.15}O$ system, crystallite sizes between 36 and 42 nm where obtained. The obtained lattice parameters can be considered as remain invariant within an error of 0.02 Å. Figure 5 shows SEM images obtained for the sample calcined at 200°C.

According to the resolution obtained by SEM, the left image of Fig. 5 shows different particle sizes ranging from 200 nm up to about 30 μm. In the write image is clearly noted the tendency of particles to agglomerate. According to these values and

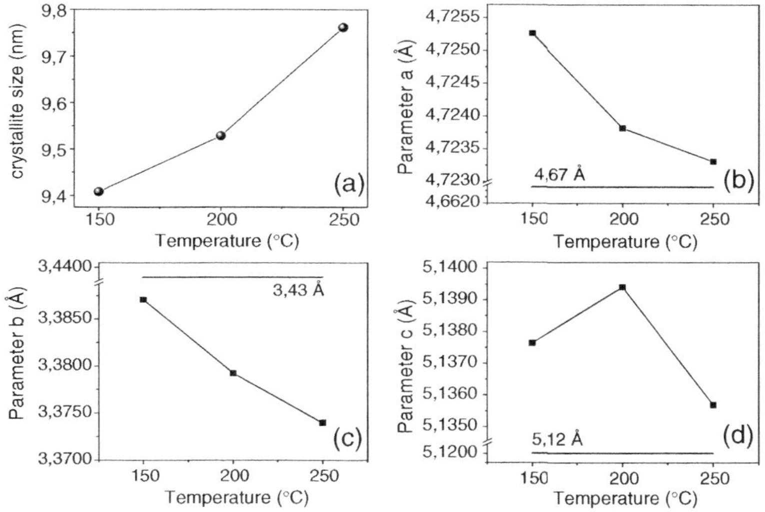

Fig. 4 **a** Crystallite sizes, **b**, **c** and **d** lattice parameters a, b and c, respectively, obtained by XRD of the $Cu_{0.94}Fe_{0.06}O$ system for different calcination temperatures

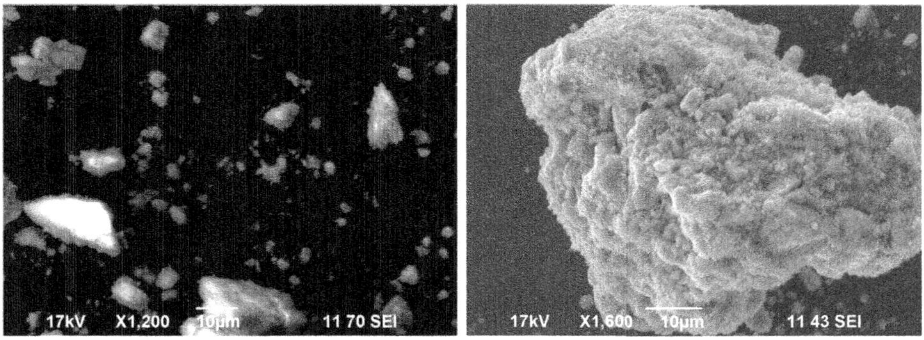

Fig. 5 SEM images for $Cu_{0.94}Fe_{0.06}O$ system for the sample calcined at 200°C

those obtained by XRD for the crystallite size it can be conclude that the particles must be polycrystalline.

4 Conclusions

The diffraction patterns obtained by XRD, and then refined by the Rietveld method, show a characteristic pattern of CuO, where Fe atoms replace Cu atoms, with a crystallite size which decreases with the calcination temperature. Transmission Mössbauer spectroscopy results show that there is a disordered paramagnetic behavior due to the large value taken of the line half width of the quadrupole splitting. XRD results indicate that the material is nanostructured, since the obtained crystallite sizes are between 10 and 42 nm. SEM results show a strong tendency of the particles to agglomerate.

Acknowledgement The authors thank the financial support of Universidad del Valle.

References

1. Gmelin, E.: Indian J. Pure Appl. Phys. **30**, 596 (1992)
2. Samokhvalov, A.A., Arbuzova, T.I., Viglin, N.A.: Fiz. Tverd. Tela (St. Petersburg) **40**(2), 295 (1998) [Solid State Phys. 40, 268 (1998)]
3. Frietsch, M., Zudock, F., Goschnick, J., Bruns, M.: Sens. Actuators, B, Chem. **65**, 379 (2000)
4. Sukhorukov, Y.P., Loshkareva, N.N., Samokhvalov, A.A., Naumov, S.V., Moskvin, A.S., Ovchinnikov, A.S.: J. Magn. Magn. Mater. **183**, 356 (1998)
5. Carnes, C.L., Klabunde, K.J.: J. Mol. Catal., A Chem. **194**, 227 (2003)
6. Dai, P.C., Mook, H.A., Aeppli, G., Hayden, S.M., Dogan, F.: Nature **406**, 965 (2000)
7. Hong, Z., Cao, Y., Deng, J.: Mater. Lett. **52**, 34 (2002)
8. Schmidt, H., Jonschker, G., Goedicke, S., Mennig, M.: J. Sol-gel Sci. Technol. **19**, 39 (2000)
9. Kodama, R.H., Makhlouf, S.A., Berkowitz, A.E.: Phys. Rev. Lett. **79**, 1393 (1997)
10. Pechini, M.P.: Method of preparing lead and alkaline earth titanates and niobates and coating method using the same to form a capacitor. United States Pattent Office 3.330.697 (1967)
11. Stewart, S.J., Borzi, R.A.: J. Appl. Phys. **87**, 9 (2000)
12. Makhlouf, S.A., Parker, F.T., Spada, S., Berkowitz, A.E.: J. Appl. Phys. **81**, 5561 (1997)
13. Young Ahn, G., Park, S., Shim, I., Kim, C.S.: J. Magn. Magn. Mater. **282**, 166 (2004)

Hyperfine Interact (2011) 202:145–151
DOI 10.1007/s10751-011-0343-3

Nanocrystalline Fe/Zr alloys: preparation by using mechanical alloying and mechanical milling processes

V. A. Peña Rodríguez · J. Medina Medina · J. Quispe Marcatoma · Ch. Rojas Ayala · C. V. Landauro · E. M. Baggio-Saitovitch · E. C. Passamani

Published online: 25 August 2011

Abstract Nanocrystalline Fe/Zr alloys have been prepared after milling for 9 h the mixture of elemental Fe and Zr powders or the arc-melting produced Fe_2Zr alloy by using mechanical alloying and mechanical milling techniques, respectively. X-ray and Mössbauer results of the Fe and Zr powders, mechanically alloyed, suggest that amorphous Fe_2Zr phase and α-Fe(Zr) nanograins have been produced with relative concentrations of 91% and 9%, respectively. Conversely, the results of the mechanically milled Fe_2Zr alloy indicate that nanograins of the Fe_2Zr alloy have been formed, surrounded by a magnetic inter-granular phase that are simultaneously dispersed in a paramagnetic amorphous phase.

Keywords Nanostructured materials · Mechanical alloying · Mechanical milling · ^{57}Fe Mössbauer spectroscopy

1 Introduction

Mechanical attrition, which is a well-established processing material method [1, 2], can be performed starting either from a mixture of chemical elemental powders (mechanical alloying, MA) or from an ordered crystalline state (mechanical milling,

V. A. Peña Rodríguez (✉) · J. M. Medina · J. Q. Marcatoma · Ch. R. Ayala · C. V. Landauro
Facultad de Ciencias Físicas, Universidad Nacional Mayor de San Marcos, P.O. Box 14-149, Lima 14, Peru
e-mail: vpenar@unmsm.edu.pe

E. M. Baggio-Saitovitch
Centro Brasileiro de Pesquisas Físicas, Rua. Dr. Xavier Sigaud 150, Urca, CEP 22290-180, Rio de Janeiro, Brazil

E. C. Passamani
Depto. de Física, Universidade Federal do Espírito Santo, CEP 29075-910, Vitória, Espírito Santo, Brazil

MM). In the former case, a great advantage is gained since solid amorphous phases from immiscible components can be produced, while in the MM process the advantage comes from a controlled crystalline–amorphous (C–A) phase transition that may occur during the milling process. Therefore, the study of C–A transitions can lead to a better understanding of their transformations, since it is common to investigate the opposite process.

Laves phase alloys, such as Fe_2Zr, have attracted a special interest in the recent decades due to their potentials for hydrogen absorption capability. Particularly, the Fe_2Zr Laves phase has a face-centered cubic lattice, a $MgCu_2$-type structure (C15), with Fe-atoms located at the 16d sites and Zr-atoms occupying the 8a sites. In fact, the composition range for occurrence of this Laves phase is within the 27.5–34.4 at. % range of Zr [3, 4]. At this Zr-range, the compound behaves as a ferromagnetic material with Curie temperatures decreasing from 733 to 588 K, as Zr-content is increased [3, 5].

In this work, it is reported results from the MA technique of elemental Fe and Zr powders and the MM method of the Fe_2Zr Laves-phase alloy, using similar milling conditions (ball/mass ratio; milling time; atmosphere; milling tools and container; etc). X-ray diffraction (XRD), which detects long-range-crystalline order (LRO), has been applied to follow the structural changes in milled materials. ^{57}Fe-Mössbauer spectroscopy (MS), which senses the chemical short-range order (SRO) around Fe-sites, was used to understand the local Fe-environments in the alloys obtained from both processes: MA and MM.

2 Experimental

The preparation of the samples was done in a cylindrical stainless steel vial containing five steel balls sealed in Ar atmosphere. The vial was mounted on a vibratory mixer ball mill, a SPEX 8,000 unit and the milling process lasted up to 9 h. Two series of samples were investigated. The first series was produced by MA an initial mixture of the Fe and Zr chemical elemental powders, with a nominal composition of 66.7 at% of Fe and 33.3 at% of Zr (Samples I). The second series, hereafter called as Sample II, was prepared by MM the Fe_2Zr intermetallic compound, previously obtained by arc-melting and annealed at 900 C for 2 days. In both preparations, high purity powders of Fe (99.5 + %, below 10 μm) and Zr (99.7 + %, below 44 μm) were used.

At pre-defined milling times small portions of the milled powders were collected from the vial, which was opened inside of glove-box with a positive Ar pressure, for X-ray and Mössbauer analysis. X-ray diffraction patterns (XRD) were recorded by using a Bruker diffractometer, model D8-FOCUS, employing Cu-K_α radiation. Microstructures of Fe-sites were studied at room temperature (RT) by Transmission Mössbauer Spectroscopy (TMS), using a 25 mCi ^{57}Co/Rh radioactive source that moved in a constant acceleration mode. The Mössbauer spectra were fitted using NORMOS program [6]. The isomer shift (δ) values are given relative to α-Fe at RT.

3 Results and discussion

The XRD pattern of the Sample I (Fig. 1a) milled for 3 h displays: (a) absence of several Bragg peaks seen in the blended powders (zero-hour powders) and

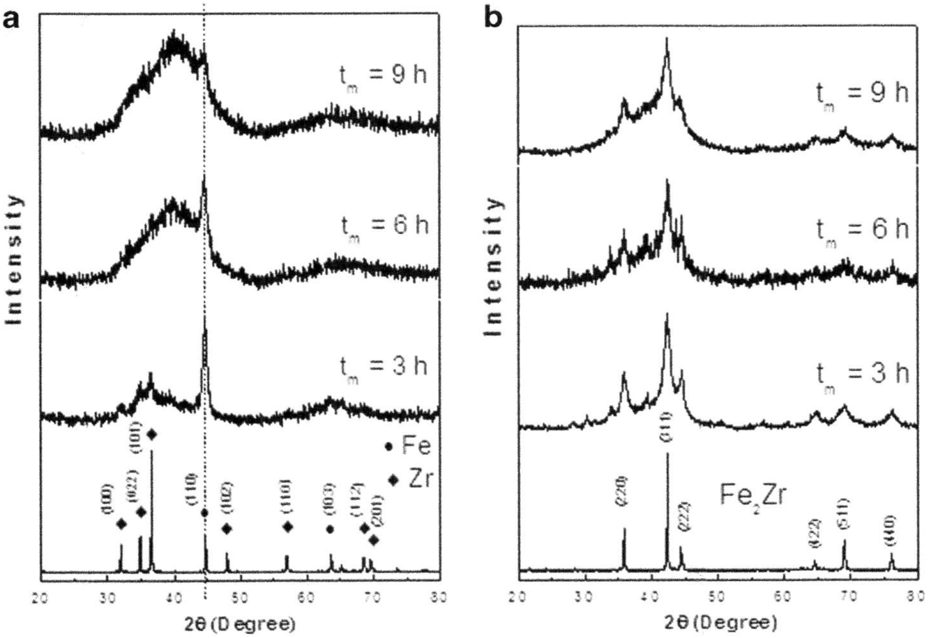

Fig. 1 X-ray diffraction patterns of the starting and ball-milled powders for the MA Samples I (**a**) and MM Samples II (**b**)

(b) broadening of the initially sharp diffraction lines of the Fe and Zr powders. For longer milling times (t_m), the most important feature of the resulting X-ray patterns is the broad diffraction line at $2\theta \approx 40°$. In addition to this, the main Fe-Bragg peak (at $2\theta = 45°$) can still be observed after milling the Sample I with times up to 9 h. Thus, the presence of a broad Bragg peak at about $2\theta = 45°$ suggests the formation of a crystalline phase with short-range order, probably immersed in the amorphous state characterized by the broad peak at $2\theta \approx 40°$. On the other hand, Fig. 1b presents XRD patterns of the Samples II MM treated. The as-prepared powder shows Bragg peaks due to the C15-type structure [lattice parameter $a = (7.0771 \pm 1)$ Å]. Bragg peaks of the characteristic Fe_2Zr alloy become broad after milling it for 3 h, and LRO may still be present up to 9 h of milling. It is important to note that for $t_m \geq 3$ h an additional set of Bragg peaks appears in the XRD patterns after, which may be associated with an Fe/Zr-segregated phase. It is also observed that most of the Bragg peaks of the C15-type structure broaden for $t_m \geq 3$ h; similar results are observed in the XRD patterns of the Sample I. In other words, after milling the alloy for 9 h, there exist broad Bragg peaks due to the Fe_2Zr phase superimposed to a broad shoulder centered at $2\theta \approx 40°$, probably associated with a disordered phase. Thus, whereas in the MA process (Sample I) the contributions to the XRD pattern are due to an amorphous-like phase and a solid phase with short-range-crystalline order, in the MM process, performed under similar milling conditions, the contributions to the XRD pattern may also come from two species: nanograins of Fe_2Zr-type phase and probably the amorphous-like one. In fact, the MM process first favors a segregation process of the components and by increasing the t_m-value, the milling

Fig. 2 Effective crystallite sizes and root mean square strains (rms) of the Samples I and Samples II as a function of the milling time

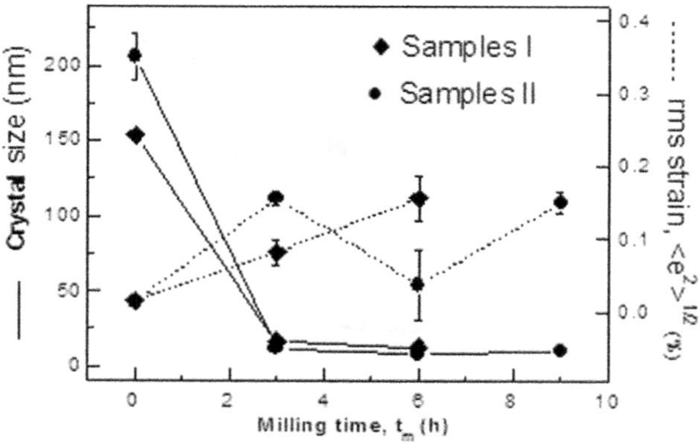

Fig. 3 a RT Mössbauer spectra for the MA Samples I at indicated milling times. b Quadrupole splitting distribution P(Δ) curves obtained from the fittings of the corresponding Mössbauer spectra shown in (a)

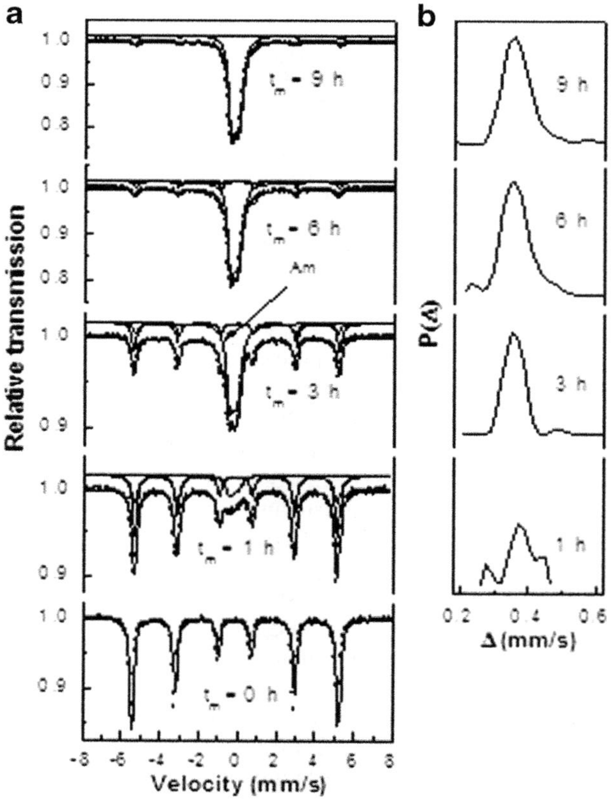

process induces a recombination of the components yielding an amorphous-like phase, as also reported in $ErFe_2$ compound [7].

Figure 2 shows the average crystal size and the root mean square strain (rms) versus milling time (t_m) calculated using the single-line method employed by Keijser et al. [8]. For Samples II, a slight reduction of the crystallite size is seen when one

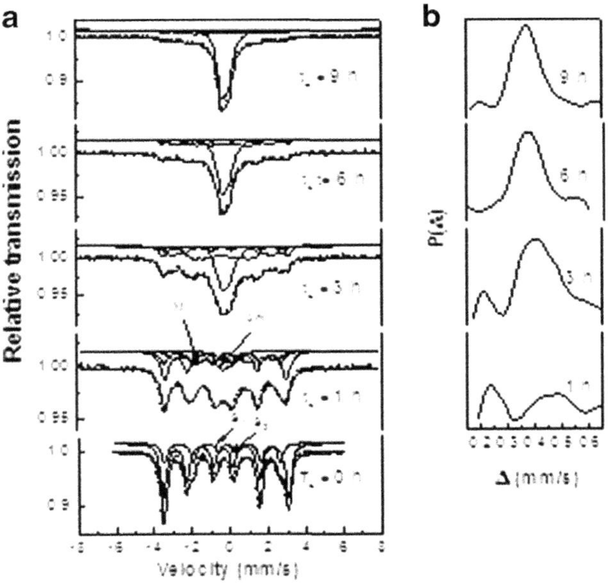

Fig. 4 **a** RT Mössbauer spectra for the MM Samples II at indicated milling times. **b** Quadrupole splitting distribution P(Δ) curves obtained from the fittings of the corresponding Mössbauer spectra displayed in (**a**)

compares with that of the Samples I. It is worth noting that for both-final milled-samples (9 h-samples), the crystallite sizes of the Fe (Sample I) and Fe_2Zr (Sample II) are in a nano-metric scale. Then, it can be inferred that the milled materials are composed by nano-crystals embedded in Fe/Zr amorphous matrixes.

In contrast to the results of the X-ray diffraction technique that measures modifications on the structural LRO, the MS senses changes in the chemical SRO. Figures 3a and 4a display RT Mössbauer spectra of milled powders for Samples I and II, respectively.

The Mössbauer spectra of Samples I (Fig. 3a) were fitted assuming three components: (1) a sextet with hyperfine parameters close to those of the α-Fe phase ($S_{\alpha\text{-}Fe}$), (2) another sextet (hereafter labelled as S') with hyperfine parameters: $\delta = 0.01(1)$ mm/s and a hyperfine magnetic field (B_{hf}) of 32.4(1) T, which could be associated with an α-Fe(Zr) crystalline solid solution, and (3) an electric field distribution component [P(Δ)], which can be attributed to an amorphous FeZr phase (*Am*) [9, 10]. Figure 3b plots the corresponding P(Δ) curves obtained from the fittings of the Mössbauer spectra shown in Fig. 3a. After milling the elemental powders for 1 h, the Mössbauer spectrum consists of two sextets arising from the $S_{\alpha\text{-}Fe}$ and S' phases as the dominant products, where 41% is due to the S' sub-spectrum, which grows at the expense of the $S_{\alpha\text{-}Fe}$ sextet as the t_m-value is increased. For $t_m > 1$ h, the *Am* phase fraction dominates the Mössbauer spectra of milled powders. It is worth noting that the *Am* phase fraction increases at the expenses of the sextet ones, $S_{\alpha\text{-}Fe}$ and S', respectively, and after 9 h of MA, 91% of the RT Mössbauer spectrum is due to the *Am* phase. The residual nano-crystals of α-Fe(Zr), also seen by XRD data, has only 9% of total Mössbauer spectrum absorption area.

Figure 4a presents the RT Mössbauer spectra of Samples II. The Mössbauer spectrum of the un-milled Fe_2Zr compound is fitted with two sub-spectra labeled as S_1 and S_2. These two sub-spectra are due to the fact that Fe-magnetic moments in

Table 1 Hyperfine parameters obtained from the fittings of the Mössbauer spectra for the Sample I and Sample II milled for 9 h

Sample	Site	Hyperfine parameters	
Sample I after 9 h of MA	S'	δ (mm/s)	0.009 (7)
		Γ (mm/s)	0.37 (1)
		B_{hf} (T)	32.6 (1)
		Area (%)	9
	Am	<δ> (mm/s)	−0.097 (8)
		Γ (mm/s)	0.26 (1)
		<Δ> (mm/s)	0.37 (1)
		Area (%)	91
Sample II after 9 h of MM	S_2	δ (mm/s)	−0.174 (1)
		Γ (mm/s)	0.45 (1)
		2ε	0.18 (1)
		B_{hf} (T)	19.8(2)
		Area (%)	12
	M	<δ> (mm/s)	−0.066 (9)
		Γ (mm/s)	0.34 (1)
		<B_{hf}> (T)	12.0 (1)
		Area (%)	20
	Am	<δ> (mm/s)	−0.091 (8)
		Γ (mm/s)	0.36 (1)
		<Δ> (mm/s)	0.38 (1)
		Area (%)	68

C15-type structure may have two non-equivalent directions on the 16a site relative to [111] easy direction of magnetization ($\theta = 0°$ and $\theta = 73°32'$) [10, 11]. The area ratio of these sub-spectra (S_1:S_2) is 1:3 and their B_{hf} values are 18.8(1) T and 20.3(1) T, respectively. For $t_m \geq 1$ h, the two sextets S_1 and S_2 become less resolved due to a line-broadening effect, which is probably associated to an enhancement of lattice defects and crystallite size reduction of the Fe_2Zr Laves-phase. Its area ratio (S_1:S_2) decreases concomitantly with the appearance of two new components, in accordance with XRD data that suggest the segregation effect. The first component is characterized by a broad sextet and a multi-peaked hyperfine magnetic field distribution (*M*), with a relative content of about 27% and an average B_{hf} value of 12.5(1) T. The second component is a broad paramagnetic doublet (*Am*) with 13% of the total absorption area and having a δ value similar to that observed for the amorphous phase of Sample I. Then, this paramagnetic doublet, which was analyzed with an electric field distribution component [P(Δ)], is associated with amorphous Fe_2Zr phase [a-Fe_2Zr].

Figure 4b displays the P(Δ) curves obtained from the fittings of the Mössbauer spectra shown in Fig. 4a. Reiteratively, the *Am* phase fraction increases at the expenses of the magnetic components: S_1, S_2 and *M*. After MM the Fe_2Zr alloy for 9 h, its RT spectrum still has the S_2-component, but now with a line width (Γ) broader than those obtained at shorter t_m-values; an effect which is probably due to lattice defects in the nano-crystalline Fe_2Zr alloy (n-Fe_2Zr). Its relative content is 12% of the total area. On the other hand, 68% of absorption area is due to the *Am* phase and 20% of the spectrum is related to the hyperfine magnetic field distribution, which can be attributed to ordered nano-crytallites of the Fe/Zr alloy [n-(Fe/Zr)].

This ordered component probably behaves as an interface between the a-Fe_2Zr and the n-Fe_2Zr phases.

It has been included Table 1 in order to resume the Mössbauer results of the MA and MM processes applied to the Fe–Zr system, giving the values of the hyperfine parameters of the resulting components, such as: isomer shift (δ), line width (Γ), quadrupole shift (2ε), hyperfine magnetic field (B_{hf}), average quadrupole splitting ($\langle\Delta\rangle$), average hyperfine magnetic field ($\langle B_{hf}\rangle$) and relative areas of each component as detected in Samples I and II milled for 9 h. The uncertainties are giver in parenthesis.

4 Conclusions

Nanocrystalline materials with nano-crystals of α-Fe(Zr) solution and (n-Fe/Zr) alloys are embedded in the amorphous Fe_2Zr phase (a-Fe_2Zr) which can be produced by either MA or MM under similar milling conditions. After milling the elemental Fe and Zr powders for 9 h (MA process), the a-Fe_2Zr phase and nano-crystals of α-Fe(Zr) solution, with 91% and 9% relative contributions, respectively, are produced. On the other hand, milling the arc-melting produced Fe_2Zr alloy for 9 h (MM process), the a-Fe_2Zr phase was produced with only a 68%. The remaining 32% consisted of nanostructured Fe species (Fe_2Zr nano-crystals and a magnetic intergranular phase at room temperature).

Acknowledgements The authors would like to acknowledge the financial support provided by CSI-UNMSM under Project No. 041301011, FAPERJ-Brazil (Cientista do Nosso Estado), FINEP, FAPES and Latin American Center of Physics.

References

1. Benjamín, J.S.: Metall. Trans. **1**, 2943 (1970)
2. Koch, C.C.: Ann. Rev. Matter. Sci. **19**, 1121 (1989)
3. Stein, F., Sauthoff, G., Palm, M.: J. Phase Equilibria **23**(6), 480 (2002)
4. Okamoto, H.: J. Phase Equilibria **14**, 652 (1993)
5. Kiss, L.F., Huhn, G., Kemény, T., Balogh, J., Kaptás, D.: J. Magn. Magn. Mater. **160**, 229 (1996)
6. Brand, R.A., Lauer, J., Herlach, D.M.: J. Phys. F. **12**, 675 (1983)
7. Passamani, E.C., Larica, C., Santos, W.R., Alves, K.M.B., Biondo, A., Nunes Filho, E.: J. Phys. Condens. Matter **11**, 1147 (1999)
8. Keijser, T.H., Langford, J.L., Mittemeijer, E.J., Vogels, A.B.P.: J. Alloys Compd. **297**, 199 (2000)
9. Bionducci, M., Navarra, G., Bellissent, R., Concas, G., Congiu, F.: J. Non-Cryst. Solids **250–252**, 605 (1999)
10. Concas, G., Congiu, F., Spano, G., Bionducci, M.: J. Magn. Magn. Mater. **279**, 421 (2004)
11. Wertheim, G.K., Jaccarino, V., Wernick, J.H.: Phys. Rev. **135**(1A), A151 (1964)

Hyperfine Interact (2011) 202:153–157
DOI 10.1007/s10751-011-0362-0

γ-Fe phase plasma-induced on the surface of thin S3A alloy ribbons

A. Cabral-Prieto · I. García-Sosa · N. Nava · E. Camps · Luis Escobar · R. López-Castañarez · O. Olea-Cardoso

Published online: 25 August 2011

Abstract Amorphous alloy ribbons of $Fe_{77}Cr_2B_{16}Si_5$ were exposed to cold plasmas of N_2 and $Ar–N_2$ at temperatures lower than $T_x = 808$ K. The conversion X-ray Mössbauer spectra of the plasma-exposed ribbons consist of a singlet and a broadened magnetic sextet. The singlet with isomer shift $\delta = -0.11$ mm/s can be assigned to γ-Fe austenite phase. Minor bulk magnetic changes in the alloy were measured as a consequence of these treatments; e.g. the relative intensities A23 of the transmission Mössbauer spectra of the untreated and treated samples, were 3.22 and 3.56, respectively, the B_{hf} values changed from 22.9 T (untreated sample) to

A. Cabral-Prieto (✉) · I. García-Sosa · E. Camps · L. Escobar
Departamento de Química, Instituto Nacional de Investigaciones Nucleares,
Apartado postal No. 18-1027, Col. Escandón, Del. Miguel Hidalgo, 11801, México D. F., México
e-mail: agustin.cabral@inin.gob.mx

I. García-Sosa
e-mail: irma.garcia@inin.gob.mx

E. Camps
e-mail: enrique.camps@inin.gob.mx

L. Escobar
e-mail: luis.escobar@inin.gov.mx

N. Nava
Instituto Mexicano del Petróleo, Eje Central Lázaro Cárdenas, 152,
Col. San Bartolo Atepehuacan, Del. Gustavo A. Madero, 07730, México D. F., Mexico
e-mail: tnava@imp.mx

R. López-Castañarez · O. Olea-Cardoso
Facultad de Química, Universidad Autónoma del Edo. de México,
Toluca Edo. de México, México

R. López-Castañarez
e-mail: rlc@anuies.mx

O. Olea-Cardoso
e-mail: olc@anuies.mx

22.4 T (plasma treated samples). Unexpectedly, the γ-Fe phase can also be produced by simply heating the alloy ribbons under N_2 flux at temperatures as low as 423 K. Mössbauer data of the crystallized samples are also reported, and a qualitative assessment on the mechanical properties of the $Fe_{77}Cr_2B_{16}Si_5$ alloy associated with the plasma and/or temperature surface induced γ-Fe phase is given.

Keywords S3A alloy · Crystalline state · Iron-based alloy · Amorphous alloy

1 Introduction

Amorphous alloys with soft magnetic properties have been the subject of many research studies given their technological and scientific interests [1]. It is well known that the mechanical properties of thin ribbons or bulk amorphous iron-based alloys deteriorate when they transform from the amorphous to the crystalline state. In the latter state, these alloys are quite brittle [1–5] where the crystallized phases contribute to the brittleness problem [5, 6]. On the other hand, the concentration of volumetric defects in the crystalline state is also higher than in the amorphous state contributing as well to the brittleness problem [5, 6]. The brittleness problem has been reduced while changing the elemental composition of the alloys without losing the important magnetic properties. Despite the scientific research carried out throughout six decades, the brittleness problem still remains an open research area, and as a consequence of this, the working temperature of these alloys tends to be much lower than that of the crystallization (T_x).

Thus, the aim of this work is to study surface changes on the alloy while treating it with cold plasmas of N_2 or N_2–Ar, followed by heat treatments under N_2 gas, and trying to relate these changes with the mechanical properties of the alloy. Conversion X-ray Mössbauer Spectroscopy (CXMS), Transmission Mössbauer Spectroscopy (TMS) and qualitative mechanical results are presented.

2 Experiment

Amorphous alloy ribbons, 20 μ thick, of $Fe_{77}Cr_2B_{16}Si_5$ (Metglas code: 2605S3A) were donated to us by Dr. R. Hasegawa from Metglas Inc. Co. in the USA. The plasma process (PP) was carried out in a microwave electron cyclotron resonance (ECR) plasma device, described in detail elsewhere [7]. Either pure N_2 gas (PP1) or a mixture of Ar and N_2 gases in a proportion of 60/40, (PP2) was used to produce the cold plasma under a pressure of 8×10^{-4} Torr [7] and an input power (kept constant) of 450 Watts [7]. Under these conditions the plasma density of 2×10^{11} ions/cm^3 was established. Two temperature conditions were studied, at 473 K (PP1) and 423 K (PP2). The dull side of the ribbon was exposed to the plasma only. Samples exposed to the PP1 are named as PP1AFe1 and PP1AFe2 and those PP2 treated are named as PP2AFe3 and PP2AFe4. The latter sample was exposed on both sides of the ribbon. The plasma treated samples were then crystallized at 808 K under N_2 gas flux by using a heating rate of 4°C/min. The crystallization process at 808 K lasted for 20 min. The crystallized samples are named as PP1CFe1, PP1CFe2, PP2CFe3, PP2CFe4, respectively.

A constant acceleration Mössbauer spectrometer with a $^{57}Co/Rh$ source was used to record the CXMS and TMS spectra. The Ar (90%)–CH_4 (10%) and Kr gases proportional counters were used to obtain the CXMS and TMS spectra, respectively. The isomer shift (δ) is related to that of α-Fe. Micro-hardness measurements on the plasma treated and heated 2605S3A alloy were unsuccessful due to its brittleness problem. In order to complement these results micro-hardness and Mössbauer measurements on untreated and thermally treated 4140 stainless steel were carried out. The Shimadzu apparatus tester with a diamond pyramid indenter was used to measure the Vickers number (H_V) of the alloy and 4140 stainless steel by using a load of 50 g/15 s.

3 Results and discussion

3.1 Mössbauer results

Whereas the bulk TMS spectrum of the untreated amorphous alloy (UAA) sample shows the characteristic broad sextet of an amorphous alloy (Fig. 1a) with a hyperfine magnetic field of $B_{hf} = 22.9$ T and area ratio A23 = 3.22 [8, 9], the bulk TMS spectrum of any plasma treated sample (PP1AFe1, PP1AFe2, PP2AFe3, PP2AFe4) is also similar to that in Fig. 1a but with slight different parameters such as $B_{hf} = 23.4$ ($\pm$0.20)T and A23 = 3.56, i.e., there are slight changes in the hyperfine magnetic parameters as a result of the plasma processes PP1 and PP2. Similarly, the surface CXMS spectrum of the UAA ribbons, also consists of the same broad sextet, with a $B_{hf} = 22.6$ ($\pm$0.20)T and A23 = 3.0. In this sense, the surface CXMS spectrum of the UAA sample presents the lowest magnetic Mössbauer parameters (Fig. 1b). The isomer shift and quadruple splitting of these magnetic patterns practically do not change with the plasma treatments; they are $\delta = -0.03$ ($\pm$0.01) mm/s and $\Delta = -0.04$ ($\pm$0.01) mm/s. On the other hand, the surface CXMS spectra of any plasma treated ribbon are practically the same, consisting of a singlet with isomer shift $\delta = -0.11$ mm/s (relative to α-Fe), superimposed to a broad sextet of relative intensity lower than that of the singlet (Fig. 2a). The singlet can be associated with the γ-Fe austenite structure [10, 11]. The magnetic pattern may have the characteristics of the untreated or crystallized sample (Figs. 1 or 2, respectively), depending on the treatment conditions of the alloy. The relative intensity of these spectra ranges from 60% to 68% for the singlet and 40% to 32% for the sextet (or multi-sextet), respectively. From the CXMS point of view, the PP's produces an iron phase which is different to that of the bulk material at temperatures as low as 423 K. As far as the authors know, it is the first time that the γ-Fe phase on the surface of the S3A alloy is reported.

The bulk TMS spectra of any plasma treated and then crystallized alloy ribbon are among them quite similar as well, they consist of four superimposed sextets with average hyperfine magnetic fields of 22, 28, 31 and 33 T, suggesting the formation of four iron phases (Fig. 2b). Each Mössbauer spectrum of the crystallized alloy may be related to amorphous material and/or Fe–Si, Fe–B, Fe–Cr, and pure Fe phases, respectively [6, 9].

A final comment is in order. The surface CXMS spectra of the non-plasma exposed bright-side of the ribbons consisted of the typical broad sextet characteristic

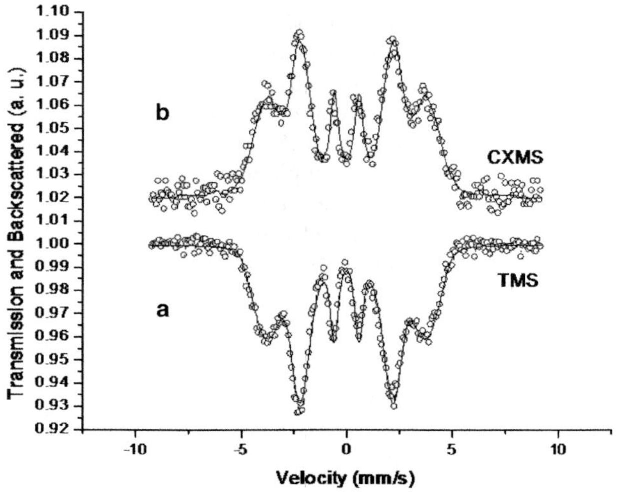

Fig. 1 Bulk (**a**) TMS and (**b**) CX MS spectra of the untreated amorphous alloy sample

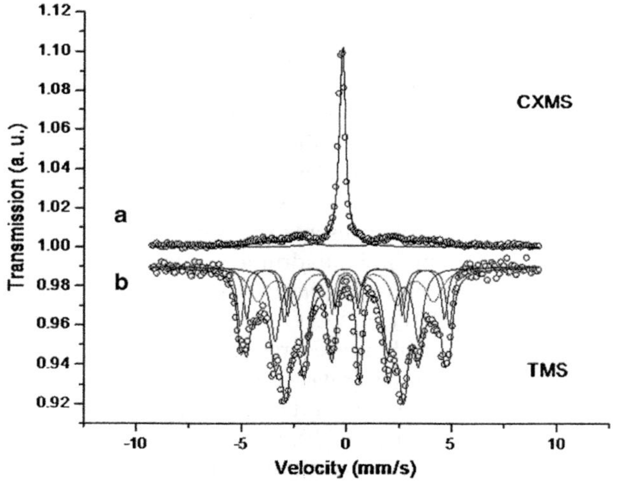

Fig. 2 Surface (**a**) CXMS and bulk (**b**) TMS spectra of the treated ribbon

of the amorphous state (Fig. 1a). However, the surface CXMS spectra of this bright-side of the ribbons after heating them at 808 K under N_2 gas unexpectedly showed the same surface CXMS spectra of the plasma treated samples, i.e., the singlet, (Fig. 2a). That is, the γ-Fe phase is not uniquely plasma induced but by simply heating the 2605S3A ribbons such a phase can be produced at temperatures as low as 423 K. That is, the plasma process has anything to do with the formation of this γ-Fe phase.

3.2 Mechanical properties

Micro-hardness measurements on the plasma treated and crystallized ribbons were carried out unsuccessfully because of their brittleness problem and deformation in the ribbons. The brittleness is, however, still high after these treatments. The presence of the γ-Fe phase on the surface of the alloy ribbons might change this brittleness of the alloy to some extent. Its mechanical effect has, however, been indirectly inferred while heating 4140 stainless steel up to 773 K/2 h. The Vickers number of the unheated and heated 4140 stainless steel is about 400 and 200, respectively. Whereas the amount of γ-Fe phase in the heated stainless steel is 46% in the unheated steel is about 30%. Similar tendencies have been reported for white cast iron where the higher the content of the austenitic phase, the lower the hardness in cast iron [12]. Whether this behavior applies to the present alloy is a subject matter of further study.

4 Conclusions

The 2605S3A alloy was treated by using pure N_2 and N_2–Ar cold plasmas at 473 and 423 K and subsequently crystallized at 808 K under N_2 gas. The main effect of these treatments on the alloy was the formation of the γ-Fe phase on the surface of the alloy ribbons as detected by CXMS unambiguously. Notice that this non-magnetic γ-Fe phase appearing at temperatures as low as 423 K, is the working temperature of the alloy as reported by Metgals, and this phase is undesirable from the magnetic point of view. On the other hand, the amount of γ-Fe phase plasma-induced on the surface of the alloy by either the PP1 or PP2 or by heat treatments is about the same as measured from CXMS results. The plasma treated alloy ribbons are still brittle and a mechanical evaluation of the presence of the γ-Fe phase on the surface of the alloy is still in process. Magnetization measurements would be required to completely asses the effect of γ-Fe phase.

Acknowledgement We thank to Dr. R. Hasegawa from Metglas Inc. Co., USA for his donation of a long ribbon of the 2605S3A amorphous alloy.

References

1. Inoue, A.: Acta Mater. **48**, 297–306 (2000)
2. Bang, J.Y., Lee, R.Y.: J. Mater. Sci. **26**, 4961–4965 (1991)
3. Huang, D.-R., Li, J.C.M.: Mater. Sci. Eng. **A133**, 209–212 (1991)
4. Hasegawa, R., Ramanan, V.R.V., Fish, G.E.: J. Appl. Phys. **53**(3), 2276–2278 (1982)
5. Cabral-Prieto, A., García-Santibañez, F., López, A., López-Catañares, R., Olea-Cardoso, O.: Hyperf. Interact. **161**, 69–81 (2005)
6. López-Morales, A., Cabral-Prieto, A., García-Santibañez, F., Ramírez, J., López-Monroy, J.: Internet. Electron. J. Nanociencia et Moletronica **8**(1), 1445–1458 (2010)
7. Camps, E., Muhl, S., Alvarez-Fregoso, O., Juárez-Islas, J.A., Romero, S.: J. Vac. Sci. Technol. A **17**, 2007–2014 (1999)
8. Brand, R.A.: Wissenschaftlich Electronik GMBH, Würmstr 8, D-8130 Starnberg 3, Germany
9. López-Morales, A., Cabral-Prieto, A., García-Santibañez, F., Ramírez, J., López-Monroy, J., Orozco-Velazco, J.: Internet Electron. J. Nanociencia et Moletronica **8**(1), 1473–1488 (2010)
10. Fall, I., Génin, J.-M.R.: Hyperf. Interact. **69**, 513–516 (1991)
11. Schaaf, P., Illgner, Ch., Niederdrenk, M., Lieb, K.P.: Hyperf. Interact. **95**, 199–225 (1995)
12. Mulaba-Bafubiandi, A.F., Waanders, F.B., Jones, C.: Hyperf. Interact. **139–140**, 455–462 (2002)

Hyperfine Interact (2011) 203:1–8
DOI 10.1007/s10751-011-0363-z

Nanostructuration of i-$Al_{64}Cu_{23}Fe_{13}$ quasicrystals produced by arc-furnace

J. Quispe-Marcatoma · C. Rojas-Ayala · C. V. Landauro · M. A. de Sousa · F. Pelegrini · M. Taquire · V. A. Peña Rodriguez · E. M. Baggio-Saitovitch

Published online: 25 August 2011

Abstract The $Al_{46}Cu_{23}Fe_{13}$ quasicrystal phase, prepared by arc-melting and nanostructured using high energy ball milling technique, was studied employing X-ray diffraction, transmission electron microscopy, ferromagnetic resonance and electric transport measurements. Fe local environments were studied by Mössbauer spectroscopy. The resistivity ratio R(4.2 K)/R(300 K) is within the expected values commonly observed in an icosahedral phase. In general, the experimental results show that an appropriate heat treatment of the as-cast alloy prepared by arc-melting makes possible to obtain good quasicrystal samples. On the other hand, for milling time longer than five hours, the average grain-size of quasicrystal phase reduces, but it preserves Fe local atomic orders. It is also observed that the quasicrystalline sample decomposes in an iron rich nano-quasicrystalline phase and a ε-Al_2Cu_3 phase. Electric transport measurements show that at low temperatures the nano-quasicrystalline samples behave strongly different to their solid counterparts, an effect attributed to a long-range order reduction and an increasing of grain boundary regions. The presence of local magnetic moments in the nanostructured sample is also discussed.

Keywords $Al_{64}Cu_{23}Fe_{13}$ · Quasicrystals · Crystalline system · Magnetic moment · Nanostructuration

J. Quispe-Marcatoma (✉) · C. Rojas-Ayala · C. V. Landauro · M. Taquire · V. A. Peña Rodriguez
Facultad de Ciencias Físicas, Universidad Nacional Mayor de San Marcos, P.O. Box 14-0149, Lima 14, Peru
e-mail: jquispe@cbpf.br

J. Quispe-Marcatoma · C. Rojas-Ayala · E. M. Baggio-Saitovitch
Centro Brasileiro de Pesquisas Físicas, Rua Dr. Xavier Sigaud 150, Urca, Rio de Janeiro, 22290-180, Brazil

M. A. de Sousa · F. Pelegrini
Instituto de Física, Universidade Federal de Goinia, Campus Samambaia, 131, Goiania, Brazil

1 Introduction

Since its discovery about 25 years ago [1], quasicrystals (QCs) opened a new class of matter in the solid state between the amorphous and crystalline systems. The interest to understand their physical properties and to employ these materials for diverse technological applications was growing worldwide. Interesting properties for industrial applications are low coefficient of friction [2, 3], high hardness [4], low surface energy [5], good wear-resistance [6], unusual electric and thermal properties [7], and thermal stability at high temperatures [8].

Among hundreds of alloys with quasicrystalline symmetry, the stable icosahedral i-Al-Cu-Fe QC, discovered in 1987 by Tsai et al. [9], is currently one of the best studied icosahedral QCs due to its excellent thermal stability. The nominal composition $Al_{64}Cu_{23}Fe_{13}$ is considered to represent the ideal icosahedral composition [10]. On the other hand, recently nanostructured materials are intensively studied because they provide the possibility to improve the physical properties of their solid counterparts. One of the most important technique used to prepare the nanostructure materials we can say that is the ball milling method, which has been receiving a special attention since it first application by Benjamin and Koch [11, 12] for material processing, and specially for reducing the grain size to the nanometer scale [12–15]. The challenge of this method lies in controlling these material properties as a function of their grain sizes. Hence, the study of the nanostructured counterpart of this promising new material is of relevant importance. The main goal of the present work is to investigate the structural and micro-structural characterization of a nanostructured $Al_{64}Cu_{23}Fe_{13}$ quasicrystal as well as its magnetic and electrical transport properties.

2 Experimental

We, in this work, systematically studied a quasicrystalline samples prepared in two routes: a high quality i-$Al_{64}Cu_{23}Fe_{13}$ icosahedral quasicrystal, obtained after heating the as-prepared sample, (hereafter we refer to this sample as i-QC) and the same sample after 5 h of grinding (hereafter we refer to this sample as n-QC). The preparation procedure of i-$Al_{64}Cu_{23}Fe_{13}$ was described in a previous paper [16]. The structural characterization was made by means of x-ray diffraction (XRD) and transmission electron microscopy (TEM). A high resolution X-ray experiment was carried out in the beam-line XRD2 on Brazilian Synchrotron Light Laboratory (LNLS) using 11.2714 keV bending-magnets. Complementary X-ray diffraction measurements were performed in an X'Pert Pro Panalitical diffractometer using Cu $K_{\alpha 1}$. TEM measurements were done employing a Philips CM 300 transmission electron microscope. The local structure around of Fe sites was analyzed by transmission Mössbauer spectroscopy (TMS) employing a ^{57}Co/Rh radioactive source. The spectra analyzed were recorded at 4.2 K and at room temperature (RT). Ferromagnetic resonance (FMR) spectra were recorded at RT employing a Bruker (ESP300C) spectrometer (X-band with 300 KHz field modulation). The electric resistivity measurements were carried out as a function of temperature from 4 K to RT using a standard DC four probe method.

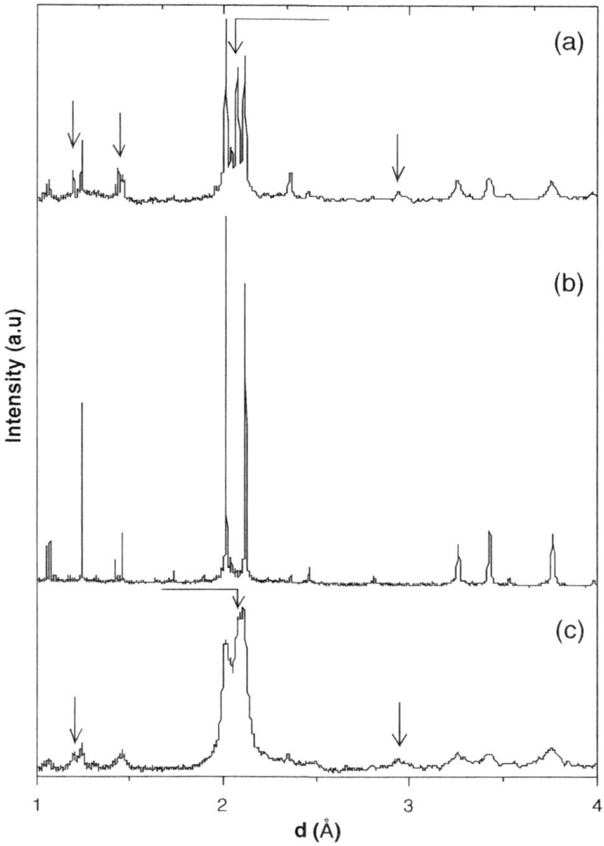

Fig. 1 X-ray diffraction pattern of the i-$Al_{64}Cu_{23}Fe_{13}$ sample **a** as-prepared, **b** after heat treatment, i-QCs, and **c** after grinding for 5 h, n-QCs

3 Results and discussion

Figure 1 shows the XRD pattern of i- $Al_{64}Cu_{23}Fe_{13}$ (a) as-prepared sample by arc-melting, (b) after an isothermal heat treatment at 700°C and (c) after ball milling the sample for 5 h. As can be seen in Fig. 1a, b, in order to remove the spurious phase from the as-prepared sample, indicated by arrows in Fig. 1a, it was necessary a heat treatment. The metastable phase was identified as ε-Al_2Cu_3 [17]. Following Weisbecker et al. [18], and according to the icosahedral scheme proposed by Cahn et al. [19], we determine the six dimensional lattice constant of the i-QC (a_{6D} = 6.309Å), which is in good agreement with those values reported in previous works [18, 20]. Figure 1c shows a reduction of the intensity and a broadening of the diffraction peaks, which is associated with the formation of the n-QC. The average grain-size of n-QC, determined from the Scherrer equation, was estimated to be of the order of 10 nm. It is also observed that after 5 h of milling, the ε-Al_2Cu_3 phase precipitates from the i-QC.

The selected area diffraction (SAD) pattern of the i-QC sample is shown in Fig. 2a–c. The SAD patterns exhibit the five-fold, three-fold and two-fold symmetry, which are typical symmetries for icosahedral phases [21]. The diffraction spots in

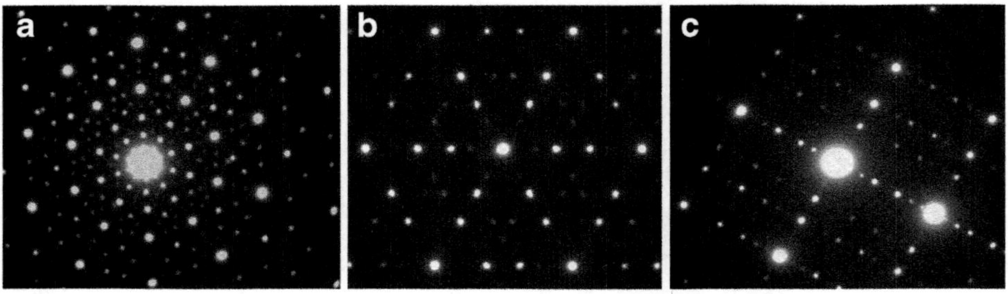

Fig. 2 The selected area diffraction pattern from i-QCs: **a** five-fold zone axis, **b** three-fold zone axis and **c** two-fold zone axis

Fig. 3 Normalized resistance R(T)/R(300 K) versus temperature plot for samples: i-QCs (*filled circles*) and n-QCs (*unfilled circles*)

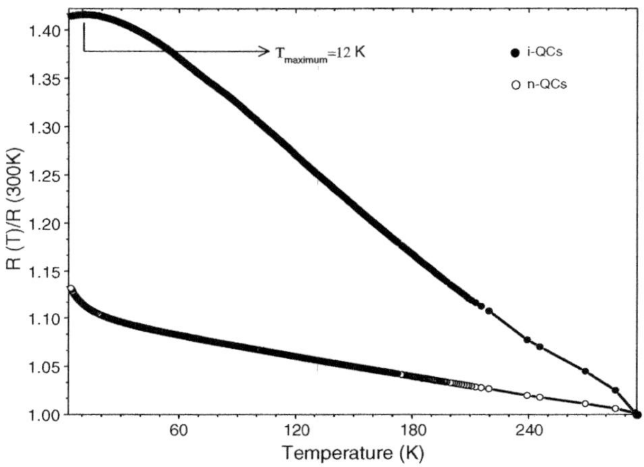

these figures are rather sharp and arranged at fixed and symmetrical position as expected in a high quality icosahedral quasicrystal.

The normalized electrical resistance, R(T)/R(300 K), versus temperature for the i-QC and n-QC samples are presented in Fig. 3. Both measurements show a negative temperature coefficient of resistivity (TCR). Moreover, the i-QC sample exhibits a maximum in the resistance around 12 K. The ratios R(4.2 K)/R(300 K) for the i-QC and n-QC were 1.5 and 1.1, respectively. Similar results for i-QC are reported by other research groups [10, 22–24]. At low temperatures, both samples behave very different. In this temperature region magnetic effects could be important, as was reported in [10]. In such case, the behavior of the resistivity is a result of competing Kondo and RKKY interactions. However, in the case of n-QC sample this behavior could also be associated with a decrease of the quasi-periodic long-range order, due to the grain-size reduction concomitantly with the increase of the grain boundary regions and due the appearance of magnetic moments in the n-QC lattice [24].

The results of the study by TMS of the local order around Fe atoms at RT and 4.2 K are show in Fig. 4a–d. The Mössbauer spectra were fitted employing a quadrupole splitting, Δ, distribution, and assuming a linear relation between the isomer shift, δ, and Δ (for the elementary Lorentzian doublets of full line width at half-maximum Γ [25]). The parameters, obtained from the fitting procedures, are

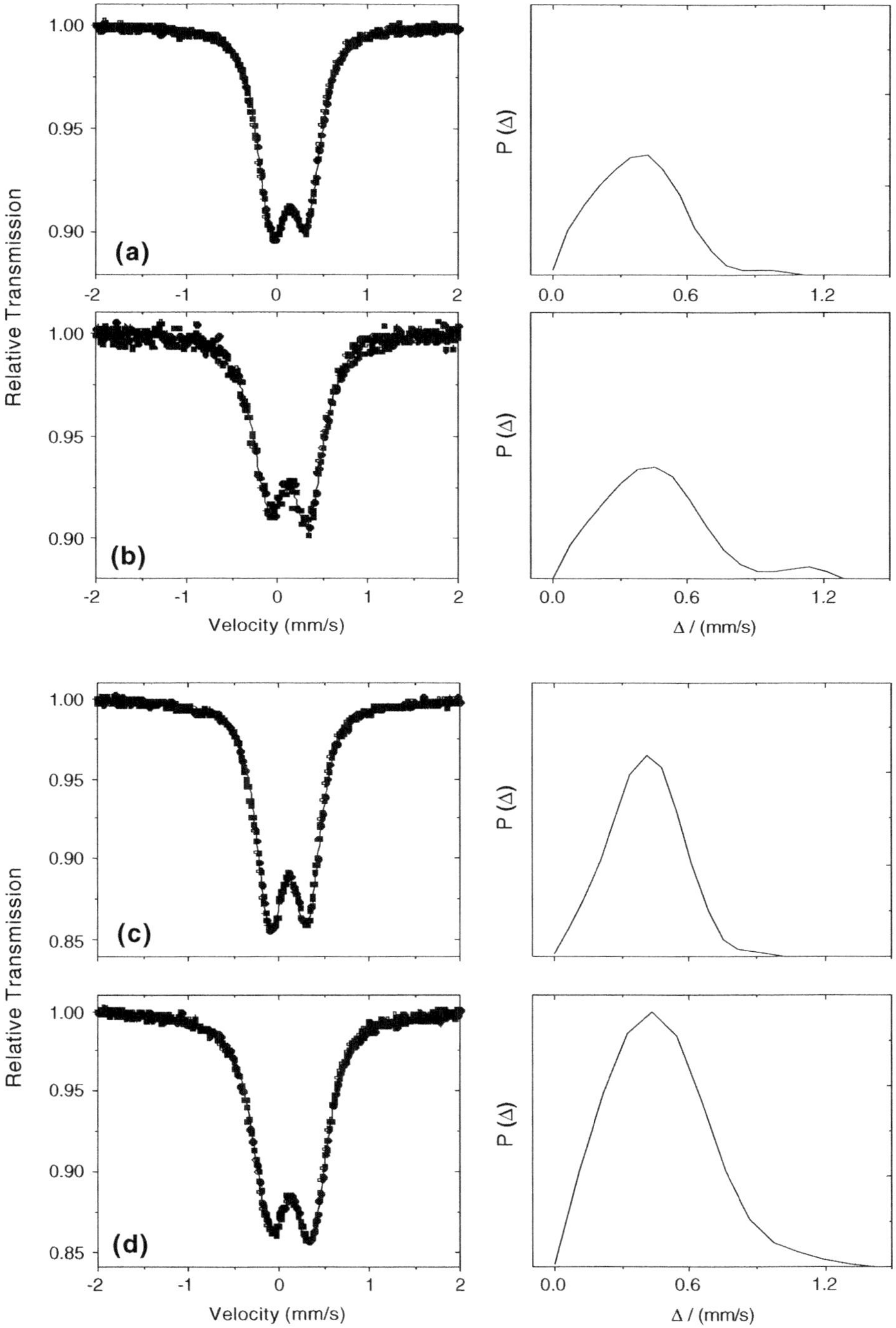

Fig. 4 ^{57}Fe Mössbauer spectra at room temperature **a** i-QCs and **b** n-QCs. ^{57}Fe Mössbauer spectra at 4.2 K **c** i-QCs and **d** n-QCs. The *right panels* correspond to the distributions P(Δ) for each case of the *left* (see text for details)

Table 1 Hyperfine parameters and their corresponding standard deviations (σ_δ, σ_Δ) obtained from the Mössbauer spectra of QCs samples

Parameters	Room temperature		He (4.2 K)	
	i-QCs	n-QCs	i-QCs	n-QCs
$<\delta>$ (mm/s)	0.229(3)	0.231(3)	0.206(2)	0.229(2)
σ_δ (mm/s)	0.003(1)	0.010(2)	0.002(1)	0.009(2)
$<\Delta>$ (mm/s)	0.381(2)	0.461(3)	0.411(2)	0.490(2)
σ_Δ (mm/s)	0.184(2)	0.243(5)	0.172(2)	0.274(4)
Γ (mm/s)	0.248(2)	0.249(3)	0.257(4)	0.267(4)
χ^2	1.200	0.987	1.690	1.270

The $<\delta>$ values are relative to α-Fe at RT
$<\delta>$ average isomer shift, $<\Delta>$ average quadrupole splitting, Γ linewidth

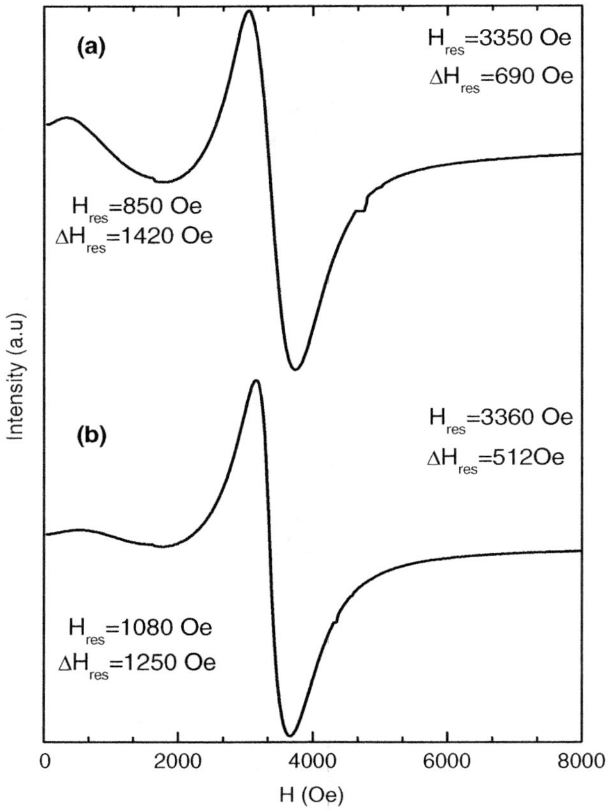

Fig. 5 First-derivate X-band FMR spectra for **a** i-QCs and **b** n-QCs. ΔH_{res} and H_{res} are the line width and magnetic field resonance respectively

shown in Table 1. These values, interpreted in the framework proposed by Stadnik [25], indicate the presence of a quasi-continuous distribution of local environments in the quasicrystal [16, 25–29]. Moreover, the results indicate that after five hours of milling the milled sample does not present remarkable changes in the distribution of the Fe-sites. We only observe a broadening of the quadrupole splitting distribution for the n-QC with respect to the i-QC sample. An effect already expected because of several parameters like as: (1) the intrinsic disorder, (2) the reduction of the quasiperiodic long-range order, (3) the reduction of the grain-size and increase of the grain boundary region and (4) an increasing number of the Fe-sites in n-QC sample; due to the precipitation of the ε-Al_2Cu_3 phase from the i-QC lattice (as indicated by XRD).

Figure 5, shows the ferromagnetic resonance spectra of the i-QC and n-QC samples. In quasicrystals, the lattice periodicity is lacking so that there exist significant local environment variations with no translational periodicity. Moreover, the magnetic properties are sensitive to small perturbations of the atomic environment and the inter-atomic distances. This fact is observed by broad resonances lines in FMR measurements of i-QC [30]. However, n-QC sample exhibits narrower resonance lines, which could be associated with a decreasing of the quasi-periodic long-range order. Furthermore, the high intensity of one resonance line in n-QC could be explained due to the increase of different Fe-sites in the n-QC lattice [31]. These results are supported by X-ray diffraction that shows the formation of the ε-Al_2Cu_3 phase together with the n-QC sample after 5 hours milling. This result agrees with those obtained by TMS measurements.

4 Summary

We have shown that high quality polycrystalline i-QC can be obtained after appropriate heat treatments of the as-cast quasicrystalline alloy prepared by arc-furnace. After five hours of milling, the i-QC sample reduces its average grain-size (approximately to 10 nm), preserving the Fe local order. We also observed that the i-QC (after five hours of milling) can be decomposed in an iron-rich nano-quasicrystalline phase (n-QC) and a ε-Al_2Cu_3 phase. These results are supported by XRD, TEM and TMS measurements. FRM, XRD and TMS measurements also indicate the presence of strain and extrinsic and intrinsic disorder in the studied samples, which are responsible for the high electronic resistivity in the n-QC sample as compared to the i-QC sample (not shown in the present work). The next step is to obtain the dependence of the transport properties with the grain-size in order to study the transport mechanism at low and high temperatures in these nanostructured quasicrystals.

Acknowledgements We are grateful to Alfredo Tolley for his support during the TEM measurements, Mariela Alzamora and Magda Bittencour for their support during the electron transport measurements. This work was supported by the CSI-Project No. 101301011 funded by the National University of San Marcos.

References

1. Shechtman, D., Blech, I., Gratias, D., Cahn, J.W.: Phys. Rev. Lett. **53**, 1951 (1984)
2. Dubois, J.M.: Mater. Sci. Eng. A **4**, 294–296 (2000)
3. Brunet, P., Zhang, L.M., Sordelet, D.J., Besser, M., Dubois, J.M.: Mater. Sci. Eng. A **74**, 294–296 (2000)
4. Wolf, B., Bambauer, K.O., Paufler, P.: Mater. Sci. Eng. A **298**, 284 (2001)
5. Belin-Ferré, E., Dubois, J.M., Fournée, V., Brunet, P., Sordelet, D.J., Zhang, L.M.: Mater. Sci. Eng. A **818**, 294–296 (2000)
6. Sordelet, D.J., Besser, M.F., Logsdon, J.L.: Mater. Sci. Eng. A **54**, 255 (1998)

7. Macía, E.: Phys. Rev. B **80**, 205103 (2009)
8. Yoshimura, M., Tsai, A.P.: J. Alloys Compounds **342**, 451 (2002)
9. Tsai, P., Inoue, A., Masumoto, T.: Jpn. J. Appl. Phys. **26**, L1505 (1987)
10. Dolinšek, J., Vrtnik, S., Klanjšek, M., Jagličič, Z., Smontara, A., Smiljanič, I., Bilušič, A., Yokoyama, Y., Inoue, A., Landauro, C.V.: Phys. Rev. B **76**, 54201 (2007)
11. Benjamin, J.S.: Metall. Trans. **1**, 2943 (1970)
12. Koch, C.C.: Ann. Rev. Mater. Sci. **19**, 121 (1989)
13. Zoz, H.: Mater. Sci. Forum. **419**, 179–181 (1995)
14. Basset, D., Matteazzi, P., Mani, F.: Mater. Sci. Eng. A **168**, 149 (1993)
15. Barua, P., Murty, B.S., Mathur, B.K., Srinivas, V.: J. Appl. Phys. **91**, 5353 (2002)
16. Quispe-Marcatoma, J., Landauro, C.V., Taquiere, M., Rojas Ayala, C., Peña Rodriguez, V.A.: Hip Int. **195**, 105 (2010)
17. Lin, C.M., Lin, S.T.: Chinese Journal of Physics **31**, 297 (1993)
18. Weisbecker, P., Bonhomme, G., Bott, G., Dubois, J.M.: J. Non-Cryst. Solids **351**, 1630 (2005)
19. Cahn, J.W., Shechtman, D., Gratias, D.: J. Mater. Res. **1**, 13 (1986)
20. Cornier-Quiquandon, M., Quivy, A., Lefebvre, S., Elkaim, E., Heger, G., Katz, A., Gratias, D.: Phys. Rev. B **44**, 2071 (1991)
21. Tiwari, R.S., Yadav, T.P., Mukhopadhyay N.K., Shaz, M. A. : Z. Kristallogr. **224**, 26 (2009)
22. Smiljanić, I., Bilušić, A., Bihar, Ž., Lukatela, J., Leontić, B., Dolinšek, J., Smontara, A.: MTAEC9 **41**(6), 265 (2007)
23. Singh, K., Bahadur, D., Radha, S., Nigam, A.K., Prasad, S.: JALCOM **257**, 57 (1997)
24. Bigg, B.D., Li, Y., Poon, S.J.: Phys. Rev. B **43**, 8747 (1991)
25. Stadnik, Z.M., Stroink, G., Ma, H., Williams, G.: Phys. Rev. B **39**, 9797 (1989)
26. Chien, C.L, Lu, M.: Phys. Rev. B **45**, 12793 (1992)
27. Stadnik, Z.M., Takeuchi, T., Tanaka, N., Mizutani, U.: J. Phys., Condens. Matter **15**, 6365 (2003)
28. Brand, R.A., Pelloth, J., Hippert, F., Calvayrac, Y.: J. Phys. Condens. Matter **11**, 7523 (1999)
29. Brand, R.A., Hippert, F., Frick, B.: J. Phys. Condens. Matter **21**, 045405 (2009)
30. Bahadur, D.: Prog. Cryst. Growth Charact. Mater. **34**, 287 (1997)
31. Misra, S.K., Kahrizi, M., Singh, K., Venkatramani, N., Bahadur, D.: JMMM **150**, 430 (1995)

Hyperfine Interact (2011) 203:9–15
DOI 10.1007/s10751-011-0364-y

Mössbauer analysis of high-energy mechanical-milled sand fraction of a magnetic soil developing on basalt

José Flávio Marcelino Borges · Marlon Luiz Hneda · André Maurício Brinatti · João Batista Marimon da Cunha · Jadir Aparecido Rosa · José Domingos Fabris

Published online: 25 August 2011

Abstract A sample of the coarse sand fraction from the soil material of the A-horizon (0–0.2 m from the soil surface) of a dusky red magnetic Oxisol was submitted to high-energy mechanical milling for different times. This assay aimed mainly at (a) monitoring the individualization of strongly aggregated mineral particles, and (b) measuring the effect of the milling pressure on the mineralogy changes of the material. These data are also intended to experimentally subside any physical model describing the mechanical behavior of the superficial soil layer that is subjected to intensive machine management, in agriculture fields. Powder X-ray data reveal that some mineralogical phases, notably gibbsite, disappear soon after the first few hours milling. The 298 K-transmission Mössbauer spectrum for the non-milled sand sample shows a qualitatively typical pattern for the sand fraction of basalt derived soils, with magnetically ordered sextets, assignable mainly to hematite and maghemite, and an intense central (super)paramagnetic Fe^{3+} doublet. For the milled samples, spectra revealed progressive spectral reduction of the magnetic hyperfine structure, with concomitant increase of relative subspectral areas due to (super)paramagnetic phases, as the milling time increased. This result is consistent with the reduction

J. F. M. Borges · M. L. Hneda · A. M. Brinatti
Department of Physics, State University of Ponta Grossa,
Av. Gal. Carlos Cavalcanti 4748, CEP 84030-900 Ponta Grossa, PR, Brazil

J. B. M. da Cunha
Institute of Physics, Federal University of Rio Grande do Sul,
Av. Bento Gonçalves 9500, CEP 91501-970, Porto Alegre, RS, Brazil

J. A. Rosa
Agronomy Institute of Paraná, Polo Regional de Ponta Grossa,
BR 376, km 496, CEP 84001-970, Ponta Grossa, PR, Brazil

J. D. Fabris (✉)
Federal University of Jequitinhonha and Mucuri Valleys (UFVJM),
Campus JK, Building 2 office 4, CEP 39100-000 Diamantina, Minas Gerais, Brazil
e-mail: jdfabris@ufmg.br

of measured saturation magnetization, from 4.96(8) J T^{-1} kg^{-1}, for the non-milled sample, to 3.26(7) J T^{-1} kg^{-1}, for the sample milled for 8 hours.

Keywords Iron oxide · Magnetic soil · Mineral aggregates

1 Introduction

There has been a surge of soil studies addressing the many agriculture management systems in view to increase crop yields and prevent soil degradation. This would also sustain natural amounts of many essential nutrients to plants, supporting an environmentally more rational agriculture system [1]. Many reported studies have been concerning on developing a conservation-based practice, mainly characterized by requiring less revolving the soil of agriculture lands [2, 3], in a management known as no-tillage system [4]. This system greatly relies on protecting the soil from water- or wind-driven erosion by maintaining a soil coverage with a layer of biomass remaining from previously cultivated crops in the same land area, with, eventually, admixed animal manures [5].

This work was designed to focus on changes of the mineralogy of the coarse sand fraction of a magnetic soil developing on basalt, when the material is submitted to high-energy mechanical milling for different times. High-energy mechanical alloying is commonly used as an affordable alternative to produce non-equilibrium phases, such as amorphous materials, nanocomposites, and rapid solid solutions [6]. This preparation technique has drawn quite a lot of attention, as it allows the study of metal–metal, oxide–metal, and oxide–oxide solid solutions, which are sometimes found to be metastable [7]. The final milling product is often in the nano-sized scale, being characterized by displaying high surface area and high density of structural flaws. Indeed, the so-milled materials very often become chemically much more reactive. This increase in the reactivity is thought to occur in consequence of increasing strains in the crystalline framework of minerals after milling, after the material is subject to impacts that produce structural fractures and deformations [8]. In this study, the coarse sand fraction separated from the magnetic pedomaterial has been submitted to different times under high-energy mechanical milling, aiming primarily at monitoring the individualization of the strongly aggregated mineral particles, containing quartz grains, probably coated with magnetic iron oxides, and, secondarily, measuring the effects of the impact pressure on the mineralogy of the milled pedomaterial. This would subside any physical model eventually proposed to describe the behavior of the superficial soil layer under intensive agriculture management with heavy machinery.

2 Material and methods

The sample of an umbric, dystrophic, clay-textured, alic red Oxisol (LVd4), of a subtropical field phase, geomorphically characterized by a slightly undulated relief, was collected from the experimental area of the Agronomy Institute of Paraná (IAPAR), Regional Pole of Ponta Grossa, state of Paraná, Brazil. The sampling site (geographic coordinates 25°13′S 50° 01′W) is located at an altitude of 875 m above the sea level, under a Cfb climate—mesothermic, subtropical, wet, according to the Köppen classification, with cool summers, absence of dry seasons, and severe frost. A corresponding sample of this pedomaterial was earlier analyzed in the Laboratory of Soil Physics and Environmental Sciences, of the Physics Department at the State University of Ponta Grossa (UEPG). The aim at that time was to evaluate differences in mineralogy [9] and the organic matter composition [10] of samples from agricultural sites under no-tillage system, by comparing them with those under traditional agriculture practices. This soil is particularly rich in iron oxides, specifically hematite and goethite [9], in its mineral fraction.

The soil sample was first dried at 60°C for 24 h on a heating plate; a 16-mesh sieve was used to separate roots and gravels. The resulting soil material was then mixed with sodium hydroxide in water (NaOH 1 mol L^{-1}), for 24 h, to disperse mineralogical particles, and submitted to a 30 min-ultrasonic cleaning bath.

Only the coarse sand fraction (average particle diameter 2×10^{-3} m–2×10^{-4} m), separated by sieving the magnetic soil material with a 270 mesh-sieve, was taken for this study. Finally, the magnetic particles were separated from the sand mass by using a hand-magnet. The resulting sample from this selection was submitted to high-energy mechanical milling, at the Institute of Physics at the Federal University of Rio Grande do Sul (IF-UFRGS), using a Spex 500 mill [10] with a 2 g soil relation on a 9.5 g steel ball used for impact. Four milling processes were run for 2, 5, 8 and 12 h in air atmosphere. The temperature change during milling was lower than 20°C.

Mössbauer spectroscopy measurements were also carried out at the IF-UFRGS, in transmission geometry, at room temperature (~298 K), and a radioactive ^{57}Co source in Rh-matrix. The X-ray patterns were obtained at the Multipurpose Laboratory of UEPG, with a Shimadzu Lab-X 6000.

The saturation magnetization was measured with a portable magnetometer under a fixed magnetic field of ~0,3 T [11, 12].

3 Results and discussion

XRD data for the non-milled sample revealed intense reflections of quartz (SiO_2; Qz), gibbsite ($Al(OH)_3$; Gb), and many other less intense reflection peaks for no iron-containing compounds. The following oxidic phases could be identified: hematite (αFe_2O_3; Hm), and maghemite (γFe_2O_3; Mh), as shown in Fig. 1. Less intense reflections can be assigned to other iron-silicon compounds (SF), namely FeSi and Fe_2Si (JCPDS # 76-1748, 83-1259 respectively), also observed in Fig. 1. Figure 2 shows XRD patterns for samples milled for different milling times. It can be seen that there was prompt collapse of the gibbsite reflections within the first two milling hours. On the other hand, the magnetic mineralogical phases occurring in the original sample were found to be somehow preserved. It can be also observed that there was no relevant crystallographic–structural change of iron oxides even by increasing the milling times.

The Mössbauer spectra showed relatively complex patterns, with somewhat intense and asymmetrical resonance lines. From Fig. 3, a slight increase in the contribution of non-magnetic phases relatively to the magnetically ordered structures, namely hematite and maghemite, could be observed as the milling time increased.

Fig. 1 Powder XRD pattern for the sand fraction of the magnetic soil material, before milling (Qz quartz; Gb gibbsite; Hm hematite; Mh maghemite; SF iron–silicon compounds)

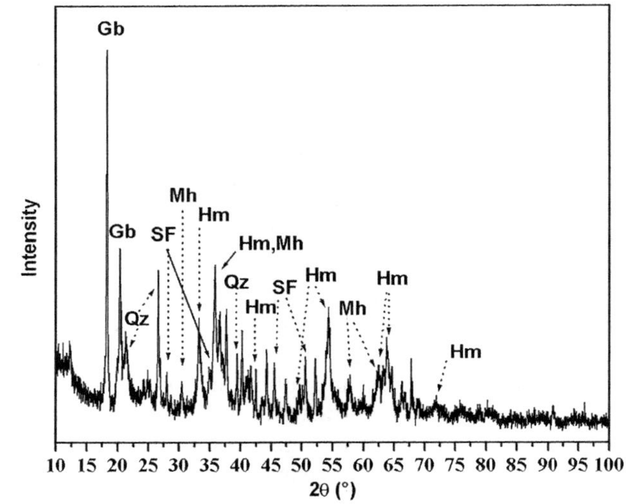

Fig. 2 Powder XRD patterns for the sand fraction of the magnetic soil material, milled for different times. Reflections of quartz are no longer observed, comparatively to the pattern (Fig. 1) for the sample before milling. (Qz quartz; Hm hematite; Mh maghemite)

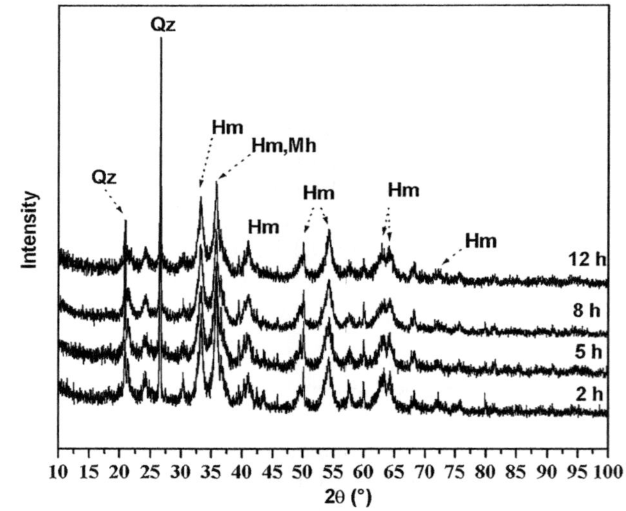

From Figs. 4a and b, it is possible to follow the evolution of the mineralogical phases proportion, at different milling times, considering the calculation of Mössbauer subspectral areas.

Regarding the non-milled sample, the subspectrum assigned to the hematite contributes with approximately 28% of the relative spectral area. This contribution decreases to almost 15% for the 12 hour-milled sample. The Fe^{3+} (super)paramagnetic doublet due to the non-milled sample contributes with approximately 48% of the relative subspectral area. This contribution is probably related to the iron silicates in the coarse sand grains, as it is observed for the non-milled sample X-ray pattern, and increases the subspectral area to about 74% for the 12 hour-milling sample. The disappearance of X-ray reflections assignable to gibbsite suggests that the mineral is

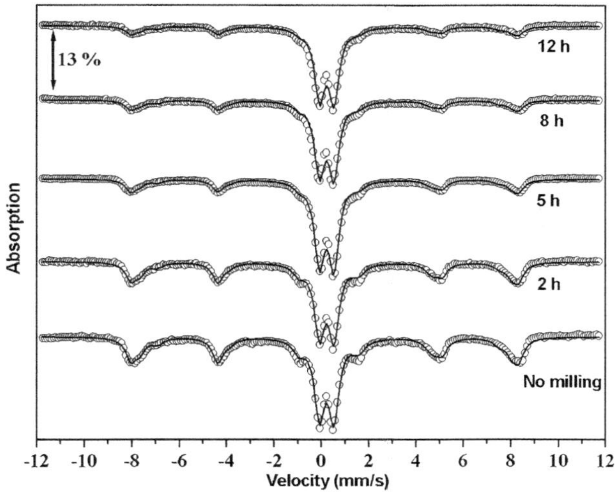

Fig. 3 298 K-Mössbauer spectra for milled samples (corresponding fitted hyperfine parameters are presented in Table 1)

Table 1 Fitted Mössbauer parameters

Time (h)	Subspectra	B_{Hf} (T)	Γ (mm/s)	δ (mm/s)	Δ or ε (mm/s)
0	Hematite	50.8	0.55	0.26	−0.11
	Maghemite	48.7	0.55	0.20	−0.01
	Maghemite	44.9	0.55	0.31	−0.01
	Paramagnetic		0.52	0.23	0.30
2	Hematite	50.8	0.55	0.26	−0.11
	Maghemite	48.7	0.55	0.20	−0.01
	Maghemite	44.9	0.55	0.31	−0.01
	Paramagnetic		0.52	0.23	0.30
5	Hematite	50.8	0.55	0.26	−0.11
	Maghemite	48.7	0.55	0.20	−0.01
	Maghemite	44.9	0.55	0.01	−0.01
	Paramagnetic		0.52	0.23	0.30
8	Hematite	50.8	0.55	0.26	−0.11
	Maghemite	48.7	0.55	0.20	−0.01
	Maghemite	44.9	0.55	0.31	−0.01
	Paramagnetic		0.52	0.23	0.30
12	Hematite	50.8	0.55	0.26	−0.11
	Maghemite	48.7	0.55	0.20	−0.01
	Maghemite	44.9	0.55	0.31	−0.01
	Paramagnetic		0.52	0.23	0.30

The corresponding relative spectral areas for subspectra are graphically represented in Fig. 4
B_{hf} hyperfine field; Γ resonance line width; δ isomer shift relative of αFe; Δ quadrupole splitting and ε quadrupole shift.

destroyed under high-energy mechanical milling; probably, the increase of the (super)paramagnetic Mössbauer fraction with milling time reflects the higher intensity of the central doublets of spectra—which is likely due to the increasing number of finer particles. Confirmation for this model requires, nevertheless, measurements at lower temperatures.

Magnetization measurements pointed to a decrease of the saturation magnetization value, from 4.96(8) J T^{-1} kg^{-1} (non-milled sample) to 3.26(7) J T^{-1} kg^{-1} (8 hour-

Fig. 4 Proportions of fractions of the milled soil–sand **a** according to their magnetic behavior and **b** relatively to all mineralogical phases, as they can be directly determined from Mössbauer subspectral areas

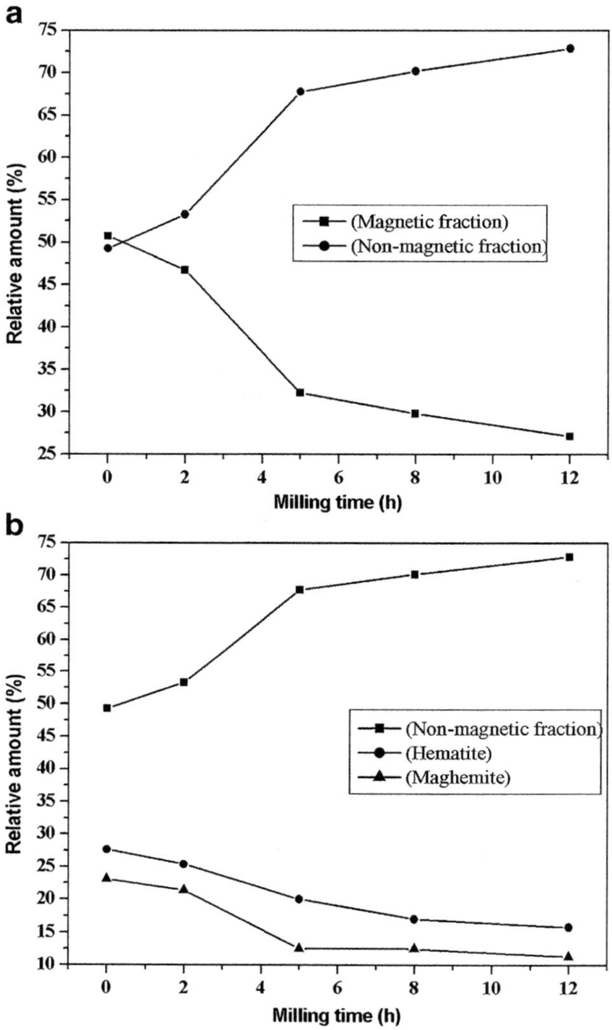

milled sample) (values in parentheses are standard deviations over the corresponding means of 10 readings). Considering the spectral pattern that can be assigned to maghemite (non-stoichiometric spinel ferrite; γFe_2O_3), for the milled sample, the most probable hyperfine fields being at 44.9 T (tetrahedral coordination iron site) and 48.7 T (octahedral coordination), the estimated area is 11% (Fig. 4). Although the spectral areas do not directly correspond to the phase proportion, due to the differences in the recoil-free fractions for iron sites, it is possible to roughly compare their alterations throughout the milling process with the magnetic variation. Yet, any further refinement in the spectral analysis will require Mössbauer measurements taken under different experimental conditions, like low temperatures or strong magnetic fields. Based solely on the current results, the areas of sites assigned to maghemite are 18% of the total spectrum, for example, an 8-hour milling sample. The ratio for maghemite areas (samples before and after 8-hour milling) would then be (23%) / (18%) = 1.3. This result is reasonably consistent with what is found as the ratio of the saturation magnetization values: $\frac{4.96\ \mathrm{J\,T^{-1}kg^{-1}}}{3.26\ \mathrm{J\,T^{-1}kg^{-1}}} = 1.5$

4 Conclusions

Gibbsite was readily converted and no characteristic reflection appeared in powder X-ray diffraction (XRD) patterns for the sand samples subjected to high-energy mechanical milling, in air atmosphere. Although XRD patterns did not show significant structural changes of the two main iron oxides, hematite and maghemite, Mössbauer spectroscopy measurements showed a clear variation of the proportion of magnetic subspectra with increased milling time, although their hyperfine structure are not significantly changed on milling. Data obtained from the used analytical techniques, XRD and ^{57}Fe-Mössbauer spectroscopy, were not enough to confirm any encapsulation of particles of quartz of the coarse sand with magnetic material.

Acknowledgements Work supported by CNPq and Araucária Foundation in Paraná (Brazil). Araucária also supported an undergraduate studentship to MLH and CAPES (Brazil) grants the Visiting Professor PVNS fellowship to JDF at UFVJM. FAPEMIG (state of Minas Gerais, Brazil) supported (grant # PPM-00217/08) the travel of JDF to Lima, Peru, to attend LACAME 2010.

References

1. Moreti, D., Alves, M.C., Valério Filho, W.V., Carvalho, M.P.: Atributos Químicos de um Latossolo Vermelho sob Diferentes Sistemas de Preparo, Adubações e Plantas de Cobertura. R. Bras. Ci. Solo. **31**, 167–175 (2007) (in Portuguese)
2. Schaefer, C.E.G.R., Souza, C.M., Vallejos M.F.J. et al.: Características da porosidade de um argissolo vermelho-amarelo submetido a diferentes sistemas de preparo de solo. R. Bras. Ci. Solo. **25**, 765–769 (2001) (in Portuguese)
3. Silva, M.L.N., Curi, N., Blancaneaux, P.: Sistema de Manejo e Qualidade Estrutural de Latossolo Roxo. Pesqui. Agropecu. Bras. **35**, 2485–2492 (2000) (in Portuguese)
4. Paterniani, E.: Agricultura sustentável nos trópicos. Estud. Av. **15**, 303–326 (2001) (in Portuguese)
5. Huggins, D.R., Reganold, J.P.: Plantio Direto, uma Revolução na Preservação. Sci. Am. Bra. **75**, 56–65 (2008) (in Portuguese)
6. Suryanarayana, C.: Mechanical Alloying and Milling. Prog. Mater. Sci. **46**, 1–184 (2001)
7. Cótica, L.F., Zanatta, S.C., Rocha, M.A., Santos, I.A., Paesano, A., Da Cunha, J.B.M., Hallouche, B.: Phase evolution and magnetic properties of a high-energy ball-milled hematite-alumina system. J. Appl. Phys. **95**, 1307–1314 (2004)
8. Calka, A., Wexler, D.: Mechanical milling assisted by electrical discharge. Nature **419**, 147–151 (2002)
9. Gonçalves, D., Leite, W.C., Brinatti, A.M., Saab, S.C., Iarosz, K.C., Mascarenhas, Y.P., Carneiro, P.I.B., Rosa, J.A.: Mineralogia de um Latossolo Vermelho Distrófico submetido a diferentes manejos por 24 anos. R. Bras. Ci. Solo, **32**, 2647–2652 (2008) (in Portuguese)
10. Murty, B.S., Ranganathan, S.: Novel material synthesis by mechanical alloying/milling. Int. Mater. Rev. **43**, 101–141 (1998)
11. Coey, J.M.D., Cugat, O., McCauley, J., Fabris, J.D.: A portable soil magnetometer. Rev. Fis. Apl. Instrum. **7**, 25–30 (1992)
12. Fabris, J.D., Coey, J.M.D.: Espectroscopia Mössbauer do ^{57}Fe e Medidas Magnéticas na Análise de Geomateriais. In: Alvarez, V.V.H., Schaefer, C.E.G.R., de Barros, N.F., de Mello, J.W.V., da Costa, L.M. (eds.) Tópicos em Ciências do Solo. Viçosa, Sociedade Brasileira de Ciência do Solo, vol. 2, pp. 47–102 (2002) (in Portuguese)

Hyperfine Interact (2011) 203:17–23
DOI 10.1007/s10751-011-0365-x

Characterization of the Carancas–Puno meteorite by energy dispersive X-ray fluorescence, X-ray diffractometry and transmission Mössbauer spectroscopy

María L. Cerón Loayza · Jorge A. Bravo Cabrejos

Published online: 25 August 2011

Abstract We report the results of the study of a meteorite that impacted an inhabited zone on 15 September 2007 in the neighborhood of the town of Carancas, Puno Region, about 1,300 km south of Lima. The analysis carried out by energy dispersive X-ray fluorescence, X-ray diffractometry and transmission Mössbauer spectroscopy (at room temperature and at 4.2 K), reveal the presence in the meteorite sample of magnetic sites assigned to taenite (Fe,Ni) and troilite (Fe,S) phases, and of two paramagnetic doublets assigned to Fe^{2+}, one associated with olivine and the other to pyroxene. In accord with these results, this meteorite is classified as a type IV chondrite meteorite.

1 Introduction

On 15 September of 2007 it was reported the impact of a meteorite in the neighborhood of the town of Carancas, Desaguadero, Chucuito Province, Puno Region, near the Bolivian border. One of the authors traveled to this site 4 days later to gather in situ samples of the meteorite and soil inside and outside the crater. The purpose of this work was to characterize the structural properties of these samples applying the analytical techniques available in our Faculty of Physical Sciences-UNMSM such as energy dispersive X-ray fluorescence (EDXRF), X-ray diffractometry (XRD) and particularly transmission Mössbauer spectroscopy (TMS) for being an isotopically selective technique. This impact caused a great explosion and the formation of a crater with approximate dimensions of 20 m in diameter and 3 m in depth. It was also reported the emission of sulfurous gases that irritated the eyes and respiratory organs of the people in the area. Newspapers announced that the fallen object contained radioactive material. To confirm the news a Geiger counter model Radalert 50

M. L. Cerón Loayza (✉) · J. A. Bravo Cabrejos
Laboratorio de Análisis de Suelos, Facultad de Ciencias Físicas,
Universidad Nacional Mayor de San Marcos, Aptdo. 14-0149, Lima 14, Perú
e-mail: malucelo@hotmail.com

was carried to the site and gave very low readings of 0.030 μS. At the site, the underground water table lies at a depth of about 3 m; therefore, the crater lies partially filled with water.

2 Materials and methods

The collection of the sample used in this study was undertaken 5 days after the meteorite impact. The meteorite sample is grayish and of fine texture, with small surface striates probably related to the impact with the ground. About 180 mg of loose material that separated from the sample was used to prepare the analyte for analysis. This material was grinded in an Agata mortar to a fine 160 μm mesh powder for analysis by XRD and TMS at RT; a pellet was prepared for EDXRF analysis. The TMS spectrum at 4.2 K was taken at CBPF in Brasil using an analyte with a mass of 100 mg from another section of the sample.

2.1 Analysis by energy dispersive X-ray fluorescence (EDXRF)

The elemental composition analysis was performed using a portable EDXRF AMPTEK instrument. This instrument uses an X-ray tube with an Ag anode, that operated at 30 kV and about 30 μA. This instrument allows the identification of elements with $Z > 12$ (magnesium). Figure 1a shows the experimental spectrum and the results of the quantitative analysis are given in Table 1. This was done by fitting the experimental EDXRF spectrum with a simulated XRF spectrum based on the fundamental parameters model. The simulation of the spectra is done by using a program written in FORTRAN It simulates all the physical processes that affect X-rays; production, scattering, absorption and photoelectric production, that take place in an XRF measurement with the used experimental geometry. It uses X-ray cross sections data provided by NIST and atomic characteristic X-ray parameters provided by IAEA. The X-ray distribution function, which includes continuous and discrete components, has been calibrated performing scattering of these X-rays by known samples. The performance of the program has been checked using reference samples. The estimated uncertainty in these measurements of the elemental concentrations is about 10%. Figure 1b shows the fitting of this spectrum of the meteorite sample together with its simulated version in semi logarithm scale (Dr. J.A. Bravo C., private communication) [1].

2.2 Analysis by X-ray diffractometry (XRD)

For the structural analysis of the minerals present in the sample, the XRD technique was applied using a RIGAKU diffractometer, model Miniflex, using Cu–Kα radiation with $\lambda = 1.54178$ Å and a vertical goniometer. The scanned angle interval was $4° < 2\theta < 70°$ and the 2θ advance was of 0.02°/step with a time interval of 3 s per step.

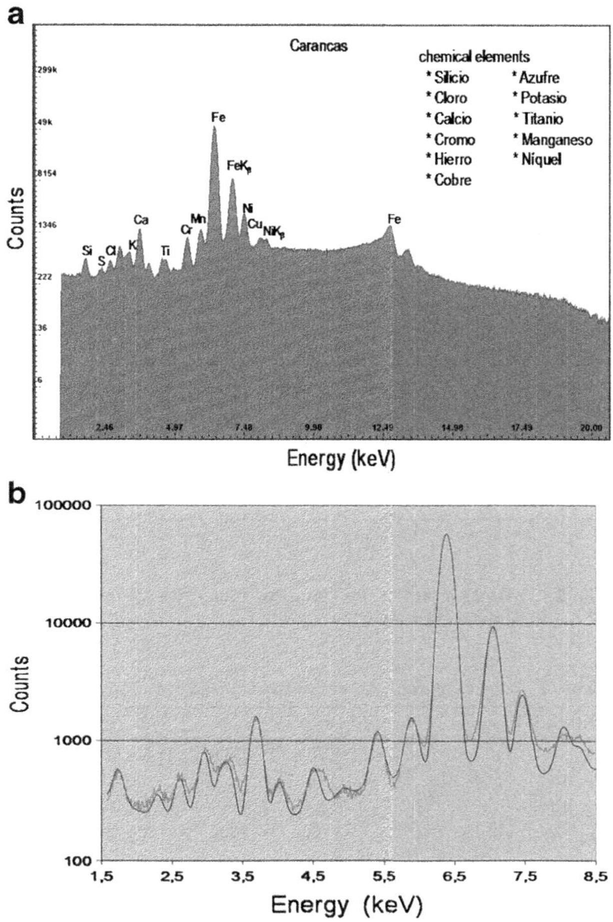

Fig. 1 **a** EDXRF experimental spectrum of the meteorite sample from Carancas- Puno, shown in semi-logarithmic scale. **b** EDXRF experimental spectrum of the meteorite sample shown together with its simulated version in semi-logarithmic scale (Bravo C., J.A., private communication)

Table 1 Quantitative elemental analysis of a meteorite sample from Carancas, Puno

Element	Al	Si	S	Cl	K	Ca	Ti	Cr	Mn	Fe	Ni	Zn	La
Concentration (% weight)	7.5	13.0	0.90	0.90	0.35	0.66	0.04	0.11	0.12	6.00	0.22	0.01	0.10

2.3 Analysis by ^{57}Fe transmission Mössbauer spectroscopy (TMS)

TMS was used to obtain more detailed information about iron containing minerals. A conventional spectrometer was used with a sinusoidal velocity modulation signal and 1024 channels. The Mössbauer spectrum at room temperature (RT) of the sample was collected at the Laboratory of Archaeometry, Facultad de Ciencias Físicas, UNMSM. TMS measurements at liquid helium temperature (LHT) were collected at the Centro Brasileiro de Pesquisas Fisicas (CBPF). A ^{57}Co source in a Rh matrix was used to collect the spectra, which were analyzed using the Normos program by Brand in its crystalline sites version (Normos Site) [2].

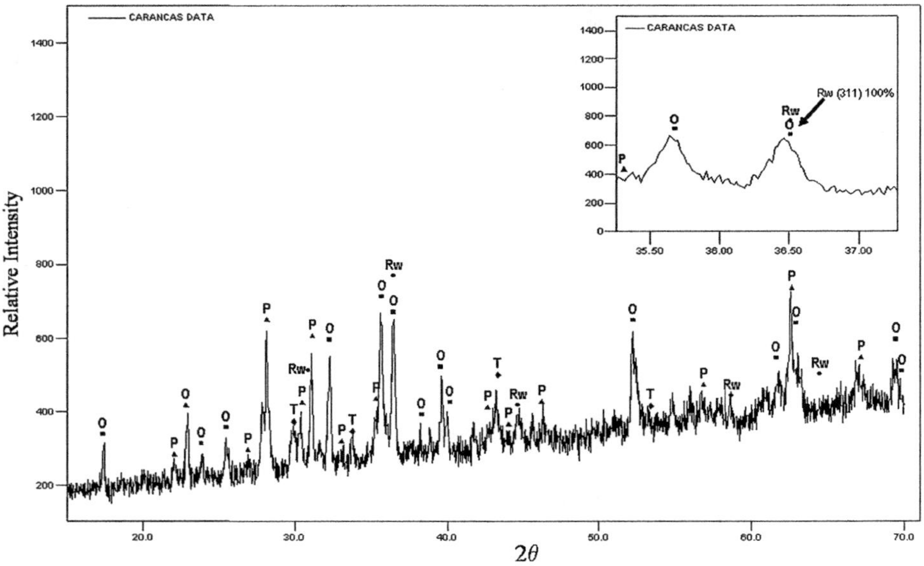

Fig. 2 Diffractogram of the Carancas meteorite sample showing: olivine (*O*), piroxene (*P*), troilite (*T*), and ringwoodite (*Rw*)

Table 2 Hyperfine parameters of the Carancas meteorite sample

MineralPhases	δ (mm/s)	2ε (mm/s)	B (T)	A (%)
	CARANCAS(RT) Total area: 0.33mm/s			
Taenite	−0.10	0.03	33.8	19
Troilite	0.65	−0.16	31.4	17
Olivine	1.04	2.95		43
Pyroxene	1.03	2.09		21
	CARANCAS (LHT) Total area: 0.28 mm/s			
Troilite I	0.73	−0.17	35.7	10
Troilite II	0.60	−1.14	38.6	9
Taenite	0.14	0.16	39.6	4
Tetrataenite	0.00	0.65	34.4	9
Antitaenite S1	−0.06			2
Olivine D1 / Fe^{+2}	1.08	2.44		5
Olivine D2	1.21	3.03		55
Pyroxene D3	1.09	3.40		6

Estimated errors are of about 0.01 mm/s for the isomer, δ, and quadrupole shifts, 2ε, of about 0.1 T for the magnetic hyperfine field, B, and of about 2 % for area values, A

3 Results and discussion

The results of the analysis by EDXRF of the meteorite sample (see Table 1) show the presence of the following elements: Si, Cl, S, K, Fe, Cr, Cu, Ca, Ti, Mn and Ni; with relatively high concentrations of Fe, as well as of S, Cl, Cr and Ni compared to a regular terrestrial soil sample (see Fig. 1a). The XRD results show the presence of

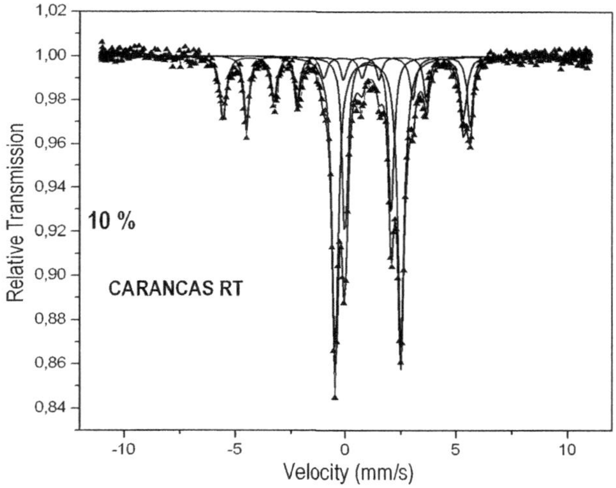

Fig. 3 Mössbauer spectrum of the Carancas meteorite sample taken at room temperature (RT)

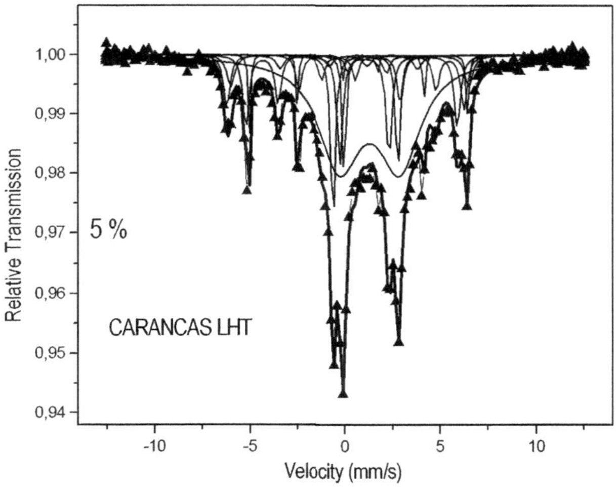

Fig. 4 Mössbauer spectrum of the Carancas meteorite sample taken at liquid helium temperature (LTH)

mineral phases such as troilite (FeS), ringwoodite (Rw, $(Mg, Fe)_2SiO_4$), olivine, and pyroxene as the main mineralogical phases (see Fig. 2).

The origin of Rw (a polymorph of olivine) may be related to the extremely high pressures generated during the impact of a meteorite moving at high speed [3]. Fused materials may be also generated by meteoritic impacts at high speed [4]. The presence of a metastable high pressure mineral, such as Rw, indicates crystallization at pressures between 13 and 23 GPa (maximum pressure generated by the shock wave) [5]. These extremely high pressures are impossible in the crust of the Earth except in the case of meteoritic impacts. Under this hypothesis, it is supposed that the mineral rich in magnesium, precursor of Rw (probably olivine), had previously been part of the meteorite. Since olivine in meteorites is present with three different crystalline structures, it is possible that rhombic olivine transformed into Rw which takes place at pressures between 6 and 16 GPa (depending on Fe content) [6]. The primary mineral quartz (SiO_2) and the Fe oxides and hydroxides are not observed.

Table 2 lists the hyperfine parameters of the different subspectra of the meteorite sample taken at RT and LHT. As shown in Fig. 3 the spectrum taken at RT was fitted by two magnetic sextets, assigned to troilite (FeS) and the Fe–Ni phase (taenite) respectively, and two paramagnetic Fe^{2+}doublets, one associated to olivine $(Fe, Mg)_2SiO_4$ and the other to pyroxene (Mg, Ca, Mn, Fe) Si_2O_6 [7]. In the results of the spectrum taken at LHT (see Fig. 4), it is observed the presence of two ordered magnetic components of Fe^{3+} associated with troilite I and troilite II, and two magnetic sextets attributed to Fe–Ni phases, which may be associated to crystalographically ordered taenite (tetrataenite [8]), with Bhf = 34.54 T, and to taenite in a disordered phase, Bhf = 38.83 T [8]. A singlet, S1, is assigned to antitaenite. It is known that tetrataenite is believed to be formed when meteorites cool very slowly below 320°C. Antitaenite, as evident from Mossbauer spectroscopy, has never been observed alone in meteorites [9]. In meteorites olivine may appear in three different crystalline structures. From the obtained results, it is probable that due to this fact olivine is observed to be associated to two paramagnetic doublets, D1 and D2 (with large line width). Pyroxene is associated with an Fe^{2+} site and continues appearing as a paramagnetic doublet D3.

4 Conclusions

1. The results obtained from analytical techniques show the merits of these techniques to obtain basic information about the studied sample. This information allowed to classify this meteorite as a type IV chondrite.
2. The analysis by EDXRF allowed to identify the presence of S in relatively high concentration, which was probably the source of reported toxic gases that affected the population of Carancas.
3. The analysis by XRD allowed to identify the presence of the structural phases of olivine and Rw - with a spinel structure with its main peak at (311). The presence of these two phases, olivine and Rw, would indicate that part of the olivine transformed into Rw and that the meteoritic impact generated pressures between 6 and 16 GPa. The presence of Rw is an evidence of the high pressures generated by the meteoritic impact in Carancas-Puno.
4. The analysis by TMS at RT allowed the observation of superposed magnetic and paramagnetic phases. The presence of the magnetic phases of troilite are evident. The Fe–Ni phases may be related to taenite (crystalline and disordered); the singlet S1 may be also related to antitaenite. With these results we have shown the presence of the Fe–Ni phase in the Carancas meteorite. At low temperature it is observed that olivine, which is present as a paramagnetic phase, is associated with two doublets. With the help of XRD it can be shown that one of them is associated to ringwoodite (cubic) and the second doublet may correspond to wadleysite o forsterite. Additional Mössbauer measurements in an applied external magnetic field are required to better resolve the structural phases of taenite.

Acknowledgements The economic support of the Superior Research Council of the National University of San Marcos, through contract N° 081301091, is acknowledged. Likewise, the collaborations of the Brazilian Physics Research Center (CBPF) and of the research laboratories at the Faculty of Physical Sciences of this University are much appreciated.

References

1. Cerón, M.L., Bravo, J.A.: Caracterización Mineralógica de un Impacto Meteoritico en la Localidad de Carancas–Puno. Rev. Investig. Fís. **12** (1), 5–12 (2009)
2. Brand, R.A., NORMOS: Mössbauer Fitting Program
3. Langenhorst, F., Deutsch, A.: Advanced Mineralogy, vol. 3, Springer-Verlag, 95–110 (1998)
4. Grieve, R.A.F., Cintala, M.J.: Meteoritics planet. Science **27**, 526–538 (1992)
5. Díaz, E., Sanz, E., Fernández, C., Martínez, J.: Evidencia de un pequeño impacto meteorítico en extremadura: el "volcán" de El Gasco (Las Hurdes). Geogaceta **30**, 47–50 (2001)
6. Melgarejo, J. C.: Atlas de asociaciones minerales en lámina delgada. In: Melgarejo, J.C. (ed.) vol. 1. España (2004)
7. Stevens, J. G.: Mössbauer Mineral Handbook, Mössbauer Effect Data Center, Asheville, North Caroline (2002)
8. De Grave, E., Pollard, R.J., Vandenberghe, R.E., De Bakker, P.M.A.: The effect of high external magnetic fields on the hyperfine interactions in the Fe–Ni phases of the Santa Catharina meteorite. Hyperfine Interact. **94**, 2349–2353 (1994)
9. Yassir Ahmed Mohamed Abdu. Mossbauer Spectroscopy of Meteoritic and Synthetic Fe–Ni Alloys (2004)

Hyperfine Interact (2011) 203:25–31
DOI 10.1007/s10751-011-0344-2

Hyperfine structure of ^{57}Fe in minerals from a manganese ore deposit

C. K. Nascimento · M. C. Pereira · L. C. D. Cavalcante · A. M. Lana · E. Murad · J. P. Braga · J. D. Fabris

Published online: 25 August 2011

Abstract Investigations based on ^{57}Fe Mössbauer spectroscopy and powder X-ray diffraction report the existence of siderite in a manganese ore deposit in the city of Conselheiro Lafaiete, Minas Gerais, Brazil. Mössbauer parameters at 110 K indicated the isomorphous substitution of Fe^{2+} in siderite structure by other cations such as Ca^{2+} and Mg^{2+}, resulting in the distortion of siderite structure. The proposed mechanism suggests the formation of siderite and it is consistent with the process of carbonate formation in reducing environments.

Keywords Siderite · Mössbauer · Rhodochrosite

C. K. Nascimento (✉) · L. C. D. Cavalcante · A. M. Lana · E. Murad · J. P. Braga
Departamento de Química, Universidade Federal de Minas Gerais,
31270-901 Belo Horizonte, Minas Gerais, Brazil
e-mail: cassiuskn@yahoo.com.br

M. C. Pereira
Instituto de Ciência e Tecnologia, Universidade Federal dos Vales do Jequitinhonha e Mucuri,
39803-371 Teófilo Otoni, Minas Gerais, Brazil

J. D. Fabris
Universidade Federal dos Vales do Jequitinhonha e Mucuri, UFVJM,
39100-000 Diamantina, Minas Gerais, Brazil

C. K. Nascimento
Escola Preparatória de Cadetes do Ar, Divisão de Ensino-Laboratório de Química,
36205-058 Barbacena, Minas Gerais, Brazil

L. C. D. Cavalcante
Universidade Federal do Piauí, 64049-550 Teresina, Piauí, Brazil

1 Introduction

Intrinsic chemical elemental factors such as low geo-mobility and original abundance in the earth's crust make iron an interesting indicator of mineral genesis. Some of its chemical, crystallographic and hyperfine properties are indicative of the way minerals form and are transformed in the geo-domain. This work focuses on the characterization of structural iron in manganese minerals, particularly in rhodochrosite, $MnCO_3$, from a manganese ore, mainly using ^{57}Fe Mössbauer spectroscopy. Powder X-ray diffraction allowed the unequivocal identification of rhodochrosite, the occurrence of which suggests an association with siderite, $FeCO_3$. The studied sample corresponds to a source material of significant commercial, industrial and geochemical interest from the Precambrian "Morro da Mina" mineral deposit in Paraopeba Valley, city of Conselheiro Lafaiete, Minas Gerais, Brazil. The sample was collected from the central portion of a sampling pit; the lithology is dominated by queluzite, a siliceous-carbonate rock. First conclusive results of this mineralogical characterization will be presented and discussed in detail.

Manganese (II) carbonate ($MnCO_3$) occurs naturally in the mineral rhodochrosite that may have siderite, iron (II) carbonate ($FeCO_3$), as ancillary component. Both rhodochrosite and siderite are important minerals from commercial and industrial viewpoints, because they are widely used in metallurgical and steel industries [1]. Thus understanding the chemical processes occurring in distant ages can provide important information on the environmental conditions during formation of these minerals, thereby allowing the fundamental mechanisms of mineral genesis to be traced to the present day [2]. Intrinsic chemical elemental factors such as low geomobility and original abundance in the earth's crust make iron an interesting indicator of mineral genesis [3]. Some of its chemical, crystallographic and hyperfine properties are indicative of how the minerals were formed and transformed in the geodomain.

Iron and manganese carbonates usually form in environments that are anoxic and poor in sulfides. Although rhodochrosite can form in the presence of sulfides, siderite forms mainly in anoxic, sulfide-poor and methane-rich environments [4]. Regions rich in organic sediments are usually favorable for the production of large amounts of carbon dioxide (CO_2) due to oxidation of organic matter, which occurs partially by reduction of manganese and Fe oxyhydroxides and consequently, leads to formation of the respective carbonates.

Siderite has been generally observed in sediments and crusts formed under anaerobic conditions, typical of lake environments. These conditions are generated from the accumulation of humic substances. The process of interaction between these substances and Fe (III) minerals such as goethite (α-FeOOH) has been documented [4] whereas the mechanisms of siderite formation are not completely understood.

Here we present the first report on the existence of siderite in the Morro da Mina (Brazil) manganese ore deposit. We evaluate the different Fe sites in the mineral structures from this deposit using Mössbauer spectroscopy to provide a mechanism of siderite formation with seafloor characteristic and volcanogenic contribution.

2 Experimental

The manganese ore sample was collected from the deepest part of the pit mining the "Morro da Mina" (20° 38′ 00″ S 46° 43′ 13″ W) mineral deposit in Paraopeba, Conselheiro Lafaiete city, Minas Gerais, Brazil, which is currently the main commercial product of the mining. It is a sedimentary rock which underwent dynamic metamorphism and it is constituted mainly by siliceous-carbonate rock, dominated by queluzite (a complex metamorphic rock containing manganese in one or one more of the mineral components).

The ore sample was ground to 53 μm, dried in air and subsequently characterized using different analytical techniques.

To dissolve iron oxides, carbonates, silicates and silica verify the presence of graphite, 0.5 g of the sample were treated with a mixture of 20 ml water and 5 mL each of concentrated hydrochloric acid and concentrated sulfuric acid and heated to total evaporation. Twenty milliliters of water and 10 ml of hydrofluoric acid mixture were then added to the residue and the mixture again heated to total evaporation. The black residue was washed three times with distilled water at 100°C, dried at 60°C and analyzed by XRD.

Total nitrogen, carbon (carbonate and organic carbon) and hydrogen were determined using a Perkin Elmer PE 2400 CHN elemental analyzer.

Powder X-ray diffraction (XRD) analysis was performed using CoKα radiation with a Rigaku Geigerflex diffractometer equipped with a graphite diffracted-beam monochromator. Data were collected from 20 to 90 °2θ for 5 s per 0.02 °2θ step.

Mössbauer spectra were collected in constant acceleration transmission mode with a 50 mCi 57Co/Rh source. The spectrum was obtained at 110 K using a liquid-N_2 bath cryostat. Data were stored in a 512-channel MCS memory unit, with Doppler velocity of 4.6 mm s^{-1}. The experimental data were fitted using the NORMOS program and isomer shifts are quoted relative to α-Fe.

Saturation magnetization measurements were performed using a portable magnetometer with a fixed field of 0.3 T. XRF was performed in a energy dispersive X-ray fluorescence spectrometer, Shimadzu EDX-720, in vacuum of 40 Pa and collimators of 1, 3, 5 and 10 mm. Mineral microstructures were studied by scanning electron microscopy (SEM) using a JEOL JSM-840A instrument operated at 15 kV and 60 pA.

3 Results and discussion

Electron microprobe analysis indicates the manganese ore sample to consist essentially of Mn, Si, O, Ca, Mg, Al and Fe. XRF analysis showed the sample to be composed of 77.4 wt. % MnO_2, 5.5 wt.% CaO, 5.4 wt.% SiO_2, 4.9 wt.% MgO, 2.8 wt.% Fe_2O_3, 2.2 wt.% Al_2O_3, 1.4 wt.% SO_3, wt.% 0.2 TiO_2, 0.1 wt.% Cr_2O_3, 0.1 wt.% NiO. Elemental analysis showed carbon, hydrogen and nitrogen contents to amount to 11.3, 0 and 0.4 wt.%. The low content of sulfur and the high content of carbon make this system favorable to formation of siderite.

Fig. 1 SEM image of the manganese ore sample

SEM images (Fig. 1) show rhombohedral shapes that are characteristics for rhodochrosite which has a much higher Mn than Fe content, and is also low in Ca and Mg [5] (see XRF result given above).

The saturation magnetization of the magnetic fraction was 0.10 J T^{-1} kg^{-1}. The low magnetization is probably due to the presence of hausmannite Mn_3O_4, in the manganese ore deposit [6]

XRD of the manganese ore sample (Fig. 2) showed the presence of rhodochrosite, graphite, spessartine ($Mn_3Al_2(SiO_4)_3$), huntite ($CaMg_3(CO_3)_4$) and quartz (SiO_2).

Figure 2b shows that the residue of the manganese ore sample after treatment with the acid mixture consists mainly of graphite, indicating that rhodochrosite, spessartine, huntite and quartz have been decomposed. The high content of carbon observed by CNH analysis should be due to the graphite.

The study of the iron form in the individual minerals is very important because Fe is a pedogenetic indicator. Because the low content of Fe in the manganese

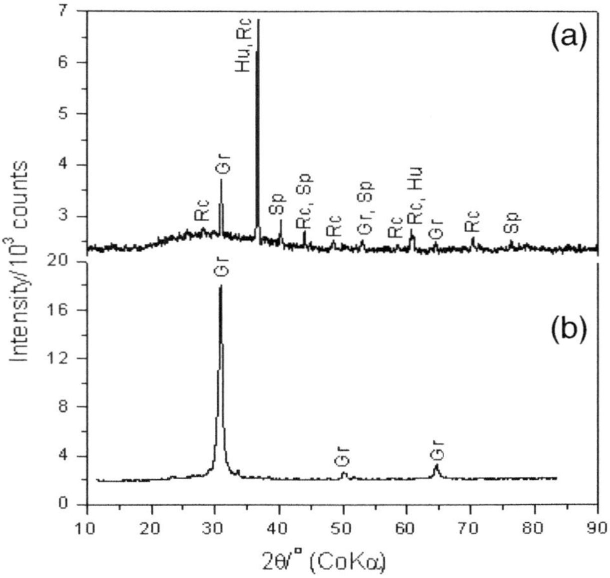

Fig. 2 XRD patterns of the manganese ore sample. **a** original sample, **b** after acid treatment. *Gr* graphite, *Rc* rhodochrosite, *Sp* spessartine and *Hu* huntite

ore sample, however, renders an evaluation of the iron forms in the individual constituents by conventional techniques very difficult [7], we used ^{57}Fe Mössbauer spectroscopy to determine the distribution of iron in the individual minerals. The Mössbauer spectrum of the manganese ore sample recorded at 110 K (Fig. 3a) was fit with a quadrupole splitting distribution for Fe^{2+} , whereas the Fe^{3+} component was fit with a Lorentzian doublet (Fig. 3b). The hyperfine parameters (Table 1) [8] show a high spin Fe^{2+} doublet with $\Delta = 2.04$ mm s^{-1}, corresponding to siderite. Isomorphous substitution of Fe^{2+} in the siderite structure by others cations such as Ca^{2+} and Mg^{2+} results in a distortion of the siderite structure. This produces a second Fe^{2+} doublet with a higher larger quadrupole splitting ($\Delta = 2.24$ mm s^{-1}) than that of pure siderite, thereby leading to an asymmetric line shape. A third Fe^{2+} doublet with $\Delta = 3.16$ mm s^{-1} can be assigned to Fe in the spessartine structure. The Fe^{3+} doublet could result either from a superparamagnetic iron oxide or Fe^{3+} in the spessartine structure.

Mössbauer spectroscopy shows most of the iron to be present as Fe^{2+}, which confirms the anoxic environment favorable for siderite formation.

4 Mechanism of siderite formation

$$\text{Oxyhydroxides}\left[\begin{array}{c}\text{goethite}\ (\alpha-\text{FeOOH})\ \text{lepidocrocite}\\ (\gamma-\text{FeOOH}),\text{etc.}\end{array}\right]\ \xrightarrow[\text{Reduction}]{\text{Organic sediments}}\ \begin{array}{c}\text{Siderite}\\ (\text{FeCO}_3)\end{array}$$

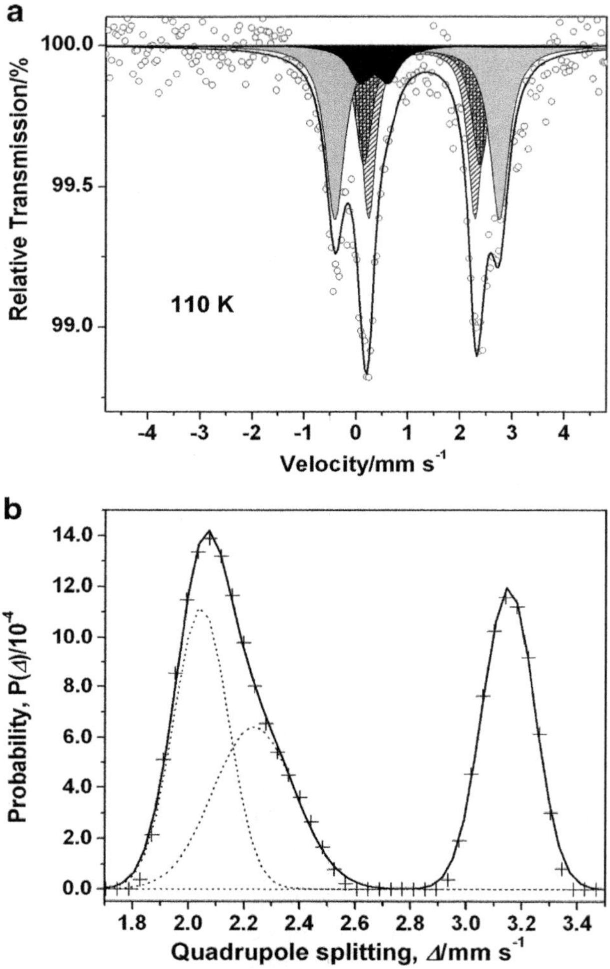

Fig. 3 **a** Mössbauer spectrum of the sample at 110 K and **b** quadrupole splitting distribution profile for the manganese ore sample

Table 1 Mössbauer parameters of the sample at 110 K

	^{57}Fe site	δ/mm s^{-1}	Δ/mm s^{-1}	Γ/mms^{-1}	RA/%
1	Fe^{2+}	1.37(1)	2.04(1)	0.33(1)	30(1)
2	Fe^{2+}	1.37(1)	2.24(1)	0.38(2)	24(1)
3	Fe^{2+}	1.28(2)	3.16(1)	0.43(2)	38(1)
4	Fe^{3+}	0.45(1)	0.55(2)	0.50(3)	8(1)

5 Conclusions

The low total content of iron is distributed over siderite, distorted siderite and spessartine. The anoxic environment and low content of sulfur and the relatively high carbon content make this system favorable for siderite formation. The identification of siderite is of importance because its presence has not been reported for this of manganese ore deposit. Formation of the iron and manganese carbonate was observed to have been simultaneous.

The proposed mechanism is consistent with the process of carbonate formation in reducing environments. The presence of significant amounts of graphite can be attributed to organic sediments that are characteristic for lake or ocean bottoms.

Acknowledgements Work financially supported by CNPq, FAPEMIG (Brazil) and Brazilian National Laboratory of Synchrotron Light (ref LNLS proposal # D10B - XPD-9920). CNPq also supported the undergraduate scholarship to AML. CAPES (Brazil) grants the Visiting Professor PVNS fellowship to JDF at UFVJM. FAPEMIG also supported the travel of JDF to Lima, Peru, to attend LACAME 2010 (grant # PPM-00217/08) and the academic visit of EM to UFMG, Brazil (grant # BPV-00068/10).

References

1. Cornell, R.M., Schwertmann, U.: The Iron Oxides. VCH, Germany (2003)
2. Harold, A., Taylor, J.: Manganese Minerals. Society of Mining, Metallurgy, and Exploration (1994)
3. Fabris, J.D., Coey, J.M.D.: Tópicos em Ciências do Solo, 47–102 (2002)
4. Berner, R.A.: J. Sediment. Petrol. **51**, 359 (1981)
5. Rad, U., von Botz, R.: Deep sea drilling project **93**, 1061 (1987)
6. Souza, J., Fabris, J.D., Mussel, W.N., Garg, V.K.: Hyp. Interact. **2**, 47 (1997)
7. Melchior, D.C., Wildeman, T.R., Williamson, D.L.: Fuel **61**, 516 (1982)
8. Vandenberghe, R.E.: Mössbauer Spectroscopy and Applications in Geology. ITC, Belgium (1991)

Hyperfine Interact (2011) 203:33–38
DOI 10.1007/s10751-011-0345-1

Mössbauer spectroscopy and X-ray diffraction analyses of clayey samples used as ceramic sourcing materials, in Peru

Rubén Quille · Ángel Bustamante · Ybar Palomino

Published online: 25 August 2011

Abstract The ceramic industry is an important area of economic activity in the Ayacucho Region, in particular in the District of Quinua. As a consequence, there is a huge demand for clay to produce ceramic pastes in that region. This paper reports on results concerning the mineralogical characterization of four clayey samples, which were collected MAA and SPQA from the area Pampa de La Quinua with geographic coordinates 13° 02′ 49″ S 74° 08′ 03″ W, CE1M and CE2M from the Quinua locality 13° 03′ 07″ S 74° 08′ 31″ W, both in the District of Quinua, Province of Huamanga, Ayacucho, Peru. The chemical and mineralogical characterization of these samples was carried out with powder X-ray diffraction detecting quartz, albite, montmorillonite, kaolinite and glauconite mineral phases, Mössbauer spectroscopy detected iron in kaolinite, glauconite and montmorillonite minerals. Chemical analysis was performed through scanning electron microscopy and energy dispersive X-ray spectroscopy. Data obtained from the combination of these techniques provided relevant information about the morphology, chemical composition, and the mineralogy of samples.

Keywords Clay · Ayacucho · X-ray diffraction · Mössbauer spectroscopy · SEM-EDX

R. Quille (✉) · Á. Bustamante
Laboratory of Ceramics and Nanomaterials, Faculty of Physical Sciences, San Marcos National University, P.O. Box 14-0149, 14 Lima, Peru
e-mail: quilleruben@gmail.com

Y. Palomino
Experimental Center of Ceramics,
National University of San Cristobal de Huamanga, Ayacucho, Peru

Table 1 Chemical composition (mass %) of clayey samples obtained using EDAX

Element	MAA	SPQA	CEM1A	CEM2A
O	46.66	44.57	44.31	46.29
Na	0.86	0.64	0.52	0.51
Mg	0.98	1.35	1.11	1.36
Al	12.55	11.17	11.73	13.58
Si	29.25	32.27	32.44	30.32
K	1.64	2.02	2.16	1.30
Ca	0.41	1.04	0.93	1.00
Ti	0.66	0.68	–	0.59
Mn	0.56	0.13	–	–
Fe	6.43	6.14	6.80	5.05

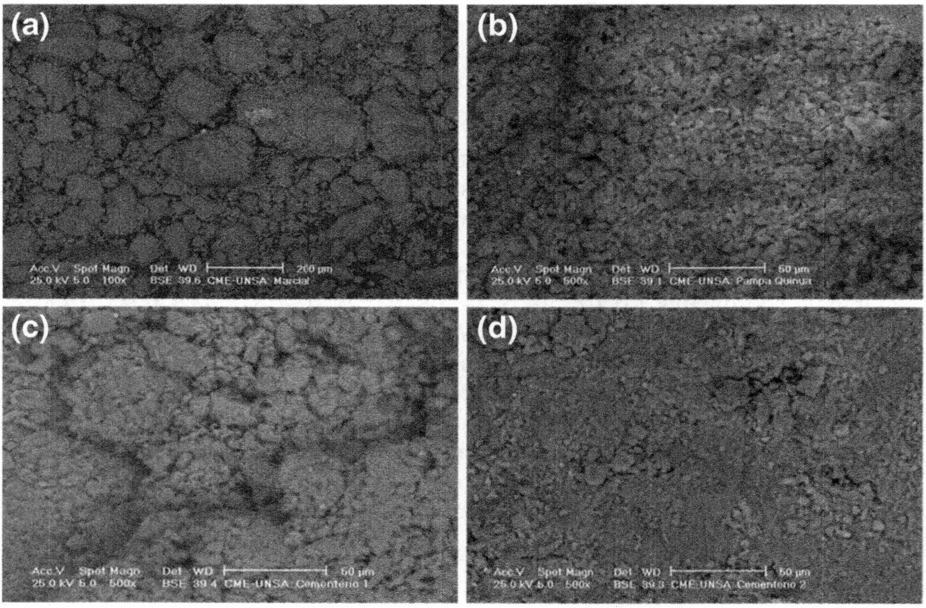

Fig. 1 SEM images of the clayey samples: **a** MAA, **b** SPQA, **c** CEM1A and **d** CEM2A with magnifications of ×100 to ×500

1 Introduction

In this work, we studied four samples of clayey materials, from the Region of Ayacucho. These materials are commonly used in the regional industry or to produce handcraft ceramic pieces of economical and cultural value. The aim of this exploratory work was to obtain detailed information about the mineralogy, morphology and chemical composition of the source clayey material, destined to the ceramic manufacture. The following analytical techniques were used: powder X-ray diffraction (XRD), Mössbauer spectroscopy (MS), scanning electron microscopy (SEM) and energy dispersive X-ray (EDX) spectroscopy. XRD data information particularly of reflection intensities and corresponding directions of

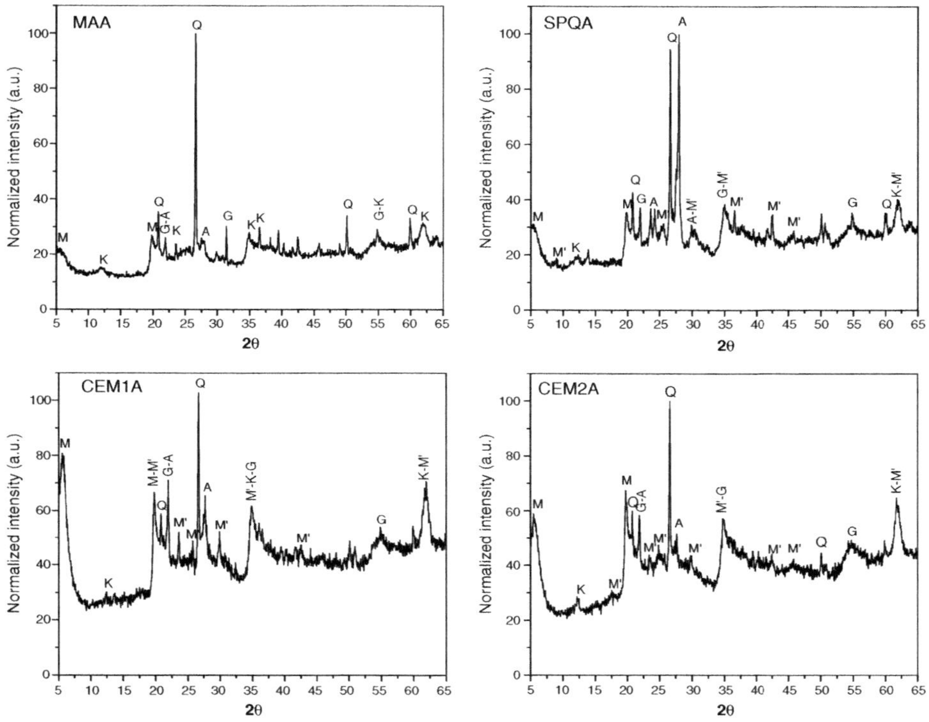

Fig. 2 Powder X-ray diffractograms of MAA through CEM2A. The peaks in the diffractograms are identified as: *M* montmorillonite, *K* kaolinite, *M′* muscovite, *Q* quartz, *G* glauconite and *A* albite

diffraction allowed an accurate interpretation of the remarkably complex crystallographic structure of selected minerals [1].

2 Experimental

Four clay samples were collected: MAA and SPQA, from the area Pampa de La Quinua and CE1M and CE2M from the locality of Quinua. The SEM and EDX analyses were performed with a scanning electron microscope Philips SEM XL 20; to assess the chemical composition, it was employed an X-ray microanalyzer based on energy dispersive EDX DX 4i spectrometry, attached to the scanning electron microscope (SEM). An XRD Rigaku diffractometer was used with the following conditions: water-cooled copper anticathode set to 30 mA and 40 kV. The XRD scanning was performed between 5° to 65° (2θ). The numerical powder diffraction data file was used [2] on phase identification. For Mössbauer spectroscopy measurements a ^{57}Co source was used in a Rh matrix. Spectra were recorded on all samples at 300 K. All fitting procedures were performed with the NORMOS program [3].

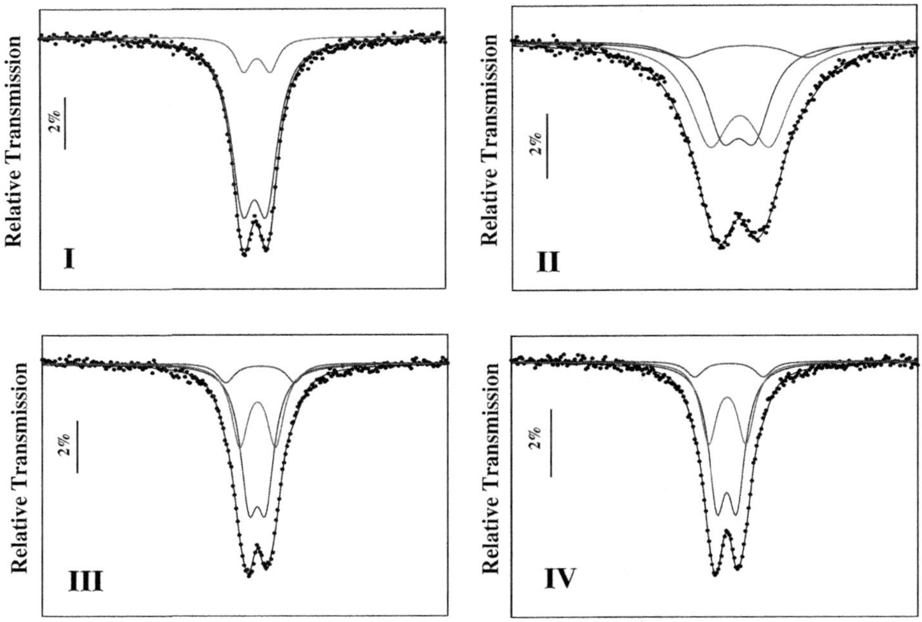

Fig. 3 Characteristic MS of clayey samples at room temperature: (*I*) MAA, (*II*) SPQA, (*III*) CEM1A and (*IV*) CEM2A

3 Results and discussion

3.1 Chemical composition

The MAA sample shows high content of Si (29.25 mass%) and Al (12.55 mass%), and Fe (6.43 mass%). The SPQA sample contains quartz, high amounts of Al (11.17 mass%), Fe (6.14 mass%) and Ca (1.04 mass%), In the CEM1A sample grains of quartz in conglomerate were identified; the sample contains Al (11.73 mass%), Fe (6.80 mass%) and accessory grains of iron oxides and some titanium. The CEM2 sample is similar to the previous sample, except that analyses reveal Al (13.58 mass%) to be higher. The amount of Ca (1.00 mass%) is slightly higher than that of Na (0.51 mass%). More data about chemical composition are presented in Table 1.

3.2 Morphology

The MAA sample shows grains of silicates, along with alumina and iron oxides (a SEM micrograph is shown in Fig. 1a). The SPQA sample shows also granular regions of quartz (Fig. 1b). The CEM1A micrograph reveals agglomerated particles of quartz with a high content of aluminum and iron oxides (Fig. 1c). That for CEM2A shows agglomerates of varying size grains of quartz containing Al and Fe (Fig. 1d).

Table 2 Hyperfine parameters for fitted Mössbauer spectra

Sample	^{57}Fe site	δ (mm/s)	Δ(mm/s)	Γ (mm/s)	A (%)	Assigned mineral
MAA	Fe^{3+}	0.351	0.48	0.55	87	Kaolinite
	Fe^{3+}	0.404	0.53	0.38	13	Glauconite
SPQA	Fe^{2+}	0.433	1.24	0.52	10	Montmorillonite
	Fe^{3+}	0.358	0.32	0.40	37	Glauconite
	Fe^{3+}	0.370	0.61	0.50	53	Muscovite
CEM1A	Fe^{2+}	0.417	1.23	0.40	09	Montmorillonite
	Fe^{3+}	0.354	0.33	0.40	55	Glauconite
	Fe^{3+}	0.362	0.70	0.40	36	Muscovite
CEM2A	Fe^{2+}	0.348	1.34	0.40	07	Montmorillonite
	Fe^{3+}	0.356	0.38	0.40	58	Glauconite
	Fe^{3+}	0.366	0.73	0.40	35	Muscovite

δ is the isomer shift relative to α iron, Δ is the quadrupole splitting, Γ is the line width, and A is the spectral fraction as obtained from the fitting

3.3 Mineralogical analysis

The mineralogical analysis of samples with powder X-ray diffraction (XRD; Fig. 2) indicates: (a) for MAA, quartz, albite, montmorillonite, kaolinite and glauconite; (b) for SPQA, quartz, albite, muscovite, montmorillonite, glauconite and kaolinite; (c) for sample CEM1A, quartz, albite, montmorillonite, glauconite and kaolinite and (d) for sample CEM2A, quartz, albite, montmorillonite, glauconite and kaolinite.

The room temperature Mössbauer spectrum for the MAA sample shows two doublets of Fe^{3+}assignable to kaolinite and glauconite [4, 5]; for the sample SPQA the spectrum was fitted with three Fe^{3+} doublets, corresponding to muscovite, glauconite and montmorillonite [4, 5].

Mössbauer parameters for samples CEM2A and CEM1A correspond to muscovite, glauconite and montmorillonite [4, 5], as it can be deduced from the three paramagnetic doublets of Fe^{3+}(Fig. 3). Corresponding Mössbauer parameters for all samples are compiled in Table 2, and XRD patterns are presented in Fig. 2.

4 Conclusion

Combining data obtained with powder X-ray diffraction and room temperature Mössbauer spectroscopy allowed us to identify the main mineralogical phases occurring in samples of clayey materials used as the raw material for ceramics in Quinua, Province of Huamanga, Ayacucho, Peru. Some of these samples, with a low proportion of minerals other than quartz or a low iron content were actually difficult to analyse and minerals (glauconite, for instance) could not be readily identified from XRD data. The two main analytical methods were confirmed to be complementary in this kind of complex mineralogical analysis. The chemical composition was based on energy dispersive X-ray spectroscopy data which was consistent with XRD patterns, as was the case for the occurrence of quartz. We identified clay minerals montmorillonite, kaolinite and muscovite. The scanning electron microscopy yielded the morphology of the clay materials. The four samples presented nearly the same set of mineral species.

Acknowledgements R.Q.R and A.B.D thank to the CONCYTEC for supporting this research, through the PROCOM 2004 program. The authors are indebted to Dr. Robert Reeve from the University of Cambridge for correcting the English grammar.

References

1. Escribano, P., Carda, J., Cordoncillo, E.: Esmaltes y Pigmentos Cerámicos. Faenza, Castellón (2001)
2. Powder Diffraction File, Publishing Joint Committee on Powder Diffraction Standard Philadelphia (1974)
3. Brand, R.A.: Normos Programs. Duisburg Universiy (1989)
4. Mössbauer Mineral Handbook: In: Stevens, J.G., Khasanov, A.M., Miller, J.W., Pollak, H.. Li. Z. (eds.) Mössbauer Effect Data Center (2002)
5. Mössbauer Spestroscopy and Applications in Geology: In: Vanderberghe, R.E. (ed.) Geological Institute Krijslaan, Belgium (1991)

Hyperfine Interact (2011) 203:39–45
DOI 10.1007/s10751-011-0346-0

Ochres from rituals of prehistoric human funerals at the Toca do Enoque site, Piauí, Brazil

Luis Carlos Duarte Cavalcante · Maria de Fátima da Luz · Niéde Guidon · José Domingos Fabris · José Domingos Ardisson

Published online: 10 August 2011

Abstract The archaeological site known as Toca do Enoque (geographical coordinates, 09° 14′ 65.3″ S 43° 55′ 62.5″ W) is a rock shelter located in the Serra das Andorinhas (Serra das Confusões National Park), rural area of the city of Guaribas, state of Piauí, Brazil. Several rupestrian paintings (anthropomorphic and zoomorphic motifs along with some pure graphisms), predominantly in red, are found on the sandstone walls. Charcoals, lithic materials, necklaces with teeth, animal bones, gastropod shells, ochres and human skeletons (dated from 6,220 ± 40 to 6,610 ± 40 years before present, BP) were identified in recent excavations in this shelter. Red and yellow ochre samples were collected from prehistoric funeral structures and

L. C. D. Cavalcante (✉) · J. D. Fabris
Departamento de Química, Universidade Federal de Minas Gerais,
31270-020 Belo Horizonte, Minas Gerais, Brazil
e-mail: cavalcanteufpi@yahoo.com.br

L. C. D. Cavalcante
Departamento de Ciências Naturais e Arqueologia, Universidade Federal do Piauí,
64049-550 Teresina, Piauí, Brazil

M. F. da Luz
Universidade Federal de Pernambuco, Recife, Pernambuco, Brazil

M. F. da Luz · N. Guidon
Fundação Museu do Homem Americano, São Raimundo Nonato, Piauí, Brazil

N. Guidon
École des Hautes Études en Sciences Sociales, Paris, France

J. D. Fabris
Universidade Federal dos Vales do Jequitinhonha e Mucuri (UFVJM),
39100-000 Diamantina, Minas Gerais, Brazil

J. D. Ardisson
Centro de Desenvolvimento da Tecnologia Nuclear, 31270-901 Belo Horizonte,
Minas Gerais, Brazil

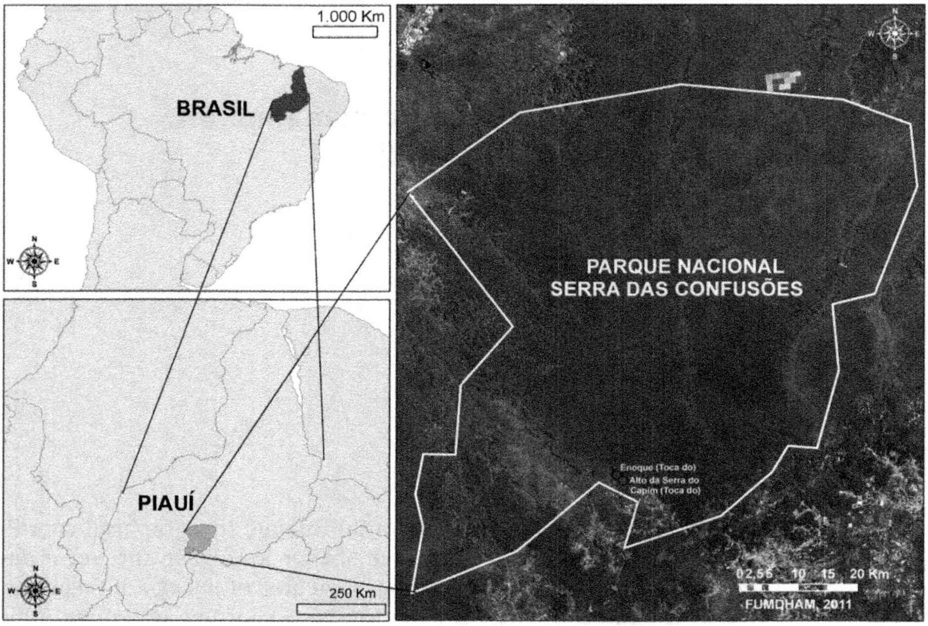

Fig. 1 Map of the Parque Nacional Serra das Confusões and Toca do Enoque site

analyzed with powder X-ray diffractometry, Fourier-transform infrared spectroscopy and ^{57}Fe transmission Mössbauer spectroscopy at 298 K and 80 K. Mössbauer data indicate that the red ochre do contain predominantly hematite (α-Fe_2O_3) whereas goethite (α-FeOOH) is the major mineral in the yellow ochre.

Keywords Archaeometry · Ochre · Funeral structures · Hematite · Goethite

1 Introduction

The Piauí state (Northeastern Brazil) hosts an extraordinary number of ancient sites, mainly prehistoric. Two of its national parks, Serra da Capivara and Serra das Confusões, draw particular interest of tourists and scientists for their palaeontological and archaeological remains, which are promptly found in caves and rock-shelters at external walls of cliffs. These rock-shelters contain an outstanding archaeological diversity of buried remains, of varied chronologies, ranging from recent ceramists groups, dating 300 years before present (BP), to human fossil skeletons, which some are recognizably older than 9,000 years BP. Data collected from some of those excavated sites have been interpreted as containing evidences of continuous human occupation since at least 100,000 years BP [1–4]. Archaeometric data have been served as important basis to characterize several of such archaeological materials [5–12].

The main objective of this work was to characterize red and yellow ochre samples collected from remaining funeral structures found in the Toca do Enoque site

Fig. 2 Funeral structure from the Toca do Enoque rock shelter

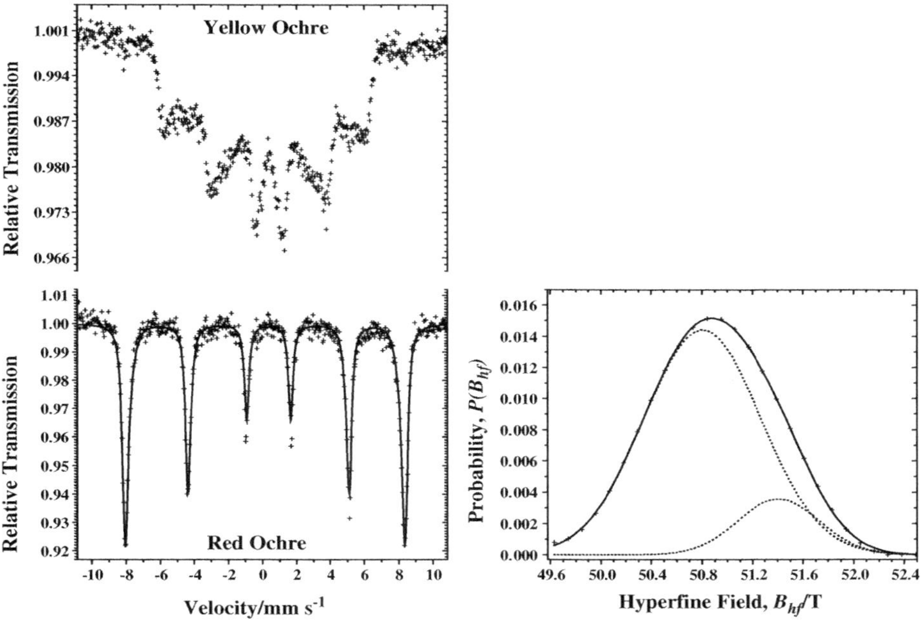

Fig. 3 298 K-^{57}Fe-Mössbauer spectra for the yellow and red ochres and corresponding profile of hyperfine field distribution for the red ochre

(Fig. 1), an archaeological rock shelter (geographical coordinates, 09° 14′ 65.3″ S 43° 55′ 62.5″ W) located in Serra das Andorinhas (Serra das Confusões National Park), rural area of the city of Guaribas, in Piauí. Several rupestrian paintings (anthropomorphic and zoomorphic motifs, along with some pure graphisms), predominantly in red, are found on the sandstone walls of this shelter. Charcoals, lithic materials, necklaces with teeth, animal bones, gastropod shells, ochres and human skeletons (dated from 6,220 ± 40 to 6,610 ± 40 years before present, BP) were identified in recent excavations in the shelter (Fig. 2) [13].

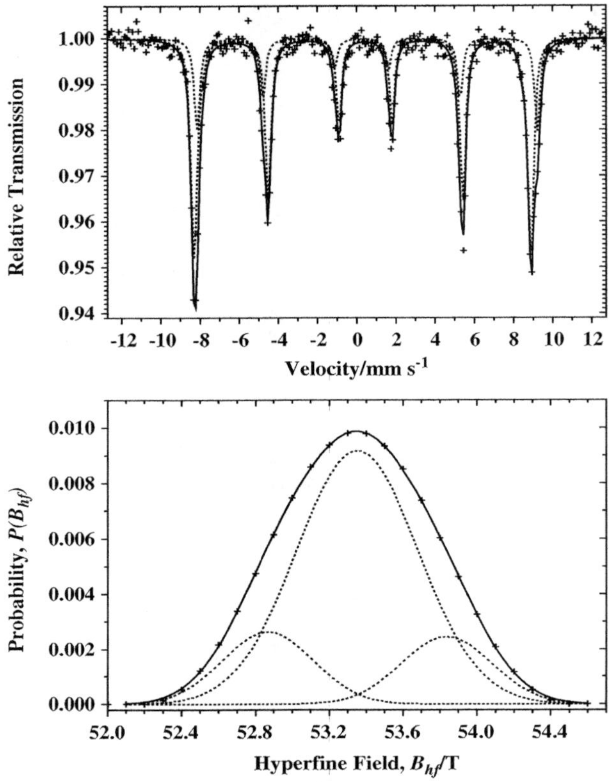

Fig. 4 80 K-^{57}Fe-Mössbauer spectrum and corresponding profile of hyperfine field distribution for the red ochre

Table 1 298 K- and 80 K-^{57}Fe-Mössbauer parameters for the red ochre

T/K	Fe site	δ/mm s^{-1}	2ε/mm s^{-1}	Γ/mm s^{-1}	B_{hf}/T or $(B_{hf})_{max}$/T	RA/%
298	Hematite	0.37(2)	–0.21(1)	0.31[a]	50.8(1)	85.3(1)
		0.37(2)	–0.21(1)	0.31[a]	51.4(1)	14.7(1)
80	Hematite	0.50(1)	0.29(2)	0.34(3)	53.9(1)	27.3(1)
	Hematite	0.47(4)	–0.17(1)	0.31[a]	52.9(2)	11.0(1)
		0.47(4)	–0.17(1)	0.31[a]	53.4(1)	51.6(1)
		0.47(4)	–0.17(1)	0.31[a]	53.8(2)	10.1(1)

δ isomer shift relative to α-Fe, 2ε quadrupole shift, Γ line width, B_{hf} magnetic hyperfine field, $(B_{hf})_{max}$ B_{hf} at maximum probability in the histogram profile of the hyperfine field distribution, RA relative subspectral area

[a]Fixed parameter during the fitting procedure. The number in parentheses are uncertainties over the last significant digit, as it was estimated from the least-squares fitting algorithm

2 Materials and methods

The Mössbauer spectra were collected at 298 K and 80 K in constant acceleration transmission mode with a ~15 mCi ^{57}Co/Rh gamma-ray source. Data were stored in a 512-channel MCS memory unit, with Doppler velocities ranging between approximately ±10 mm s^{-1}. Mössbauer isomer shifts are quoted relatively to an α-Fe foil

Fig. 5 FTIR spectra for the yellow and red ochres

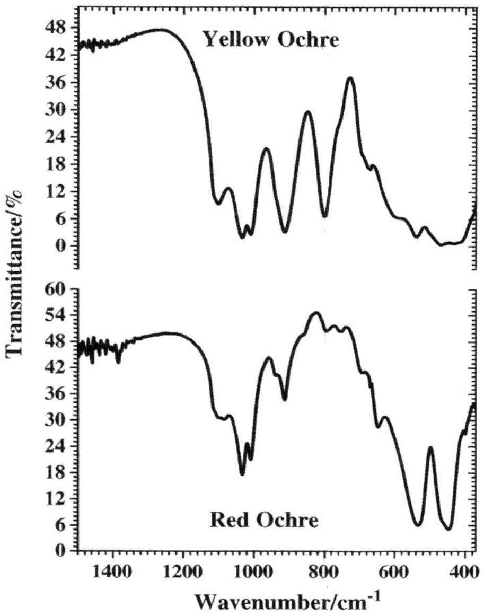

at room temperature. The experimental data were least-square-fitted with a model-independent hyperfine field distribution along with Lorentzian-shapped resonance lines, with the NORMOS™-90 computer program.

The Fourier-transform infrared absorption spectra (FTIR) were obtained with a Perkin-Elmer BX FTIR spectrometer, using KBr discs, between 4,500 and 370 cm^{-1} at a resolution of ± 2 cm^{-1}.

Powder X-Ray diffraction (XRD) measurements were performed in a Rigaku model Geigerflex diffractometer, using Cu($K\alpha$) radiation, operating at 32.5 kV and 25.0 mA, scanning from 4° to 80° (2θ) at scan rate of 4° min^{-1}, with silicon as an internal standard.

3 Results and discussion

Mössbauer spectra (Figs. 3 and 4) and corresponding hyperfine data (Table 1) indicate that the red ochre do contain predominantly hematite (α-Fe_2O_3) whereas the yellow ochre contains mainly goethite (α-FeOOH). A remarkable point is that part (27.3(1)% of the relative subspectral area) of this hematite was found to undergo the Morin transition (characteristic temperature, $T_M \sim 260$ K), as the quadrupole shift value changes from −0.21(1) mm s^{-1} (at 298 K; Fig. 3) to +0.29(2) mm s^{-1} (at 80 K; Fig. 4). This fraction is likely of bigger particles, low-isomorphically substituted and better crystallized hematite than is that not undergoing the Morin transition. The yellow ochre sample, on the other hand, rends a complex spectrum at 298 K (Fig. 3), suggesting some superparamagnetic relaxation; the corresponding spectrum 80 K (pattern not shown) reflects a well-ordered magnetic structure. New analyses are still in progress in order to get complementary data also from other techniques, as from magnetic measurements, powder XRD diffraction with synchrotron source, scanning

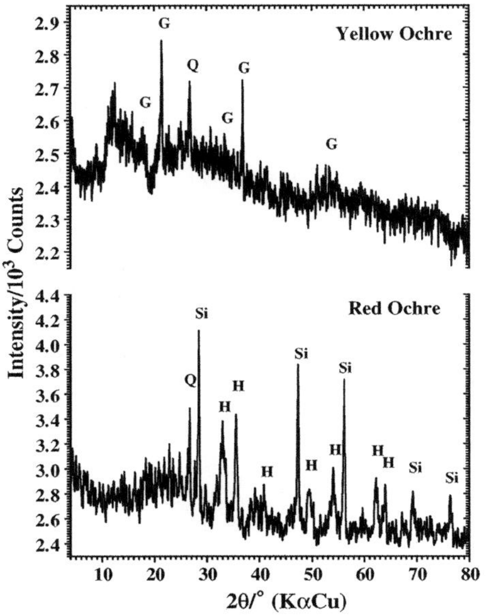

Fig. 6 XRD patterns for the yellow and red ochres. *Si* silicon, *H* hematite, *G* goethite, *Q* quartz

and transmission electron microscopy images and energy dispersive spectroscopy, in order to allow safer assignments for all of its Mössbauer spectral contributions.

The infrared spectrum of the red ochre shows prominent bands at 448 and 534 cm^{-1}, which are assignable to hematite. By contrast yellow ochre shows a band arising from ν(OH) at 3,112 cm^{-1} together with intense bands at 912 and 800 cm^{-1}, due to goethite. The IR spectroscopy data (Fig. 5) has provided a valuable way to detect and differentiate between hematite and goethite in these ochre samples [14–16].

Powder X-ray patterns are presented in Fig. 6. All three phases are identified by comparing with JCPDS [17] standards: hematite (JCPDS card # 33-664), goethite (card # 29-713) and quartz (card # 46-1045). Silicon (corresponding JCPDS card # 27-1402) was admixed to samples, as internal standard.

4 Conclusions

From these data, yellow and red ochres found among remaining structures of prehistoric burials of human funerals, from the Toca do Enoque site, were found to be constituted mainly of goethite (yellow ochre), hematite (red) and quartz. From Mössbauer data, part of the hematite in red ochre undergoes the Morin transition, revealing a bulker and better crystallized hematite, relatively to those so far reported, from similar prehistoric sites. These results also support the essential role of ^{57}Fe Mössbauer spectroscopy as central archaeometric technique to identify and characterize iron-bearing prehistoric materials from geological origin, as a basis to build safer models and hypothesis, as of anthropological nature, concerning the first human groups occupation in the American continent.

Acknowledgements Work financially supported by CNPq, FAPEMIG and FUMDHAM (Brazil). CAPES (Brazil) grants the Visiting Professor PVNS fellowship to JDF at UFVJM.

References

1. Guidon, N., Delibrias, G.: Carbon-14 dates point to man in the Americas 32,000 years ago. Nature **321**, 769–771 (1986)
2. Guidon, N., Pessis, A.-M., Parenti, F., Guérin, C., Peyre, E., Santos, G.M.: Pedra Furada, Brazil: paleoindians, paintings, and paradoxes. Athena Rev. **3**(2), 42–52 (2002)
3. Santos, G.M., Bird, M.I., Parenti, F., Fifield, L.K., Guidon, N., Hausladen, P.A.: A revised chronology of the lowest occupation layer of Pedra Furada Rock Shelter, Piauí, Brazil: the Pleistocene peopling of the Americas. Quat. Sci. Rev. **22**, 2303–2310 (2003). doi:10.1016/S0277-3791(03)00205-1
4. Valladas, H., Mercier, N., Michab, M., Joron, J.L., Reyss, J.L., Guidon, N.: TL age-estimates of burnt quartz pebbles from the Toca do Boqueirão da Pedra Furada (Piaui, Northeastern Brazil). Quat. Sci. Rev. **22**, 1257–1263 (2003). doi:10.1016/S0277-3791(03)00029-5
5. Watanabe, S., Ayta, W.E.F., Hamaguchi, H., Guidon, N., La Salvia, E.S., Maranca, S., Baffa Filho, O.: Some evidence of a date of first humans to arrive in Brazil. J. Archaeol. Sci. **30**, 351–354 (2003). doi:10.1006/jasc.2002.0846
6. Pessis, A.-M., Guidon, N.: Dating rock art paintings in Serra de Capivara National Park. Adoranten **1**, 49–59 (2009)
7. Lage, M.C.S.M.: Etude archéométrique de l'art rupestre du sud-est du Piauí—Brésil. Ph.D. thesis, Université Paris 1 Panthéon-Sorbonne (1990)
8. Lage, M.C.S.M.: Dating of the prehistoric paintings of the archaeological area of the Serra da Capivara National Park. In: Strecker, M., Bahn, P. (eds.) Dating and the earliest known rock art, pp. 49–52. Oxbow Books, Oxford (1999)
9. Cavalcante, L.C.D., Lage, M.C.S.M., Pereira, M.C., Fabris, J.D.: Estudo químico e espectroscópico dos pigmentos pré-históricos do sítio de arte rupestre Arco do Covão, Piauí, Brasil. Int. J. South Am. Archaeol. **3**, 59–66 (2008)
10. Cavalcante, L.D.C., Lage, M.C.S.M., Fabris, J.D., Etchevarne, C.A.: Análise arqueométrica de pintura rupestre do sítio Poções, Bahia, Brasil. Rev. Arqueol. **22**, 95–103 (2009)
11. Cavalcante, L.C.D., Lage, M.C.S.M., Fabris, J.D.: Análise química de pigmento vermelho em osso humano. Quim. Nova **31**(5), 1117–1120 (2008). doi:10.1590/S0100-40422008000500034
12. Alves, T.L., Brito, M.A.M.L., Lage, M.C.S.M., Cavalcante, L.C.D., Fabris, J.D.: Pigmentos de pinturas rupestres pré-históricas do sítio Letreiro do Quinto, Pedro II, Piauí, Brasil. Quim. Nova **34**(2), 181–185 (2011). doi:10.1590/S0100-40422011000200002
13. Guidon, N., Luz, M.F.: Sepultamentos na Toca do Enoque (Serra das Confusões-Piauí). Fumdhamentos **8**, 115–123 (2009)
14. Mortimore, J.L., Marshall, L.-J.R., Almond, M.J., Hollins, P., Matthews, W.: Analysis of red and yellow ochre samples from Clearwell Caves and Çatalhöyük by vibrational spectroscopy and other techniques. Spectrochim. Acta A **60**, 1179–1188 (2004). doi:10.1016/j.saa.2003.08.002
15. Marshall, L.-J.R., Williams, J.R., Almond, M.J., Atkinson, S.D.M., Cook, S.R., Matthews, W., Mortimore, J.L.: Analysis of Ochres from Clearwell Caves: the role of particle size in determining colour. Spectrochim. Acta A **61**, 233–241 (2005). doi:10.1016/j.saa.2004.03.041
16. Cornell, R.M., Schwertmann, U.: The Iron Oxides: Structure, Properties, Reactions, Occurrences and Uses. Wiley-VCH Verlag GmbH & Co. KGaA, Weinheim (2003). ISBN 3-527-30274-3
17. JCPDS—Joint Committee on Powder Diffraction Standards: Mineral Powder diffraction files data book. Swarthmore, Pennsylvania (1980)

Hyperfine Interact (2011) 203:47–50
DOI 10.1007/s10751-011-0347-z

Structural characteristics of chalcopyrite from a Cu(Au) ore deposit in the Carajás Mineral Province, Brazil

Andreza Aparecida Ribeiro · Diana Quintão Lima · Hélio Anderson Duarte · Enver Murad · Márcio César Pereira · Marcos Tadeu de Freitas Suita · José Domingos Ardisson · José Domingos Fabris

Published online: 25 August 2011

Abstract Mössbauer spectra and X-ray diffraction data show a chalcopyrite from the Cristalino Cu(Au) deposit in the Carajás Mineral Province in northern Brazil to consist of a single, tetragonal phase. This is in stark contrast to a previously described chalcopyrite from the Camaquã copper mine in southern Brazil, obviously reflecting differences in mineral (and thus ore deposit) genesis.

Keywords Mössbauer spectroscopy · Chalcopyrite · Copper deposit · Rietveld analysis

1 Introduction

The Carajás Range in the central north of the State of Pará has been termed the main mineral province in Brazil [1]. A striking feature of the Carajás Mineral Province is

A. A. Ribeiro · D. Q. Lima · H. A. Duarte · E. Murad
Departamento de Química, ICEx, UFMG, Campus Pampulha, 31270-901 Belo Horizonte, Minas Gerais, Brazil

M. C. Pereira (✉)
Instituto de Ciência e Tecnologia, Universidade Federal dos Vales do Jequitinhonha e Mucuri, 39803-371 Teófilo Otoni, Minas Gerais, Brazil
e-mail: mcpqui@yahoo.com.br

M. T. de Freitas Suita
Departamento de Geologia, UFOP, 35400-000 Ouro Preto, Minas Gerais, Brazil

J. D. Ardisson
Centro de Desenvolvimento da Tecnologia Nuclear, 31270-901 Belo Horizonte, Minas Gerais, Brazil

J. D. Fabris
Universidade Federal dos Vales do Jequitinhonha e Mucuri, UFVJM, 39100-000 Diamantina, Minas Gerais, Brazil

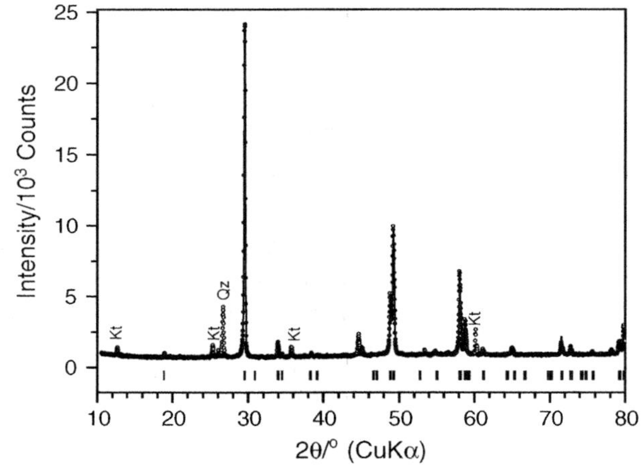

Fig. 1 XRD pattern of the chalcopyrite sample. *Qz* quartz, *Kt* kaolinite

the large number of Cu–Au mineralizations that are stratigraphically and tectonically related. Here we describe the properties of a chalcopyrite sample from the Cristalino Cu(Au) deposit, comparing its structural features with previously reported data for a sample from the Camaquã copper mine, a contrasting site in southern Brazil [2].

The Cristalino Cu(Au) deposit, located in the southeastern part of the Carajás Mineral Province, is a world class IOCG (iron oxide–copper–gold) type deposit with estimated resources greater than 500 Mt of ore averaging 1.0% Cu and 0.3 g/t Au [3]. Principal minerals constituting in this deposit are chalcopyrite, pyrite, native gold, bravoite, millerite and vaesite [3].

The main objective of the present study was to characterize a chalcopyrite from this deposit which had been used for isotope analysis [4] and compare its structural features with those of the previously mentioned chalcopyrite from southern Brazil.

2 Experimental

The studied sample was collected in the Carajás geo-province (2° 57′ 08″ S 5° 51′ 38″ W), State of Pará, Brazil. Chalcopyrite was manually separated under a binocular microscope.

A powder X-ray diffraction (XRD) pattern of this sample was obtained by slow scanning at 1/8° 2θ/min from 10° to 80° on a Rigaku Geigerflex 3064 diffractometer equipped with a Cu tube and a graphite-diffracted beam monochromator. Rietveld refinement was performed with FULLPROF 2010 [5] in order to estimate the structural parameters of the chalcopyrite. The average chemical composition was obtained by probing 13 points on seven polished sections of selected grains with a Jeol 733 scanning electron microscope equipped with an energy-dispersive X-ray spectrometer. A surface-scanning experiment was performed to assess the homogeneity of sample for Fe and Cu. Mössbauer spectra were obtained with the samples held at room temperature and at 80 K (liquid-N_2 bath cryostat) using a constant acceleration transmission spectrometer equipped with a $^{57}Co/Rh$ source of about 10 mCi. Data were stored in a 1024-channel MCS memory unit. The experimental

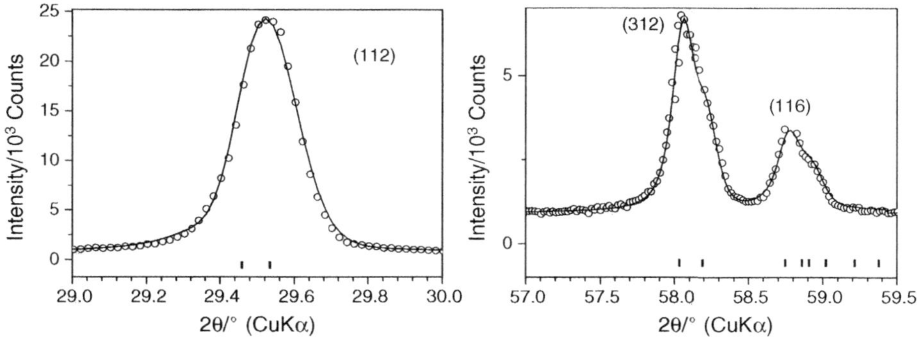

Fig. 2 XRD pattern of the chalcopyrite on an expanded 2θ-scale showing the Rietveld fits of the (*112*) reflection and (*312*) plus (*116*) reflections considering only the tetragonal chalcopyrite

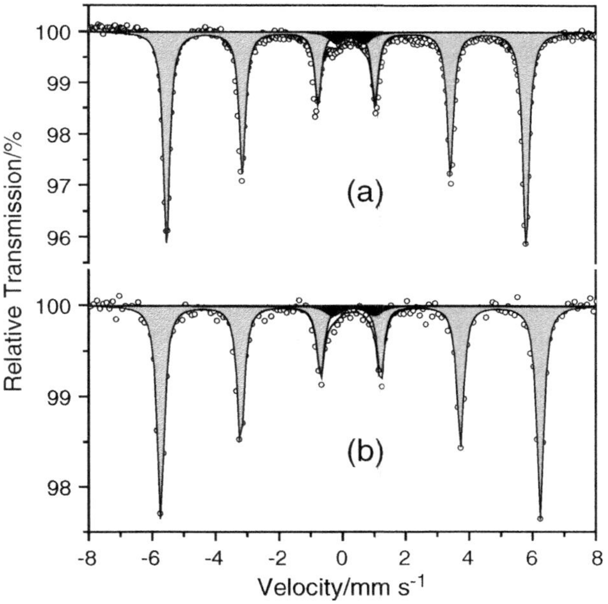

Fig. 3 Mössbauer spectra of the chalcopyrite sample **a** at room temperature and **b** at 80 K

data were fitted using Lorentzian functions with a least-squares fitting procedure based on the NORMOS program and isomer shifts are quoted relative to α-Fe.

3 Results and discussion

X-ray diffraction (XRD, Fig. 1) displays peaks at 2θ° (CuKα) of 29.46, 33.97, 48.83, 49.24, 58.04, 58.79, 71.54, 73.00, 79.15, 79.40, 79.77 and 80.00, which are characteristic of chalcopyrite. In addition to chalcopyrite, kaolinite and quartz could be identified by XRD. The XRD data were fitted using Rietveld analysis implying only one chalcopyrite with a tetragonal cell. The lattice parameter for the tetragonal

Table 1 Mössbauer parameters of the chalcopyrite sample at room temperature and 80 K

T(K)	δ/mm s^{-1}	$2\varepsilon Q$, Δ/mm s^{-1}	B_{hf}/T	RA/%	Fe site
298	0.28 (1)	0.00 (2)	35.2 (1)	96 (1)	Chalcopyrite
	0.37 (1)	0.85 (1)	–	4 (1)	Fe^{3+} in kaolinite
80	0.37 (1)	0.00 (2)	37.1 (1)	95 (1)	Chalcopyrite
	0.47 (2)	1.33 (1)	–	5 (1)	Fe^{3+} in kaolinite

chalcopyrite were a = 5.27434(9) Å and c = 10.3885(2) Å. These lattice parameters are less than the values reported for stoichiometric chalcopyrite (a = 5.2855 Å and c = 10.4117 Å) [2], and possibly indicative of oxidation of Cu^+ to Cu^{2+} and/or isomorphous substitution of Cu and Fe by other cations with smaller ionic radii. The kaolinite and quartz reflections were excluded from the Rietveld fitting using the exclusion zone. Figure 2 shows the good fitting of (112), (312) and (116) reflections, suggesting the existence of a unique phase, i.e. tetragonal chalcopyrite. The slight asymmetry is attributed to the different wavelengths of $K_{\alpha 1}$ and $K_{\alpha 2}$ radiation

Mössbauer data collected at 298 K consisted of a dominant sextet with hyperfine parameters that are typical for chalcopyrite (Fig. 3, Table 1). The sextet in Fig. 3 lacks the asymmetry characteristic for the chalcopyrite from Camaquã. A subordinate doublet in the spectrum can be ascribed to Fe^{3+} substituting for Al^{3+} in kaolinite.

4 Conclusions

The occurrence of a single crystalline phase in this chalcopyrite from the Cristalino deposit, Carajás Mineral Province in northern Brazil, presents an interesting differential characteristic compared to the chalcopyrite from the Camaquã copper mine [2]. For the chalcopyrite from Camaquã, the existence of two discrete modifications was attributed to an incomplete conversion of the high-temperature (cubic) variety to the low-temperature (tetragonal) variety, i.e. the cubic component was a remnant of an originally formed high-temperature phase. This is obviously not the case at the Cristalino deposit.

Acknowledgements CNPq, FAPEMIG and VALE (previously Companhia Vale do Rio Doce). CAPES (Brazil) grants the Visiting Professor PVNS fellowship to JDF at UFVJM. FAPEMIG also supported the travel of JDF to Lima, Peru, to attend LACAME 2010 (grant # PPM-00217/08) and the academic visit of EM to UFMG, Brazil (grant # BPV-00068/10). This work was also partially supported by the project named "Acqua Institute Mineral Resources, Water and Biodiversity", of the Brazilian National Program of Institutes of Science and Technology.

References

1. Rosière, C.A., Chemale, F. Jr.: Rev. Bras. Geociênc. **30**, 274 (2000)
2. Mussel, W.N., Murad, E., Fabris, J.D., Moreira W.S., Barbosa, J.B.S., Murta, C.C., Abrahão, W.P., de Mello, J.W.V., Garg, V.K.: Phys. Chem. Miner. **34**, 383 (2007)
3. Monteiro, L.V.S., Xavier, R.P., de Carvalho, E.R., Hitzman, M.W., Johnson, C.A., de Filho, C.R.S., Torresi, I.: Min. Dep. **43**, 129 (2008)
4. Ribeiro, A.A., Suita, M.T.F., Sial A.N., Fallick, A.E., Eli, F., Goulard, E.A.: Geochim. Bras. **23**, 159 (2009)
5. Fullprof_Suíte. Crystallographic tools for Rietveld, profile matching and integrated intensity refinements of X-ray and or neutron data 2010. http://www.ill.eu/sites/fullprof/ (2010). Accessed 7 Nov 2010

Hyperfine Interact (2011) 203:51–57
DOI 10.1007/s10751-011-0367-8

Mössbauer study of miniature vessels from the Sicán burial site of Huaca Loro

G. Cervantes · I. Shimada · W. Häusler ·
U. Wagner · F. E. Wagner

Published online: 30 September 2011

Abstract Small handmade ceramic vessels are often found in large quantities in late pre-hispanic elite tombs on the north coast of Peru. These miniature vessels are of poor artistic and technologic qualities compared with other offerings found in tombs and have not received much attention. Our hypothesis is that these miniatures were made hastily by non-elite funeral attendants, which resulted in the observed formal and material variability. Here we report on the results of a systematic archaeological analysis of the shape and of archaeometric analyses by Mössbauer spectroscopy and XRD of the firing conditions of such vessels that were conducted to test the above hypothesis.

Keywords Mössbauer spectroscopy · X-ray diffraction · Ancient ceramics · Firing of ceramics

1 Introduction

Recently, several pre-hispanic elite tombs were excavated on the north coast of Peru [1]. They usually contain large quantities of offerings, such as metal objects, decorated cloths and fine ceramic vessels. Together with these artistically and technically high quality objects, one often finds large quantities of rather crudely made and undecorated miniature ceramic vessels, which in the past did not attract much attention. We have now undertaken an in-depth examination with the aim

G. Cervantes (✉)
Pontificia Universidad Católica del Perú, Lima, Peru
e-mail: gabriela.cervantes@pucp.pe

I. Shimada
Southern Illinois University, Carbondale, IL 61901, USA

W. Häusler · U. Wagner · F. E. Wagner
Physik-Department E15, Technische Universität München, 85747 Garching, Germany

of clarifying why so many of these vessels were so often produced and how this was done. We hypothesized that these miniatures were made hastily by commoners living in diverse localities, who brought them as humble offerings to the funerals. The principal objective of this study, which is still ongoing and on which we here present first results, is to examine the formal and technological variability in the miniature vessels by means of archaeological (formal typology) and archaeometric (Mössbauer spectroscopy, X-ray diffraction) analyses, and thereby test our hypothesis.

2 Archaeological context

The archaeological setting of our study is the Middle Sicán (900–1100 A.D.) religious center situated within what is today the National Historical Sanctuary of Poma in the middle La Leche Valley. The site is dominated by a dozen monumental, truncated pyramidal temple mounds built of adobe bricks and fills, including the one designated as Huaca Loro [2]. The collection of miniature vessels under study was excavated from a major elite tomb called Central West Tomb 2 situated at the west base of Huaca Loro. This large rectangular tomb was found nearly 10 m below surface and its Principal Chamber measured 7.20 m (N–S) $\times$ 3.50 m (E–W) and was 1.5 m deep. It contained diverse objects, including ceramic bottles, pigments of different colors, camelid limbs and heads, and metal objects. In the northern part of the Principal Chamber, two small chambers, named Chamber 1 and Chamber 2, were found, one nested inside the other. Chamber 1 contained the bones of an adult woman, fine ceramic bottles and cast arsenical copper objects. Chamber 2 held the partial skeleton of a child, inferred to be the principal individual of the tomb. Miniature vessels were found in all three chambers.

3 Typological analysis

Over 1,200 miniature vessels were excavated, but only for about 1,000 the shape could be determined because they were either unbroken or because diagnostic parts were preserved. The results of the typological (shape) analysis are summarized in Fig. 1. The main criterion is the division between pots and jars, and the subdivision is the shape of the pots or of the necks of the jars. A more detailed account of the typological analysis is given elsewhere in ref. [2, 3].

4 Mössbauer spectroscopy

Mössbauer spectroscopy gives information about the chemical state of the iron, which is a tracer for the maximum temperature and the atmosphere in which the pottery was fired. Since the Mössbauer results also depend on the nature of the clay used in the production of the pottery, the ancient firing techniques can best be assessed by comparison with Mössbauer spectra of clay from the archaeological site fired in the laboratory under controlled conditions [4, 5].

So far Mössbauer measurements have been performed on 58 fragments of miniatures; 46 of these were from vessels of known shape type (Fig. 1). ^{57}Fe Mössbauer spectra were measured at room temperature (RT) and 4.2 K with sources of ^{57}Co in

	Vessel Shape	Moss Type
O (Olla=Pot) n=527 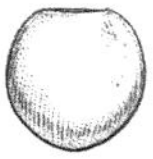	O1 spherical n=74 	T1 n=2
	O2 tubular n=180 	T1 n=5 T3 n=1
	O3 globular n=273 	T1 n=2 T2 n=1 T4 n=1
	Neck Orientation	
Ca (Cantaro=Jar) n=435 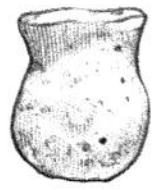	Ca1 divergent n=131 	T1 n=1 T2 n=4 T4 n=1
	Ca2 straight n=197 	T2 n=6 T3 n=2 T4 n=1 T5 n=1 T6 n=1
	Ca3 convergent n=107 	T2 n=3 T4 n=1
Other shapes n=30		

Fig. 1 Results of the typological and Mössbauer analysis of the miniature vessels. The typological classification is summarized on the left and middle columns. The diameter of the vessels is usually about 5 cm. On the right, the frequencies of the six different Mos-Types (00541 through *T*6) are given as observed for the different vessel shapes

Rh kept at the same temperature as the absorber. The spectra could be subdivided into six types (Mos-Types) according to the features of both the RT and 4.2 K spectra. Typical spectra of each type are shown in Fig. 2. Despite some variability of the spectra within each Mos-Type group, one can associate these types with different firing procedures. One specimen of each Mos-Type was refired in the laboratory in air for 12 h at temperatures increasing in steps of 50°C up to 1,150°C, and RT as well as 4.2 K Mössbauer spectra were recorded for each refiring temperature. A detailed discussion of the changes of the Mössbauer parameters would exceed the scope of this brief account. Previous work [4, 5] has shown, however, that the change in the electric quadrupole splitting of the Fe^{3+} component (Δ^{3+}) in the RT spectra during refiring is the single most helpful parameter in identifying the original firing conditions. In Fig. 2 the refiring behavior of Δ^{3+} for each Mos-Type is compared with the dependence of this parameter during the first firing of clay 39/80 from the workshop of Huaca Sialupe [6], which was excavated near to the burial site of Huaca Loro. This clay is believed to be typical for the whole region [6]. The behavior of its Mössbauer spectra observed after firing at different temperatures and in different atmospheres is described in detail in [6].

The Mössbauer spectra and the refiring behavior of Mos-Types 1 and 2 are very similar, the main difference being that there is a small amount of Fe^{2+} in Mos-Type 2. The low content of magnetically split hematite in the RT spectra suggests that these vessels were fired at rather low temperatures, between 400°C and 500°C under oxidizing conditions. The presence of some Fe^{2+} in Mos-Type 2 agrees with this conclusion, since the Fe^{2+} present in the clay oxidizes between 400°C and 500°C.

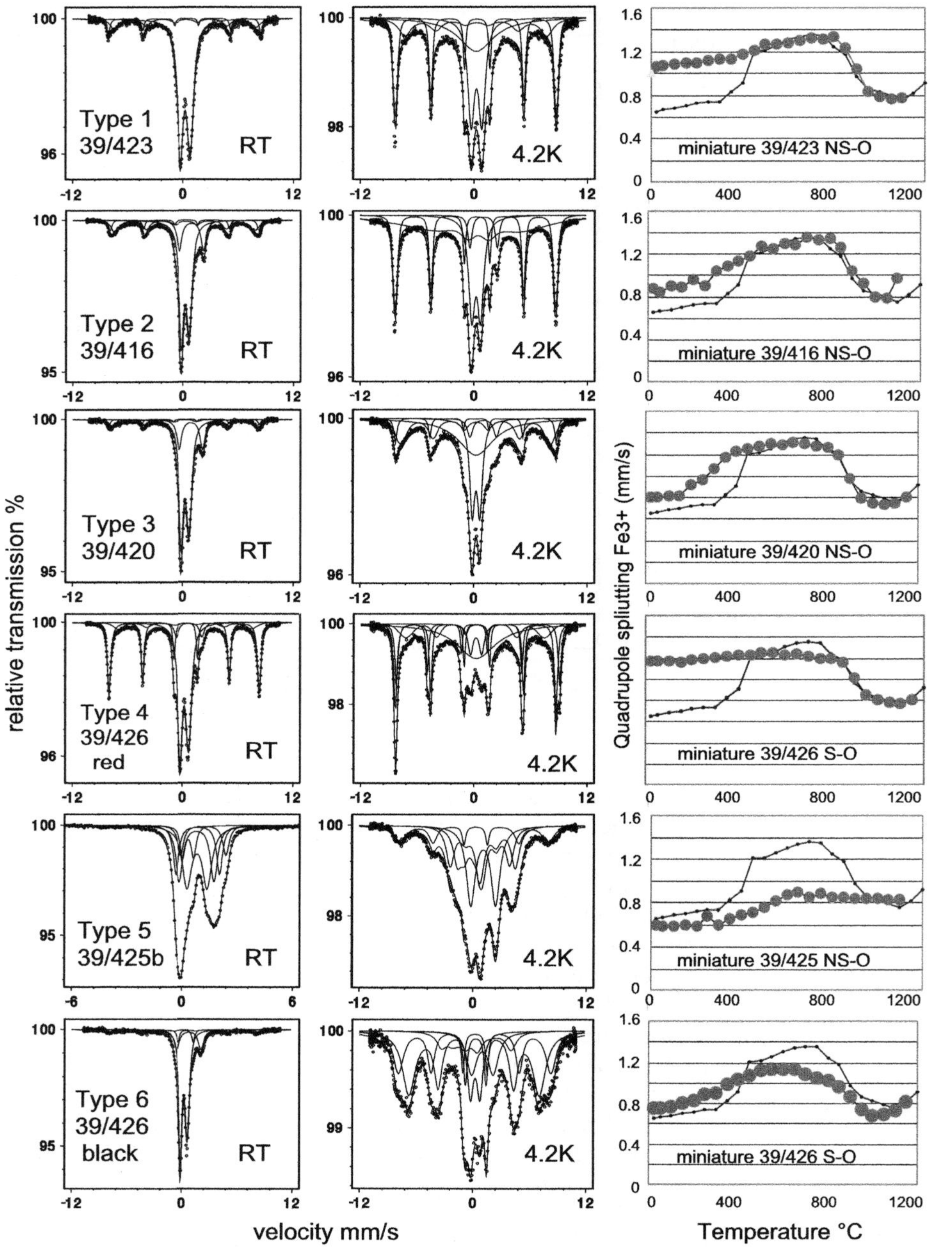

Fig. 2 Typical RT and 4.2 K Mössbauer spectra for the 6 Mos-Types into which the spectra of the miniature vessels can be classified. The codes given for each Mos-Type are the ones we used for the sample whose spectrum is shown. On the right the Δ^{3+} values obtained at RT after refiring specimens of the different Mos-Types in air at increasing temperatures (*big dots*) are compared with data obtained after firing clay 39/80 from Huaca Sialupe [6] under the same conditions (*small dots*). The scales for the quadrupole splittings are identical for all 6 Mos-Types

The absence of Fe^{2+} in Mos-Type 1 specimens indicates that these were fired at slightly higher temperatures than Mos-Type 2. The fact that Δ^{3+} in the refiring curve meets the firing curve of the clay at about 500°C confirms this conclusion. Most of the studied miniatures exhibit Mos-Type 1 or 2 (Fig. 1). The other Mos-Types are both rare and largely different from Mos-Types 1 and 2.

Mos-Type 5 contains mainly Fe^{2+}, showing that the firing as well as the cooling of the pottery occurred under reducing conditions. The Mössbauer spectra are typical for the presence of hercynite ($FeAl_2O_4$), which exhibits a magnetic hyperfine splitting at 4.2 K [7, 8], but they also show the presence of a glassy phase which still yields a quadrupole doublet at 4.2 K [4]. The formation of these phases during firing under reducing conditions requires firing temperatures in excess of about 750°C [4]. When re-oxidation of such a material occurs at the end of the firing procedure, the Fe^{2+} is oxidized to Fe^{3+} which, depending on the cooling rate and the rate of oxidation, may remain in the glassy phase and yield a magnetically split pattern or still a quadrupole doublet at 4.2 K, or it may ex-solve and form hematite as described in reference [9]. The former appears to have happened in Mos-Type 6, and the latter in Mos-Type 4. Notably, part of the hematite in Mos-Type 4 undergoes the Morin transition, which requires rather large and pure hematite crystals and is the exception rather than the rule in ceramics. The refiring behavior of Δ^{3+} of the vitreous phase of ceramic specimens still needs to be studied, but the change of Δ^{3+} of Mos-Type 6 with refiring temperature indicates a rearrangement of the local environment of the Fe^{3+} in the vitreous phase. We also note that only one instance of Mos-Type 6 has been observed, and this in a rather remarkable vessel (39/426) that is black on one side and red on the other. The Mos-Type 6 spectrum of the black side shows that this vessel was fired under reducing conditions. The black color is attributed mainly to carbon black deposited during the reducing firing phase [10]. Then, during the cooling phase, a current of air appears to have re-oxidized the vessel on both sides, but on the black side the temperature of re-oxidation must have been lower than on the red side. Thus the temperature on the black side was already too low for hematite formation and for burning off the carbon black, while on the red side the oxidation took place at a temperature high enough for burning the carbon black and for hematite formation, which resulted in a Mos-Type 4 spectrum.

The spectra of Mos-Type 3 resembles the one of Mos-Type 6 except that much of the Fe^{3+} is still yielding a quadrupole doublet at 4.2 K. This could mean that the rather rapid cooling during the final oxidizing phase of an initially reducing firing cycle did not permit the formation of iron-rich clusters that would order magnetically at 4.2 K [9]. If this is the case, Mos-Type 3 represents just another variant of firing cycles also leading to Mos-Type 4 or 6 spectra, i.e., reducing firing at or above 800°C followed by re-oxidation during cooling.

5 X-ray diffraction

XRD-scans were taken of unfired clay from the burial site of Huaca Loro (39/405), of the miniature 39/420 (Mos-Type 3) and of the black and red parts of 39/426 (Mos-Type 6 and 4). The clay contains ample amounts of kaolinite, muscovite mica and traces of 14 Å minerals like chlorite in addition to quartz and feldspars, and resembles clay 39/80 studied before [6]. The patterns obtained for the miniature

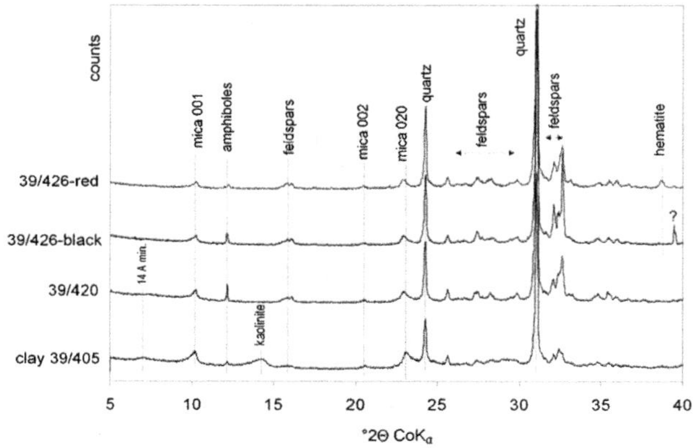

Fig. 3 XRD-scans of clay 39/405 from the burial site of Huaca Loro, of the miniatures 39/420 (Mos-Type 3) and of the *red* and *black* side of 39/426 (Mos-Types 4 and 6)

vessels still contain mica, showing that the vessels were not heated above 950°C, where mica decomposes [6]. Both kaolinite and the 14 Å minerals have disappeared in both the red and the black part of vessel 39/426; hematite is found in the red, but not in the black part, in agreement with the Mössbauer results. In the pattern of 39/420 there may still be a trace of 14 Å minerals (chlorite) visible, which would indicate the specimen was fired not above about 650°C, where chlorite decomposes [6]. This would contradict the conclusion drawn from the Mössbauer spectra that the firing temperature of this vessel was around 800°C and requires further studies (Fig. 3).

6 Conclusions

A total of 58 miniature vessels from the burial site of Huaca Loro were studied by Mössbauer spectroscopy at ambient temperature and at 4.2 K. The results allow the specimens to be subdivided into 6 groups according to the types of Mössbauer spectra they yield. These groups reflect different firing conditions and histories. Oxidizing firing at moderate temperatures of no more than 400 or 500°C is most frequently encountered. Some vessels, however, were fired under reducing conditions at or even above 800°C and, in most cases, re-oxidized at the end of the firing procedure. X-ray diffraction data confirm these findings. The results indicate that the miniatures were not fired in proper pottery kilns, but perhaps rather in bonfires or open pits, where the firing conditions were poorly controlled and where high temperatures were rarely produced, perhaps for reasons of fuel economy.

Acknowledgements We would like to thank the German Research Council and the Tokyo Broadcasting System for their support of the archaeometric work and the archaeological fieldwork, respectively.

References

1. Shimada, I., Shinoda, K., Farnum, J., Corruccini, R., Watanabe, H.: An integrated analysis of pre-hispanic mortuary patterns: a middle Sicán case study. Curr. Anthropol. **45**, 369–402 (2004)

2. Cervantes, G., Wagner, U., Perry, L.: Variabilidad de las vasijas en miniatura de Huaca Loro: un estudio multidisciplinario. In: Vetter, L., Vega-Centeno, R., Olivera, P., Petrick, S. (eds.) II Congreso Latinoamericano de Arqueometría (pp. 237–251) Lima: Editorial Universitaria, Universidad Nacional de Ingeniería (2011)
3. Cervantes, G.: El Ritualizado Proceso Funerario y el Rol de las Vasijas en Miniatura en Huaca Loro, Valle de la Leche (The ritualized funerary process and the function of miniature vessels at Huaca Loro, La Leche Valley). Unpublished Licenciatura Thesis, Faculty of Letters and Humanities, Pontificia Universidad Católica del Perú, Lima (2010)
4. Wagner, F., Wagner, U.: Mössbauer spectra of clays and ceramics. Hyperfine Interact. **154**, 35–82 (2004)
5. Wagner, U., Gebhard, R., Grosse, G., Hutzelmann, T., Murad, E., Riederer, J., Shimada, I., Wagner, F.E.: Clay: an important raw material for prehistoric man. Hyperfine Interact. **117**, 323–335 (1998)
6. Shimada, I., Häusler, W., Hutzelmann, T., Wagner, U.: Early pottery making in Northern Coastal Peru. Part I: Mössbauer study of clays. Hyperfine Interact. **150**, 73–89 (2003)
7. Dormann, J.L., Seqqat, M., Fiorani, D., Noguès, M., Souberyoux, J.L., Bhargava, S.C., Renaudin, P.: Mössbauer studies of $FeAl_2O_4$ and $FeIn_2S_4$ spin glass spinels. Hyperfine Interact. **54**, 503–508 (1990)
8. Klavius, G.M., Krimmel, A., Hartmann, O., Litterst, F.J., Wäppling, R., Wagner, F.E., Tsurkan, V., Loidl, A.: Frustration driven magnetic state of a-site spinels probed by μSR. Eur. Phys. J., B **77**, 87–100 (2010)
9. Friederich, J., Gebhard, R., Krause, R., Riederer, J., Wagner, F.E., Wagner, U.: Mössbauer study of Celtic pottery from Bopfingen, Baden-Württemberg, Germany. J. Phys. Conf. Ser. **217**, 012064 (2010)
10. Shimada, I., Häusler, W., Hutzelmann, T., Riederer, J., Wagner, U.: Early pottery making in Northern Coastal Peru. Part III: Mössbauer study of Sicán Pottery. Hyperfine Interact. **150**, 107–123 (2003)

Hyperfine Interact (2011) 203:59–66
DOI 10.1007/s10751-011-0368-7

Preparative treatment with NaOH to selectively concentrate iron oxides of a Chilean volcanic soil material to produce effective heterogeneous Fenton catalyst

Valentina Manzo · Carmen Pizarro · María Angélica Rubio · Luis Carlos Duarte Cavalcante · Vijayendra Kumar Garg · José Domingos Fabris

Published online: 25 August 2011

Abstract A Chilean volcanic Ultisol material was first size-fractionated so as to obtain the fraction with mean particle sizes $\phi < 53$ μm. This sample was then sequentially treated three or five times with 5 mol L^{-1} NaOH, in an attempt to evaluate the effectiveness of the selective chemical dissolution to concentrate iron oxides, as a preparation procedure before using the materials as heterogeneous Fenton catalysts. The effects of those treatments on the iron oxides mineralogy were monitored with Mössbauer spectroscopy. The NaOH-treated samples were tested as catalysts towards the H_2O_2 decomposition. Three or five sequential NaOH treatments were found to be comparably effective, by concentrating nearly the same proportion of iron oxides in the remaining solid phase (25.1 ± 0.4 and 23.3 ± 0.2 mass%,

V. Manzo · C. Pizarro (✉) · M. A. Rubio
Facultad de Química y Biología, USACH,
Av. L. B. O'Higgins 3363, Santiago, 7254758, Chile
e-mail: carmen.pizarro@usach.cl

V. Manzo · C. Pizarro · M. A. Rubio
Centro de Desarrollo para la Nanociencia y Nanotecnología (CEDENNA),
Av. L. B. O'Higgins 3363, Santiago, 7254758, Chile

L. C. D. Cavalcante · J. D. Fabris
Departamento de Química—ICEx, UFMG, Campus Pampulha,
31270-901 Belo Horizonte, Minas Gerais, Brazil

L. C. D. Cavalcante
Centro de Ciências da Natureza, Universidade Federal do Piauí,
64049-550 Teresina, Piauí, Brazil

V. K. Garg
Instituto de Física, Universidade de Brasília,
70919-970 Brasília, DF, Brazil

J. D. Fabris
Universidade Federal dos Vales do Jequitinhonha e Mucuri,
Campus JK, 39100-00 Diamantina, Minas Gerais, Brazil

respectively). 298 K-Mössbauer patterns were similar for both samples, with a central (super)paramagnetic Fe^{3+} doublet and a broad sextet, assignable to several closely coexisting magnetically ordered forms of iron oxides. Despite of this nearly similar effect of the two treatments, the Ultisol material treated three times with NaOH presents higher heterogeneous catalytic efficiency and is more suitable to decompose H_2O_2 than that with five treatments.

Keywords Mössbauer spectroscopy · Iron oxides · Fenton reaction · Soil minerals

1 Introduction

Iron oxides have been used as inexpensive heterogeneous catalysts in many industrial processes [1]. The Fenton reaction [2], which consists of a highly reactive mixture of hydrogen peroxide and an iron-containing catalyst, can be used to non-selectively oxidize organic materials [3]. Natural heterogeneous catalysts based on iron oxides for Fenton reaction, for their wide geological availability, as in many volcanic areas, low cost and potential chemical efficiency, have appeared as an interesting technological alternative for the scale environmental remediation of contaminated water bodies, as they often occur as freshwater or groundwater affected by industrial effluents and nearby intensively managed agricultural areas.

Some related studies dealing with natural iron oxides as catalysts for heterogeneous Fenton and Fenton-like reactions have been recently reported [4, 5]. The Fenton reaction (1) is based on the reaction of Fe^{2+} and H_2O_2 to produce highly reactive intermediate species that induce the generation of hydroxyl radical, according to the process [6]:

$$\mathrm{Fe\,(II)} + \mathrm{H_2O_2} \rightarrow \mathrm{Fe\,(III)} + \mathrm{OH^-} + {}^{\bullet}\mathrm{OH} \tag{1}$$

In some iron oxides, as in magnetite ($Fe_2^{3+}Fe^{2+}O_4$), structural Fe^{2+} and Fe^{3+} do coexist. The alternative Fenton-like mechanism (2) is based on reacting Fe^{3+} instead of starting Fe^{2+} involved in the electron transference to H_2O_2 [6]:

$$\mathrm{Fe\,(III)} + \mathrm{H_2O_2} \rightarrow \mathrm{Fe\,(II)} + \mathrm{H^+} + {}^{\bullet}\,\mathrm{OOH} \tag{2}$$

The heterogeneous Fenton or Fenton-like catalyst offers several advantages if compared with the homogeneous Fenton reaction, as no sludge is formed. Also, it operates at near-neutral pH and allows recycling the chemical state of the iron promoter. The decomposition rate of hydrogen peroxide (1) and (3) is a measure of the Fenton catalyst efficiency as it can be determined by evaluating the volume of the forming oxygen in the out coming gas.

$$2\mathrm{H_2O_2} \rightarrow \mathrm{O_2} + 2\mathrm{H_2O} \tag{3}$$

Ultisols forming on volcanic materials in Chile are characterized by having high iron oxide contents, with significant proportions of magnetic species, in their silt and clay fractions (saturation magnetization for the soil material, $\sigma = 2.40\ \mathrm{J\,T^{-1}\,kg^{-1}}$; for silt, $\sigma = 1.26\ \mathrm{J\,T^{-1}\,kg^{-1}}$, and clay, $\sigma = 1.20\ \mathrm{J\,T^{-1}\,kg^{-1}}$[7]), making them a potential natural source of iron-based Fenton catalyst.

The selective chemical dissolution of clay and other aluminum-bearing minerals is a particularly useful laboratory procedure in sample preparation, to concentrate and/or dissolves selected mineralogical phases. The 5 mol L^{-1} NaOH dissolution [8, 9] is used as chemical treatment to concentrate iron oxides in solid phases of soils, by selectively removing aluminum oxides and aluminosilicate minerals. This chemical selective dissolution was earlier used to concentrate iron oxides of silt and clay fractions from Chilean volcanic soil materials, in order to evaluate its effectiveness on the sample preparation for the Mössbauer analysis of their iron oxides [10]. However, no similar evaluation study has been made in order to confirm the more appropriate number of sequential NaOH treatments, intending to increase their efficiency on concentrating iron oxides and prepare the materials, to be used as heterogeneous Fenton catalyst.

In the present work, a sample from a Chilean Ultisol forming on volcanic materials was size-fractionated so to obtain the fraction with mean particle sizes $\phi < 53$ μm. This soil fraction was then sequentially treated three or five times with 5 mol L^{-1} NaOH, in order to promote the selective concentration of iron oxides, by removing aluminosilicates and aluminum oxides, as a preparation procedure before using these materials as heterogeneous Fenton catalysts. The dissolution effectiveness according to the number of sequential NaOH treatments was monitored with Mössbauer spectroscopy and the product was tested as catalysts for the hydrogen peroxide decomposition, in Fenton reaction, through volumetric measurement of the evolved gas.

2 Materials and method

2.1 Soil collection and preparing samples

A soil sample was collected from an Ultisol profile at a depth between 15 and 30 cm (B horizon), in a site located in southern Chile (geographical coordinates of the sampling site, 36°58′ S 72°09′ W). The collected soil material was sieved in the field, so to obtain a fraction with mean particle diameter $\phi < 2$ mm (whole soil), and packed under roughly the corresponding soil moisture as of the natural field conditions. In the laboratory, the whole soil sample was first treated with hydrogen peroxide 30 mass%, in order to remove organic matter. The resulting material was passed through a sieve so to obtain a $\phi < 53$ μm fraction, which was then chemically attacked three or five sequential times with 5 mol L^{-1} NaOH, intending to remove mainly aluminosilicates and aluminum oxides.

2.2 NaOH treatments

The treatment with 5 mol L^{-1} NaOH was based on the method first described by Norrish and Taylor [9], with minor modifications [10]: 5 g of soil sample were put to react in a Teflon® flask by adding 75 mL of 5 mol L^{-1} NaOH solution (3 or 5 additions, depending on the treatment). The mixture was left on a sand bath at 90°C for 1 hour for each of the two treatments, and then transferred to tubes to be centrifuged at 8000 rpm during 10 min. Subsequently, the supernatant was discarded and the solid phase was washed with 5 mol L^{-1} NaOH; 0.5 mol L^{-1} HCl was used

Table 1 Aluminum, iron and silicon contents of the whole Collipulli Ultisol and of its $\phi < 53$ μm fraction treated with three or five times with 5 mol L^{-1} NaOH

Sample	Al_2O_3	Fe_2O_3	SiO_2
	Mass%		
Collipulli whole soil	26.5 ± 0.3	12.2 ± 0.3	41.1 ± 0.4
Collipulli whole soil three NaOH	12.7 ± 0.1	23.3 ± 0.7	10.1 ± 1.2
Collipulli <53 μm three NaOH	9.0 ± 0.4	25.1 ± 0.2	15.8 ± 0.5
Collipulli <53 μm five NaOH	12.1 ± 0.3	23.3 ± 0.2	10.5 ± 0.3

Uncertainties are standard deviation over the mean of three independent determinations

to neutralize the remained excess of NaOH and to remove commonly formed side-products of the reaction, like sodalite. A solution of $(NH_4)_2CO_3$ 1 mol L^{-1} was used to remove side-formed NaCl and, finally, the excess of ammonium carbonate was washed out with distilled water.

2.3 Chemical and mineralogical characterizations

The chemical composition of samples was determined by digesting 0.250 g of the dry sample with HF and *aqua regia* (10:1.5 ratio), in a Parr™ bomb digester, set at 110°C, during 12 h. The resultant digestion mixture was treated with ultra-pure graded (Merck) boric acid, diluted to 100 mL with water, and stored in a plastic bottle at 5°C, prior to elemental chemical readings for Fe, Si, and Al on a atomic absorption spectroscopy (THERMO M5). Data presented in this work were expressed as averaged values of three sample-determination-replicas. Organic matter content was determined *via* the Walkley–Black method [11].

Mössbauer spectra were collected at 298 K and 80 K in constant acceleration transmission mode, with a ~30 mCi $^{57}Co/Rh$ source. Data were stored in a 512-channel MCS memory unit, with Doppler velocities ranging between approximately ±10 mm s^{-1}. The experimental data were fitted using Lorentzian functions, with least-squares fitting procedure based on the NORMOS™ computer program; isomer shifts are quoted relative to α-Fe.

2.4 Catalytic tests

The Fenton activity was evaluated through the decomposition of 30% p/p hydrogen peroxide, by measuring the evolving gaseous O_2, in a volumetric gas system, at pH 5.0, using 15 mg of the solid sample, as catalyst [5].

3 Results and discussion

This Collipulli Ultisol is typically acid and characterized by relatively low organic matter and high iron oxides contents [12]. The B horizon presents a high clay content [13]; its $\phi < 53$ μm portion encompasses the silt and clay fractions, according to the Soil Survey Staff of the US Department of Agriculture [15], and represent about 90% of the soil mass. Data of the chemical composition in aluminum, iron and silicon for the whole soil and for its $\phi < 53$ μm fraction treated with three or five times with 5 mol L^{-1} NaOH are shown in Table 1. From these chemical data, the

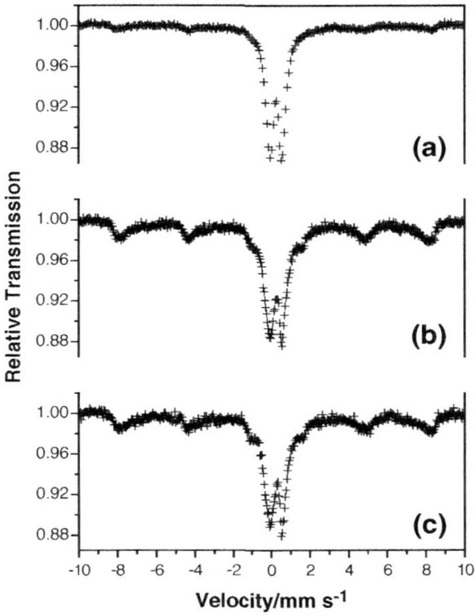

Fig. 1 298 K-Mössbauer spectra for the <53 μm fraction of the Collipulli Ultisol **a** untreated sample and after **b** three and **c** five treatments with 5 mol L^{-1} NaOH

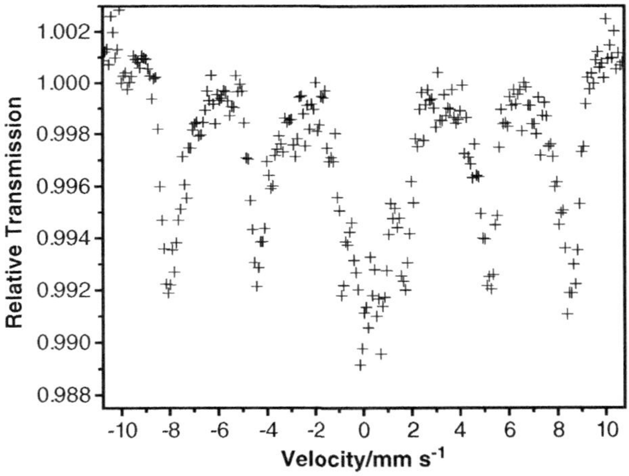

Fig. 2 80 K-Mössbauer spectrum for the <53 μm fraction Collipulli Ultisol sample obtained by treating the soil material with three treatments with 5 mol L^{-1} NaOH

treatment with NaOH was mainly effective in concentrating iron oxides: about twice as much iron oxides content is found in the remaining solid phase of the soil treated with NaOH. This procedure revealed that both, the three and five treatments, concentrated nearly the same amounts of iron oxides in the solid phase of the $\phi < 53$ μm fraction (25.1 ± 0.2 and 23.3 ± 0.2 mass%, respectively). On the other hand, a decreasing of the aluminum and silicon contents is observed, respectively, to the untreated soil samples, as it is expected, These samples, before and after the chemical treatments, were characterized with Mössbauer spectroscopy (spectra are shown in Figs. 1 and 2). The incipient magnetic hyperfine structure in 298 Mössbauer spectrum (Fig. 1a) for the untreated sample is observed and not

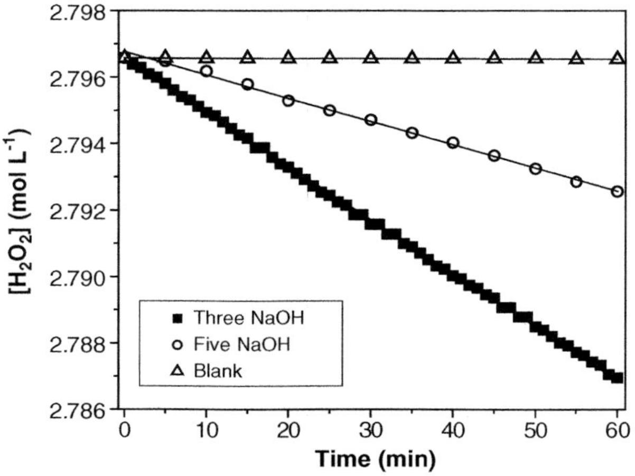

Fig. 3 Pseudo-zeroth order kinetics for the decomposition reaction of H_2O_2 with iron oxide-based heterogeneous catalysts obtained by treating the <53 μm fraction of the Collipulli Ultisol material with three and five treatments 5 mol L^{-1} NaOH

Table 2 Pseudo zeroth-order kinetic constant (k) for the ϕ < 53 μm-Collipulli Ultisol fraction, after three and five treatments with 5 mol L^{-1} NaOH, used as Fenton catalyst on the decomposition reaction of H_2O_2

Sample	$k/10^{-5}$ mol L^{-1} min^{-1}
Three NaOH	16.0 ± 0.8
Five NaOH	7.1 ± 0.7

accurately assignable to any iron oxide mineral, whereas the spectra for the sample after three and five NaOH treatments (Fig. 1b, and c, respectively) reveals a more pronounced magnetic hyperfine structure. 298 K-Mössbauer patterns were similar for samples resulting from both treatments (Fig. 1b and c). This sample is typically formed by complex mineral assemblages, as it is usually found for most silt and clay fractions from volcanic soils, rendering correspondently Mössbauer patterns mainly formed by an intense central doublet, due to (super)paramagnetic Fe^{3+} species, as structural iron in crystalline structure of phyllosilicates and/or iron (hydr)oxides in very small particles, and a broad sextet, commonly due to several closely coexisting magnetically ordered forms of iron (hydr)oxides. For the sake of accuracy and in an attempt to seek for a physically more consistent interpretation, these set of spectra were deliberately not assigned to any iron oxide. By the way, the difficulty commonly found to safely interpret Mössbauer data of volcanic soil samples was duly reported earlier [14]. However, comparing both 298 K- and 80 K-Mössbauer data can directly draw essential qualitative information. A remarkable point is that most of the central doublets of spectra are actually due to superparamagnetic iron (hydr)oxides, as the relaxation behavior is mostly blocked at 80 K (Fig. 2). From both, chemical and Mössbauer analyses, it is not possible to observe any significant differences between iron oxides, as products of three and five NaOH treatments. This means that the first three treatments with NaOH are well enough to obtain a selectively concentrated product; performing more than three chemical treatments renders no effective gain in concentrating extra amounts of iron oxides in the solid phases of this geomaterial.

The 80 K-Mössbauer spectrum for the three times NaOH-treated sample (Fig. 2) shows that the resonance signals are significantly more intense that those obtained with samples at 298 K (Fig. 1). Comparable hyperfine structures to this 80 K-spectrum were earlier reportedly found from fitting 6 K-Mössbauer spectra for ammonium oxalate-treated samples of magnetic silt and clay fractions, of the same Collipulli soil [7, 13]. From those spectra, for which all superparamagnetic relaxation effects were virtually blocked, it is plausible, by similarity, to assign the main contributions to the magnetic hyperfine structure of pattern in Fig. 2 as being due to maghemite (γFe_2O_3), goethite ($\alpha FeOOH$) and hematite (αFe_2O_3). In the present study, no corresponding spectrum was obtained for the untreated or five times NaOH-treated sample at low temperature. Even though, similar qualitative spectral patterns to that of Fig. 2 are expected for all samples.

Decomposition of H_2O_2 (Fig. 3) shows that the H_2O_2 decomposition reaction rate was of pseudo-zeroth order on the peroxide concentration. The sample treated three times with NaOH showed higher catalytic efficiency on the decomposition reaction of H_2O_2 than that observed with the sample obtained with five treatments. The determined kinetic rate constants were $16.0 \pm 0.8 \times 10^{-5}$ and $7.1 \pm 0.7 \times 10^{-5}$ mol L^{-1} min^{-1}, respectively, as shows in the Table 2. An explanation for this is that some aluminum-containing chemical specie (putatively, sodalite) was side-formed and re-precipitated on treating the sample with NaOH. Actually, the aluminum content in the solid phase increased about 25 mass% on going from three to five treatments.

4 Conclusions

Even though no outstanding differences on Mössbauer patterns for the samples after three and five sequential chemical treatments were observed, their chemical behaviors are remarkably distinctive. For providing best kinetic results, the three sequential treatments of this $\phi < 53$ μm soil fraction with NaOH were found to be well suitable and enough to effectively concentrate iron oxides, if the product is to be used as heterogeneous Fenton catalysts.

Acknowledgements Work supported by DICYT-USACH 090942-PA (Chile) and CEDENNA FB 0807 (Chile), Academic Mobility Program of the Asociación de Universidades Grupo Montevideo (AUGM), FAPEMIG (grants CEX PPM 00217/08 and PPM 00419/10), and CNPq (grants 490096/2010-7 and PROSUL 490132/2006-5) (Brazil). The Mecesup project UCH 0601 (Chile; Red Nacional de Doctorado en Quimica) supported a two week-academic visit of JDF to the University of Santiago Chile, in January 2010.

References

1. Schwertmann, U., Cornell, R.M.: Iron Oxides in Laboratory, 2nd edn. Wiley, D-69469 Weinheim (Federal Republic of Germany). ISBN 3-527-29669-7 (2000)
2. Fenton, H.J.H.: Oxidation of tartaric acid in presence of iron. J. Chem. Soc. **65**, 899–910 (1894)
3. Costa, R.C.C., Moura, F.C.C., Ardisson, J.D., Fabris, J.D., Lago, R.M.: Highly active heterogeneous Fenton-like systems based on Fe^0/Fe_3O_4 composites prepared by controlled reduction of iron oxides. Appl. Catal. B Environ. **83**(421), 131–139 (2008)

4. Pereira, M.C., Tavares, C.M., Fabris, J.D., Lago, R.M., Murad, E., Criscuolo, P.S.: Characterization of a tropical soil and a waste from kaolin mining and their suitability as heterogeneous catalyst for Fenton and Fenton-like reactions. Clay Miner. **42**, 299–306 (2007)
5. Aravena, S., Pizarro, C., Rubio, M.A., Cavalcante, L.C.D., Garg, V.K., Pereira, M.C., Fabris, J.D.: Magnetic minerals from volcanic Ultisols as heterogeneous Fenton catalysts. Hyperfine Interact. **195**, 35–41 (2010)
6. Kwan, W.P., Voelker, B.M.: Rates of hydroxyl radical generation and organic compound oxidation in mineral-catalyzed Fenton-like systems. Environ. Sci. Technol. **37**(6), 1150–1158 (2003)
7. Pizarro, C., Fabris, J.D., Stucki, J.W., Garg, V.K., Galindo, G.: Ammonium Oxalate and citrate-ascorbate as selective chemical agents for the mineralogical analysis of clay fractions of an Ultisol and Andisols from southern Chile. J. Chil. Chem. Soc. **53**(3), 1581–1584 (2008)
8. Norrish, K., Taylor, R.M.: The isomorphous replacement of iron by aluminium in soil goethites. J. Soil Sci. **12**(2), 294–306 (1961)
9. Kämpf, N., Schwertmann, U.: The 5 M NaOH concentration treatment for iron oxides in soils. Clays Clay Miner. **30**, 401–408 (1982)
10. Pizarro, C., Escudey, M., Fabris, J.D., Almeida, A.B.: Effects of sodium hydroxide selective chemical treatment on samples from some Chilean Soils. Commun. Soil Sci. Plant Anal. **31**(19&20), 3113–3119 (2000)
11. Allison, L.E.: In: C.A. Black, D.D. Evans, J.L. White, L.E. Ensminger, F.E. Clark (eds), Agronomy 9, Part 2, American Society of Agronomy, p. 1367. Madison, WI (1965)
12. Pizarro, C., Escudey, M., Fabris, J.D., Almeida, A.B.: Iron-rich spinel phases from sand fraction of three Chilean soils developing on volcanic materials. Commun. Soil Sci. Plant Anal. **32**(17&18), 2741–2754 (2001)
13. Pizarro, C., Fabris, J.D., Stucki, J., Garg, V.K., Morales, C., Aravena, S., Gautier, J.L., Galindo, G.: Distribution of Fe-bearing compounds in an Ultisol as determined with selective chemical dissolution and Mössbauer spectroscopy. Hyperfine Interact. **175**, 95–101 (2008)
14. Pizarro, C., Furet, N.R., Venegas, R., Fabris, J.D., Escudey, M.: Some caution on the interpretation of Mössbauer spectra in mineralogical studies of volcanic soils. Bol. Soc. Chil. Quím. **45**, 243–250 (2000)
15. Soil Survey Staff: Soil Taxonomy: a basic system of soil classification for making and interpreting soil surveys, 2nd edn. USDA-SCS Agric. Handb. 436 Government Printing Office, Washington, DC. ftp://ftp-fc.sc.egov.usda.gov/NSSC/Soil_Taxonomy/tax.pdf (1999). Accessed on 2 January 2011

Hyperfine Interact (2011) 203:67–74
DOI 10.1007/s10751-011-0348-y

Effect of Sn on methane decomposition over Fe supported catalysts to produce carbon

Patrícia F. Oliveira · Leandro P. Ribeiro · Marcelo G. Rosmaninho · José D. Ardisson · Anderson Dias · Rochel M. Lago

Published online: 25 August 2011

Abstract In this work, alumina-supported Sn containing Fe catalysts were investigated in CVD reactions (Chemical Vapor Deposition) using methane for carbon production. The catalysts were prepared with 10 wt.% of Fe (as Fe_2O_3) and 3, 6 and 12 wt.% of Sn (as SnO_2) supported on Al_2O_3 named hereon Fe10Sn3A, Fe5Sn6A and Fe10Sn12A, respectively. These catalysts were characterized by SEM, TPCVD, TPR, TG, Raman, XRD and ^{57}Fe and ^{119}Sn Mössbauer spectroscopy. Methane reacts with Fe10A catalyst (without Sn) in the temperature range 680–900°C to produce mainly Fe^0, Fe_3C and 20 wt.% of carbon deposition. TPR and TPCVD clearly showed that Sn strongly hinders the CH_4 reaction over Fe catalyst. ^{57}Fe Mössbauer suggested that in the presence of Sn the reduction of Fe^{+3} by methane becomes very difficult. ^{119}Sn Mössbauer showed Sn^{+4} species strongly interact with metallic iron after CVD, producing iron-tin phases such as Fe_3SnC and $FeSn_2$. This interaction Sn–Fe increases the CVD temperatures and decreases the carbon yield leading to the production of more organized forms of carbon such as carbon nanotubes, nanofibers and graphite.

Keywords Iron and tin catalysts · Carbon deposition · Chemical vapor deposition

P. F. Oliveira · L. P. Ribeiro · M. G. Rosmaninho · R. M. Lago (✉)
Departamento de Química, ICEx. Universidade Federal de Minas Gerais (UFMG), Belo Horizonte, MG, Brazil
e-mail: rochel@qui.ufmg.br

J. D. Ardisson
Laboratório de Física Aplicada, Centro de Desenvolvimento de Tecnologia Nuclear (CDTN), Belo Horizonte, MG, Brazil

A. Dias
Departamento de Química, Universidade Federal de Ouro Preto (UFOP), Ouro Preto, MG, Brazil

1 Introduction

Catalytic carbon deposition on transition metals has been the subject of intensive investigation. Previous studies have shown that the decomposition of hydrocarbons over metals can result in the formation of different carbon structures, such as sheet, fullerene, fibers or tubes. Typical metals used as catalysts for carbon deposition are Co, Fe and Ni [1–3]. Different studies have also demonstrated that the presence of a second metal, e.g. alloys such as Sn/Fe, Ni/Cu and Fe/Ni strongly affects the reactivity and the type of carbon formed [4–6]. In this work, we have studied the effect of Sn on iron catalysts supported on Al_2O_3 methane decomposition via a CVD process. The reactions were investigated by the new technique Temperature Programmed CVD (TPCVD), using a mixture of CH_4/N_2 at temperatures from 30°C to 900°C [7, 8].

2 Experimental

2.1 Preparation of Fe_2O_3, SnO_2/Al_2O_3

The composites were prepared by co-impregnation of Al_2O_3 (Aldrich) with $Fe(NO_3)_3$ and $SnCl_2$ in water, slowly dried at 60°C and calcinated in a tubular furnace (Lindberb Blue M) heat up at 10°C min^{-1} until 400°C and held in this temperature for 4 hours. The amount of iron was maintained (10 wt.% of iron oxide) and tin contents were calculated to produce systems with 3, 6 and 12 wt.% of SnO_2. These materials were called Fe10Sn3A, Fe10Sn6A and Fe10Sn12A, respectively. In addition, it was prepared an iron free system, composed by 12 wt.% of SnO_2 in Al_2O_3 (Sn12A) and tin free composite with 10 wt.% of Fe_2O_3 (Fe10A).

2.2 Temperature programmed chemical vapor deposition (TPCVD)

The reactions with methane (CVD reactions) were carried out in a tubular furnace at temperatures between 30°C and 900°C, with a heating rate of 10°C min^{-1}, using about 20 mg of catalyst and flow of 30 mL min^{-1} of 14% CH_4/N_2. The gas products were analyzed by a ChemBET3000 TPR of the Quantachrome Instruments with a thermal conductivity detector (TCD).

2.3 Characterization

The Temperature Programmed Reductions (TPR) were carried out with a flow of 5% H_2/N_2 (30 mL min^{-1}) and the hydrogen consumption was analyzed by a ChemBET3000 TPR of the Quantachrome Instruments.

^{57}Fe and ^{119}Sn Mössbauer spectra were obtained from a constant acceleration spectrometer moving a $^{57}Co/Rh$ ($CaSnO_3$) source at room temperature. The samples were analyzed at 298 and 80 K. All Mössbauer spectra were computer-fitted assuming Lorentzian line shapes. Scanning electron microscopy (SEM) analysis was done using a Jeol JSM6360 instrument. TG analyses were carried out in a Shimadzu TGA60H simultaneous TG/DTA. The Raman spectroscopy experiments were carried out in Renishaw micro Raman with the excitation wavelength of 633 nm and a laser spot size of 20 μm.

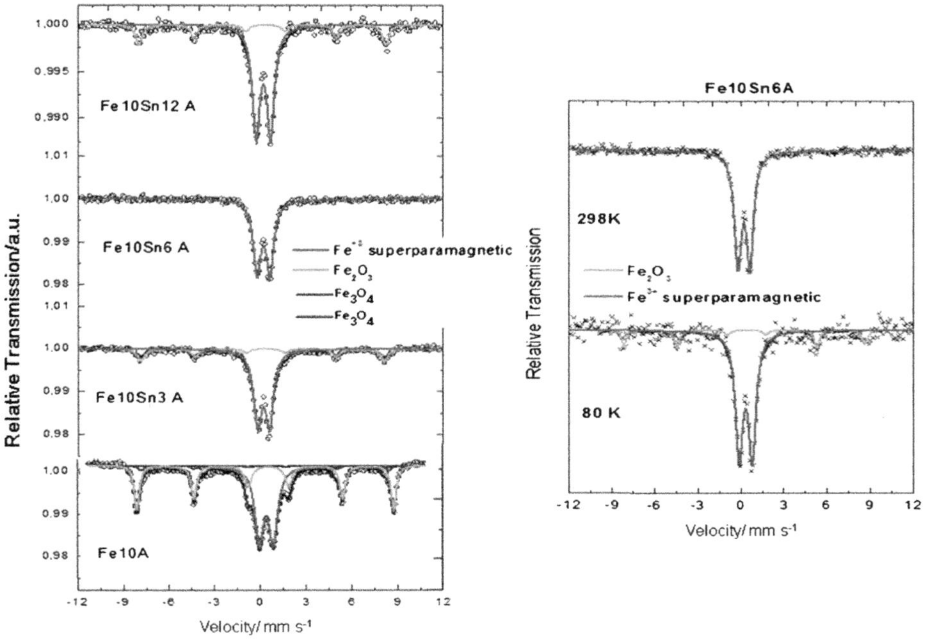

Fig. 1 ^{57}Fe Mössbauer spectra at 298 K for Fe10A and Fe–Sn supported catalysts and at 80 K for Fe10Sn6A, before CVD reaction

Fig. 2 Distribution of iron phases in Fe10A and Fe–Sn catalysts before CVD reaction, determined by ^{57}Fe Mössbauer spectroscopy

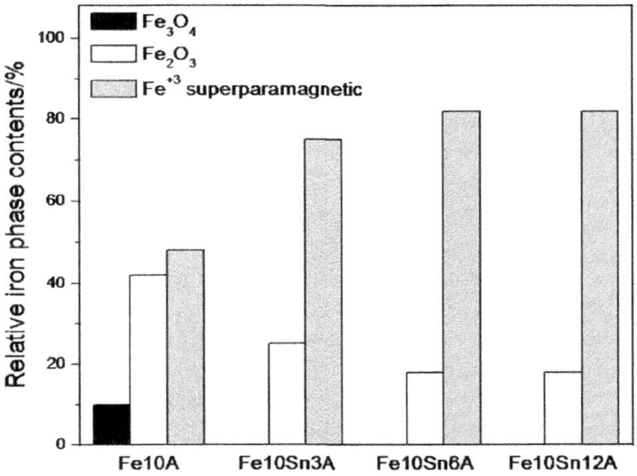

3 Results and discussion

^{57}Fe Mössbauer Spectroscopy at 298 K (Fig. 1) for Fe10A catalyst before CVD reaction showed mainly hematite and a superparamagnetic Fe^{3+} phase. After Sn addition, the ^{57}Fe Mössbauer suggested that the content of hematite decreased with the presence of tin from 42% in Fe10A to 18% in Fe10Sn12A. On the other hand,

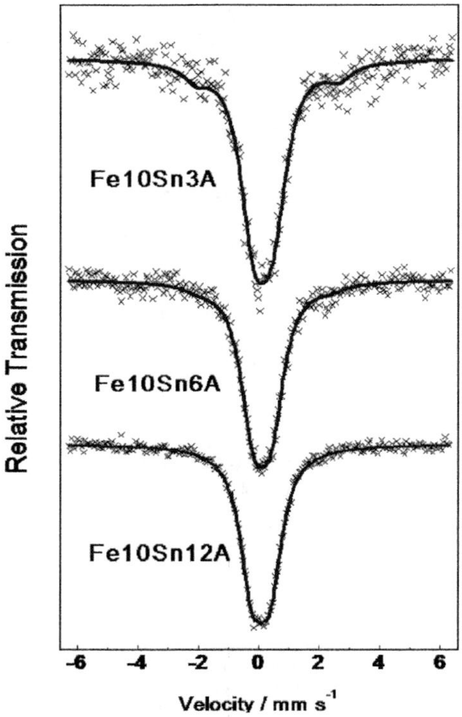

Fig. 3 ^{119}Sn Mössbauer spectra for the Fe–Sn supported catalysts at 80 K, before CVD

the presence of Sn caused an increase of Fe^{+3} superparamagnetic species from 48% in Fe10A to 82% in Fe10Sn12A (Figs. 1 and 2) [9]. ^{57}Fe Mössbauer at 80 K for Fe10Sn6A showed that the doublet observed at 298 K is a superposition of hematite with low size of crystallite and a paramagnetic Fe(III) phase (Fig. 1). ^{119}Sn Mössbauer (Fig. 3) showed only the presence of Sn^{+4}species, probably SnO_2.

TPR profiles (Fig. 4) indicated that the increase of tin contents completely shifted Fe reduction steps to higher temperatures. These results clearly show that iron is interacting strongly with Sn, probably by the formation of stable Fe–Sn compounds, avoiding the formation of common iron phases, like hematite and magnetite.

It was not possible to obtain ^{119}Sn Mossbauer spectra for the materials after TPR, since higher temperatures led the formation of Sn–H volatile compounds [10].

The catalysts were submitted at CVD reactions with methane for carbon production according to the simplified reaction (Eq. 1):

$$CH_4 \rightarrow C + 2\,H_2 \quad (1)$$

The TPCVD profiles (Fig. 5) showed that CH_4 reacts with Fe10A catalyst (without Sn) in the temperature range 680–900°C. However, the addition of tin strongly changes the Fe reactivity increasing the CVD temperatures. ^{57}Fe and ^{119}Sn Mössbauer spectroscopy after CVD reaction (Fig. 6) showed a doublet centered at isomeric shifts $\delta = 2.14$ mm s^{-1} in a ^{119}Sn Mössbauer spectra suggesting the presence of the alloy $FeSn_2$ [11]. The formation of the stable phase Fe_3SnC is suggest by a singlet in ^{119}Sn Mössbauer spectra ($\delta = 1.59$ mm s^{-1}) and a doublet in ^{57}Fe Mössbauer

Fig. 4 TPR profiles for Fe–Sn supported catalysts

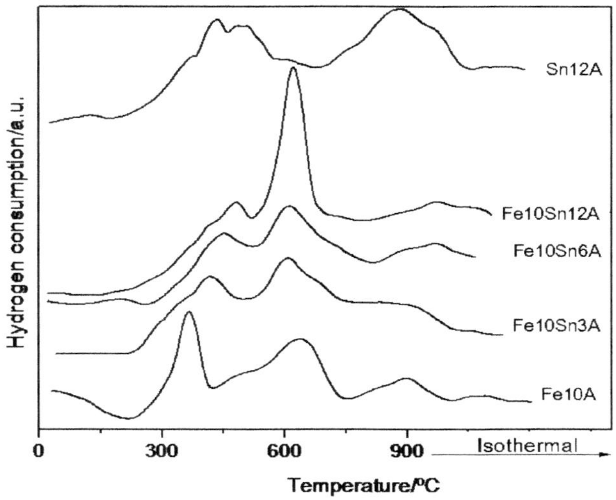

Fig. 5 Hydrogen production by methane decomposition for Fe10A and Fe–Sn supported catalysts

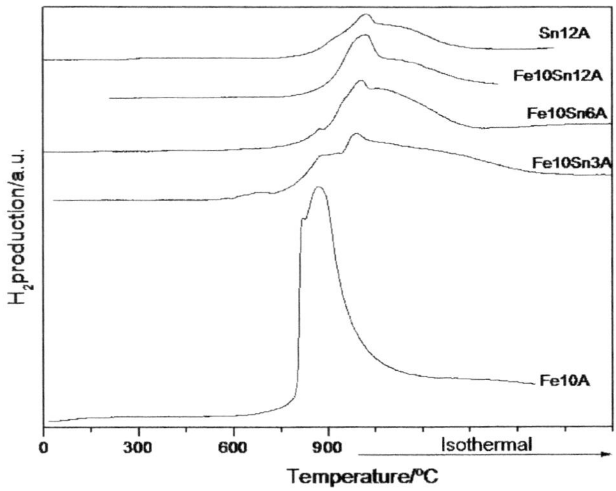

spectra ($\delta = 0.38$ mm s^{-1}, $\Delta = 1.01$ mm s^{-1}). The production of a iron-tin carbide was confirmed by XRD. The formation of this phase suggests a diffusion of carbon to the FeSn alloy during the thermal treatment [10]. These stable iron-tin phases decrease the catalytic properties to carbon deposition of the materials studied, since metallic iron or tin are required to promote methane decomposition.

The carbon deposited on the catalyst was measured using TG analysis. The amount of carbon strongly decreased with the addition of tin (Fig. 7). It is interesting to note that the pure tin catalyst Sn12A (without Fe) showed similar activity to Fe10A, leading to formation of ca. 20% of carbon. These results confirm the strong interaction between iron and tin. It was observed that the structure of deposited carbon depends of the presence of tin in the composite. TG analysis (Fig. 7) showed that the increase of tin content promotes the carbon oxidation (carbon decomposition)

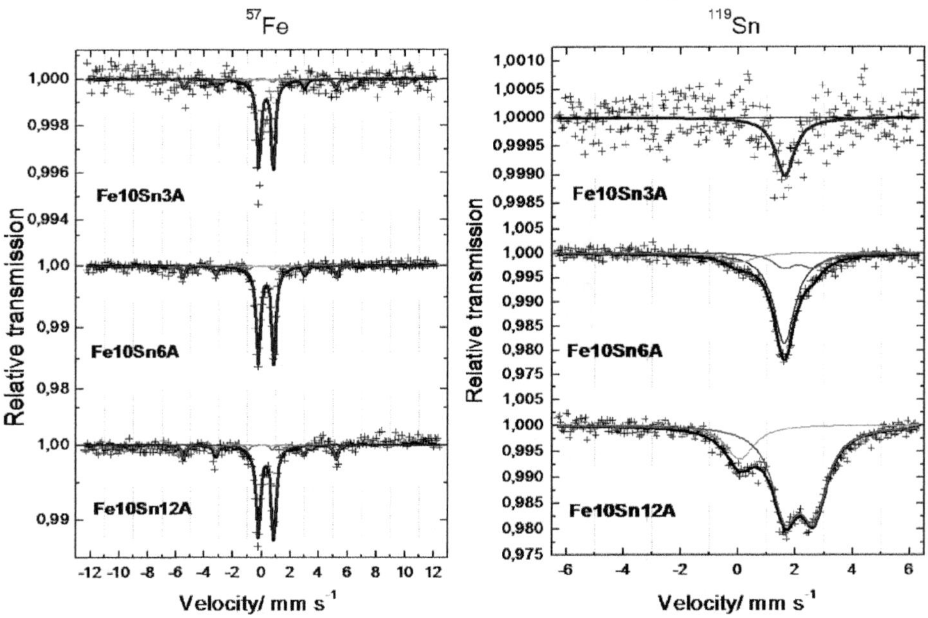

Fig. 6 ^{57}Fe and ^{119}Sn Mössbauer spectra for Fe–Sn supported catalysts after CVD at 298 and 80 K, respectively

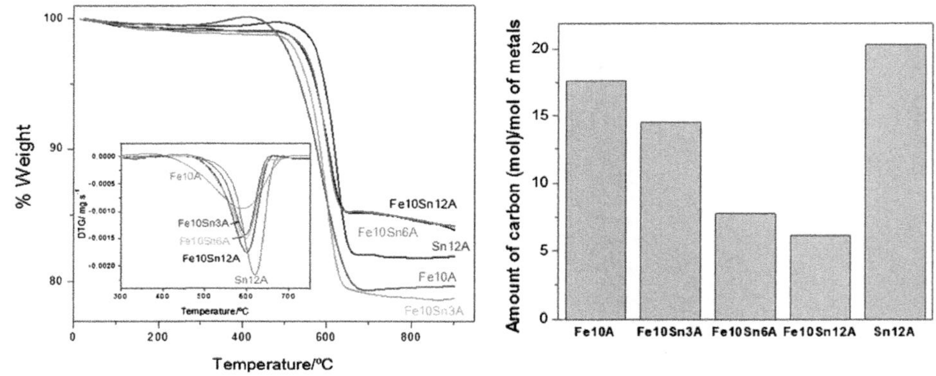

Fig. 7 Thermogravimetric profiles for Fe10A, Sn12A and Fe–Sn supported catalysts in air and the amount of carbon (mol) deposited per mol of metal by TG

at higher temperatures, suggesting that the presence of tin promotes the formation of more organized carbon structures. These results agree with Raman spectroscopy (Fig. 8), which exhibited D band more intense than G band for materials with tin. The presence of D band can be related with either to disorganized carbon or organized carbon with structural imperfection, such as nanotubes, nanofibers and graphite with imperfections.

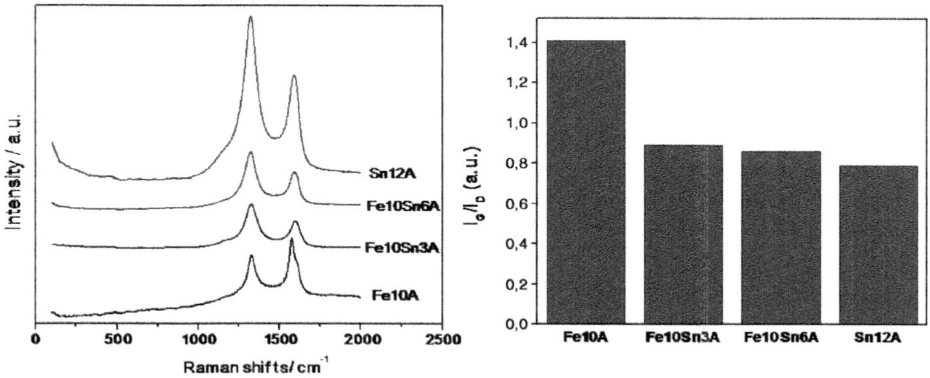

Fig. 8 Raman spectra and I_D/I_G reason for deposited carbon for Fe–Sn supported catalysts

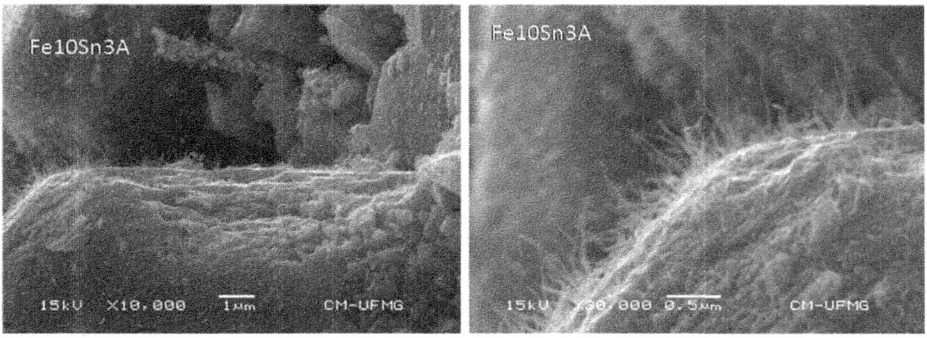

Fig. 9 SEM images for Fe10Sn3A after CVD reaction

SEM images (Fig. 9) for the tin containing materials indicated the formation of organized structures of carbon, such as filaments and graphite, which agrees with TG and Raman results.

4 Conclusions

Catalysts based on Sn and Fe supported in Al_2O_3 were studied in CVD reactions with methane to produce carbon. CVD profiles showed that the addition of tin strongly hinders the CH_4 reaction over Fe catalyst, decreasing the amount of deposited carbon with the increase of tin content. On the other hand, it was observed that the presence of tin led the formation of more organized forms of carbon, such as carbon nanotubes, nanofibers and graphite with imperfections. ^{57}Fe and ^{119}Sn Mössbauer spectroscopy after CVD suggested the formation of Fe–Sn stable phases, like Fe_3SnC and $FeSn_2$, confirming that iron and tin strongly interacts after thermal treatment.

Acknowledgements The authors are grateful to FAPEMIG and CNPq for the financial support and to Microscopy Center of UFMG for the SEM images.

References

1. Kukovitsky, E.F., L'vov, S.G., Sainov, N.A., Shustov, V.A.: CVD growth of carbon nanotube films on nickel substrates. Appl. Surf. Sci. **215**, 201–208 (2003)
2. Dupuis, A.C.: The catalyst in the CCVD of carbon nanotubes—a review. Prog. Mater. Sci. **50**, 929–961 (2005)
3. Choy, K.L.: Chemical vapour deposition of coatings. Prog. Mater. Sci. **48**, 57–170 (2003)
4. Li, D., Pan, L., Qian, J., Liu, D.: Highly efficient synthesis of carbon nanocoils by catalyst particles prepared by a sol–gel method. Carbon **48**, 170–175 (2010)
5. Okazaki, N., Hosokawa, S., Goto, T., Nakayama, Y.: Synthesis of carbon tubule nanocoils using Fe-In-Sn-O fine particles as catalysts. J. Phys. Chem., B **109**, 17366–17371 (2005)
6. Zhang, M., Nakayama, Y., Pan, L.: Synthesis of carbon tubule nanocoils in high yield using iron-coated indium tin oxide as catalyst. Jpn. J. Appl. Phys. **39**, 1242–1244 (2000)
7. Tristão, J., Moura, F.C., Lago, R.M.: Sistema RTP: uma técnica poderosa para o monitoramento da formação de nanotubos de carbono durante o processo por deposição de vapor químico. Quim. Nova **33**, 1379–1383 (2010)
8. Oliveira, A., Teixeira, I., Ribeiro, L., Tristão, J., Dias, A., Lago, R.M.: Magnetic amphiphilic composites based on carbon nanotubes and nanofibers grown on an inorganic matrix: effect on water-oil interfaces. J. Braz. Chem. Soc. **21**, 2184–2188 (2010)
9. Gaudon, M., Pailhé, N., Majimel, J., Wattiaux, A., Abel, J., Demourgues, A.: Influence of Sn^{4+} and Sn^{4+}/Mg^{2+} doping on structural features and visible absorption properties of α-Fe_2O_3 hematite. J. Solid State Chem. **183**, 2101–2109 (2010)
10. Fujinami, M., Yoshida, K., Ujihira, Y., Terasaka, M.: ^{57}Fe and ^{119}Sn conversion electron Mössbauer studies of Fe–Sn phases on tinplate. J. Mater. Sci. **20**, 4099–4102 (1985)
11. Nwokeke, U.G., Alcántara, R., Tirado, J.L., Stoyanova, R., Yoncheva, M., Zhecheva, E.: Electron paramagnetic resonance, x-ray diffraction, mössbauer spectroscopy, and electrochemical studies on nanocrystalline $FeSn_2$ obtained by reduction of salts in tetraethylene glycol. Chem. Mater. **22**, 2268–2275 (2010)

Hyperfine Interact (2011) 203:75–84
DOI 10.1007/s10751-011-0349-x

Synthesis and characterization of Cu^{2+} substituted magnetite

A. L. Morales · A. A. Velásquez ·
J. P. Urquijo · E. Baggio

Published online: 25 August 2011

Abstract Samples of magnetite, both pure and doped with divalent copper, $Fe_{3-x}Cu_xO_4$, with $x = 0, 0.05, 0.10$ and 0.20 atm.%, were synthesized hydrothermally. The samples were characterized by Atomic Absorption Spectroscopy, Mössbauer Spectroscopy, X-ray diffraction, Scanning Electron Microscopy and SQUID magnetometry. The analyses made by the above techniques showed that as the Cu^{2+} concentration increases, a simultaneous reduction in the magnetic and structural parameters takes place, namely: magnetic hyperfine interactions at octahedral sites, particle size and lattice constant. Degradation in the particles morphology as well as a distribution of their size were also observed. Our study points two important effects of Cu^{2+} in magnetite, the first one is its incorporation within the structure, replacing Fe^{2+} ions and decreasing both the magnetic hyperfine interactions at octahedral sites and the bulk magnetization, the second one is the contraction of the crystalline lattice of magnetite, because incorporation of Cu^{2+} within the structure, generation of vacancies or both simultaneous effects.

Keywords Magnetite · Divalent copper · Mössbauer spectroscopy · Morphology · SQUID magnetometry

A. L. Morales · J. P. Urquijo
Grupo de Estado Sólido, Instituto de Física, Universidad de Antioquia,
A.A. 1226, Medellín, Colombia

A. A. Velásquez (✉)
Grupo de Electromagnetismo Aplicado, Universidad EAFIT,
A.A. 3300, Medellín, Colombia
e-mail: avelas26@eafit.edu.co

E. Baggio
Centro Brasileiro de Pesquisas Físicas, CEP 22290-180,
Rio de Janeiro, Brasil

1 Introduction

Magnetite is an iron oxide present on the surface of ancient rocks, as well as in the rust formed on the surfaces of iron and steel exposed to atmospheric conditions [1–3]. In the last decades some investigations [4–6] had as main objective understanding how changes in magnetic, electrical and structural properties occur when magnetite is synthesized in presence of cations such as Al^{3+}, Co^{2+}, Cr^{3+}, Mn^{2+}, Ti^{4+}, Cu^{2+}, among others. The reason for this interest is that magnetite doped with these cations can acquire properties that it does not have in its non doped form, for example, thermal conductivity variations, electrical conductivity, crystalline structure, morphology and particles size, oxidation stability, among others. In particular, the research on the effects of copper in magnetite has two paths: the first one deals with the formation of magnetite in presence of Cu^{2+}, where copper ions can prevent the formation of other iron oxides; the second and more fundamental one deals with the substitution Fe^{2+} by Cu^{2+} and also with the consequences this has for magnetite properties. In this paper we present the results obtained from the synthesis of magnetite in presence of Cu^{2+} in different atomic concentrations, looking for a fundamental approach to the mechanism by which this cation introduces changes in structural and magnetic properties of this iron oxide.

2 Experimental procedure

The samples were synthesized by the hydrothermal method reported by Schwertmann and Cornell [7]. $FeCl_2$–$4H_2O$ was used as Fe^{2+} precursor and $NaNO_3$ was used as Fe^{2+} oxidant. Non doped magnetite was synthesized from 0.1 M (300 ml) $FeCl_2$–$4H_2O$ aqueous solution, which was constantly bubbled with $N_{2(g)}$ and heated in a thermal bath at 90°C. During a 5 min period 100 ml of an alkaline mixture composed of NaOH 3 M and $NaNO_3$ 0.06 M were added dropwise to the before solution. After that, the reaction continued at 90°C during 30 min with vigorous stirring. The black precipitate was repeatedly washed with deionized water to obtain a neutral pH in the residual supernatant. Finally the precipitate was filtered and dried in an oven at 40°C for 48 h. The doped magnetites were prepared from 300 ml of aqueous solutions 0.1 M in $FeCl_2$–$4H_2O$ and 1.662×10^{-3} M, 3.324×10^{-3} M and 6.648×10^{-3} M in $CuCl_2$–$2H_2O$ for 5, 10 and 20 atm.%-Cu^{2+}, respectively. The subsequent procedure was the same described for the non doped magnetite.

Atomic Absorption Spectroscopy measurements were made with an UNICAM 929 AA Spectrometer with copper lamp, the samples were diluted in HCl solutions for these measurements. X-ray diffraction patterns were taken with a FreibergerPrazisiosmechanik Zeiss HZG-4 powder diffractometer with copper anode and wavelength of 1.5406 Å, angular step of 0.05° and sweep velocity of 1 s/step. SEM images were taken with a Low Vacuum Scanning Electron Microscopy JEOL model JSM-6460LV, operating at voltages of 20 kV and magnifications up to ×40,000. Mössbauer spectra were taken with a Wissel Spectrometer model MR260 with ^{57}Co (Rh) radioactive source, 512 channels and triangular velocity waveform. Magnetization measurements at 300 K were taken with a Quantum Design MPSM series XL SQUID magnetometer.

Table 1 Weight percentage of copper found in the samples by atomic absorption spectroscopy

Sample	Measured copper concentration (%wt)	Expected copper concentration (%wt)
mag0cu	Non doped	Non doped
mag5cu	1.14 ± 0.02	1.62
mag10cu	2.32 ± 0.04	3.24
mag20cu	4.78 ± 0.09	6.48

3 Results and discussion

3.1 Atomic absorption analyses

The Table 1 shows the weight percentages of copper measured in the different samples, as well as the expected copper concentrations calculated from the initial solutions prepared. According to the doping percentage, the samples were labeled: mag0cu, mag5cu, mag10cu and mag20cu for $x = 0$, 5, 10 and 20 atm.%-Cu^{2+} respectively.

It can be observed that the copper concentration measured in the samples increases as the copper content increases in the synthesis process, being the weight percentage of copper close to 70% of the initial copper present in the synthesis of each doped sample.

3.2 XRD measurements

X-ray diffraction patterns of the samples are presented in Fig. 1. Only peaks of magnetite were found in all cases. The Table 2 shows the structural and spectral parameters of the samples obtained by Rietveld analysis [8]. Lattice parameter of the samples decreases as the Cu^{2+} content increases, this behavior can be seen from the shifting of the principal peak **($2\theta_{311}$)** to higher angles. Due to the ionic radius of Cu^{2+} (0.69 Å) is smaller than that of Fe^{2+} (0.76 Å) a contraction of the crystalline lattice is expected as a consequence of Fe^{2+} by Cu^{2+} substitution. The mean crystallite diameter (MCD) decreases as Cu^{2+} content increases, which is consistent with the increasing value of the FWHM parameter for the principal peak, both tendencies show that the presence of Cu^{2+} in the solution also interferes with the growth of magnetite crystals.

3.3 SEM measurements

SEM analyses of the samples are shown in Fig. 2 and Table 3. The sample mag0cu consists of nearly cubic particles of well defined planes and edges, with a particle size distribution ranging from 100 to 200 nm. In the sample mag5cu the mean particle size decreases to 120 nm, keeping a cubic shape. In the samples mag10cu and mag20cu, particles without a definite shape are observed, whose sizes are close to 80 and 50 nm, respectively.

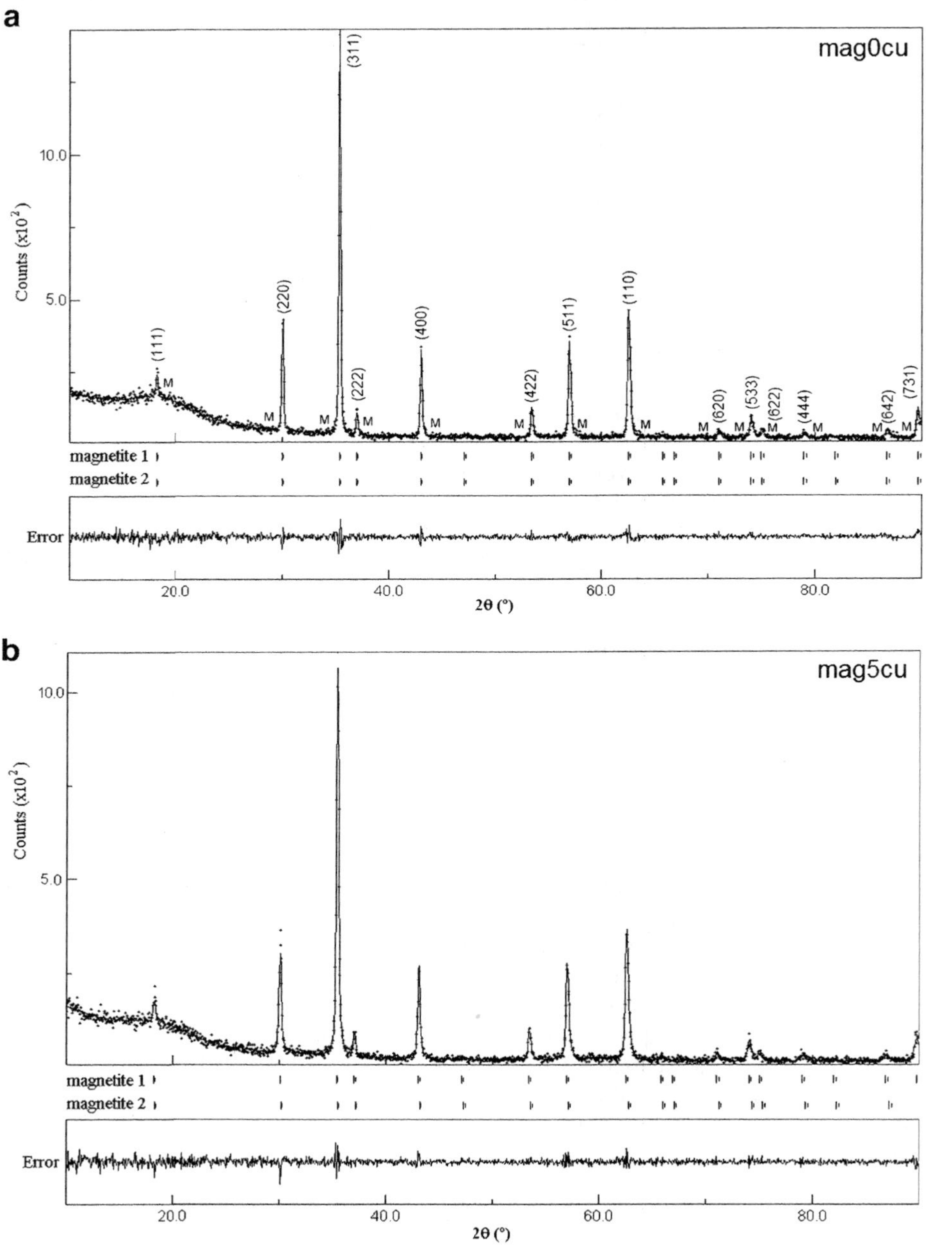

Fig. 1 X-Ray patterns of the samples

3.4 Mössbauer spectroscopy measurements

Room temperature Mössbauer spectra of the samples are presented in Fig. 3 and the hyperfine parameters obtained by least square fitting of the spectra with the

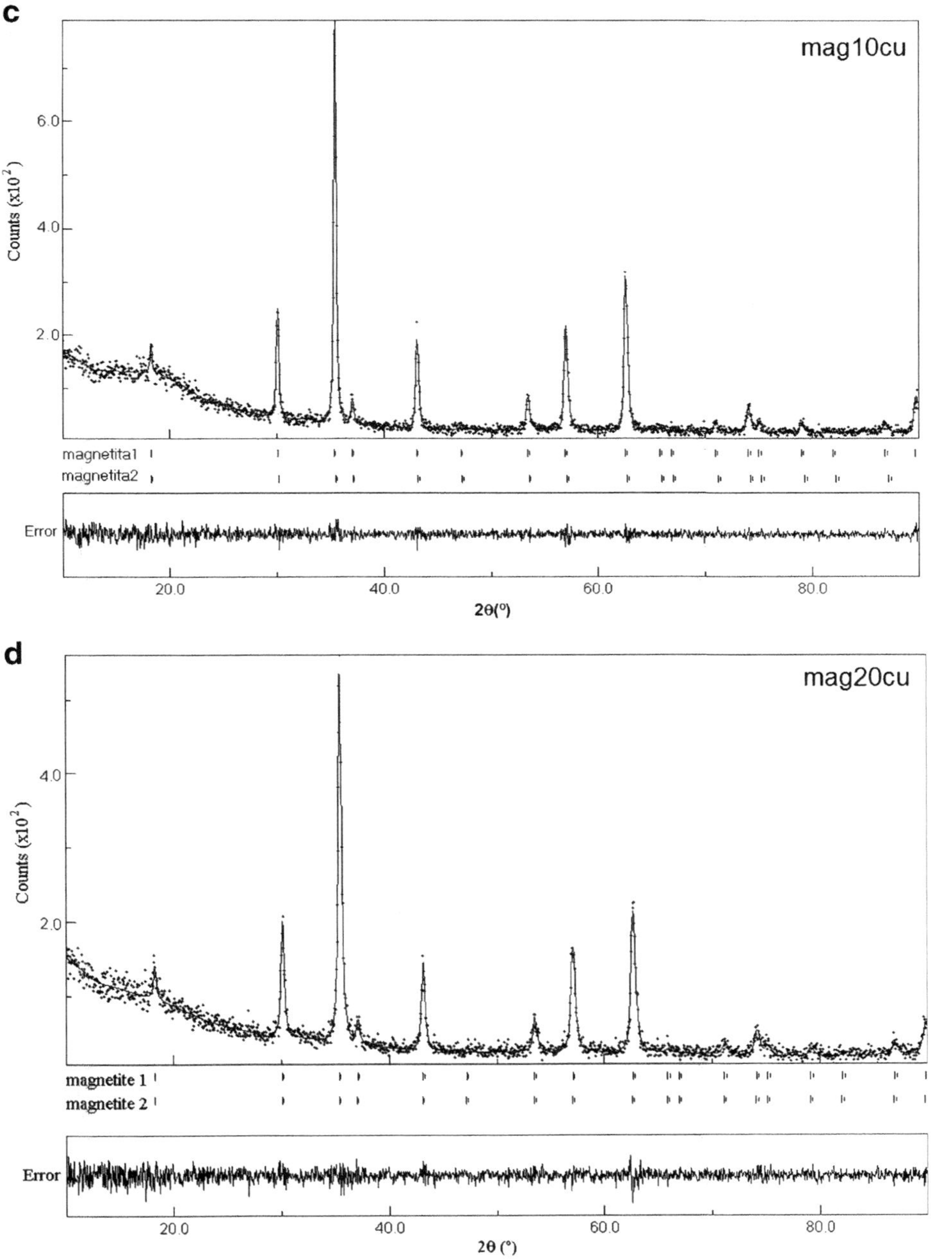

Fig. 1 (continued)

MOSF and DIST3EN software [9] are presented in Table 4. Due to the broad and asymmetrical character of the spectral lines, distributions of hyperfine magnetic fields of each spectrum were obtained first, in order to find the most probable fields

Table 2 Structural and spectral parameters of the samples obtained by Rietveld analysis

Sample	a (Å)	MCD (nm)	$2\theta_{311} \pm 0.025$	FWHM(°) $\pm$ 0.025
mag0cu	8.3852 (6)	110	35.496	0.068
mag5cu	8.3806 (6)	80	35.521	0.101
mag10cu	8.3793 (6)	80	35.550	0.118
mag20cu	8.3734 (6)	35	35.561	0.176

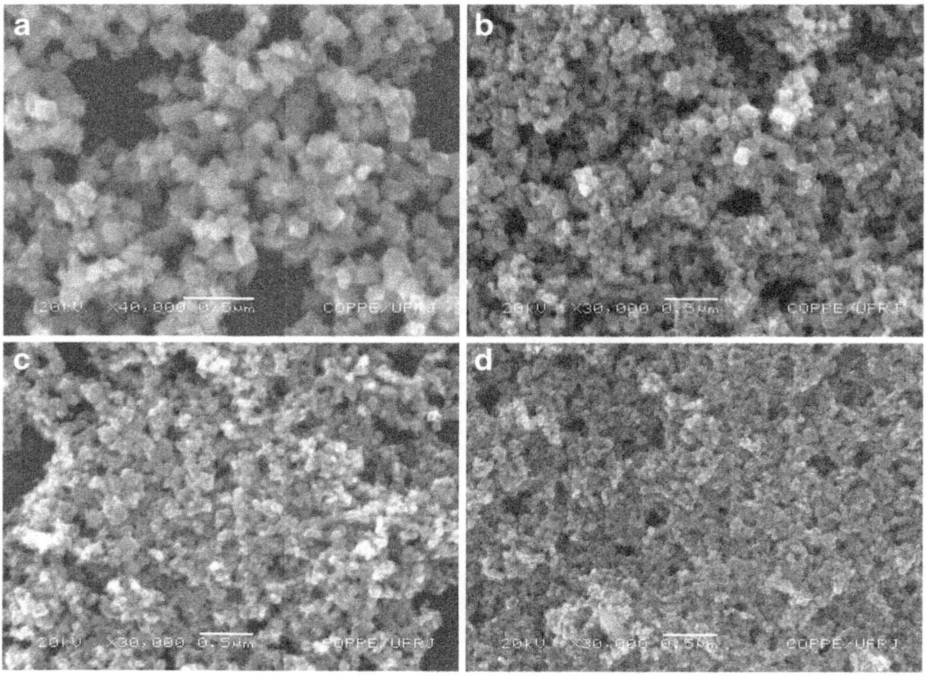

Fig. 2 SEM images of the samples

Table 3 Particle size estimated from SEM images

Sample	Particle size (nm)
mag0cu	200
mag5cu	120
mag10cu	80
mag20cu	50

of each distribution. Hyperfine magnetic field distributions are not presented here for sake of brevity.

Magnetic hyperfine field distributions of the samples mag5cu, mag10cu and mag20cu show one intermediate component of field around 45.4, 44.6 and 43.2 T, respectively. These intermediate components can be attributed to partial substitution of Fe^{2+} by Cu^{2+} at octahedral sites, due to the magnetic moment of Fe^{2+} is 4 μ_B while the magnetic moment of Cu^{2+} is 1 μ_B. A field component near to 45.8 T is present in all samples, this value is close to the hyperfine magnetic field of 46.0 T at octahedral

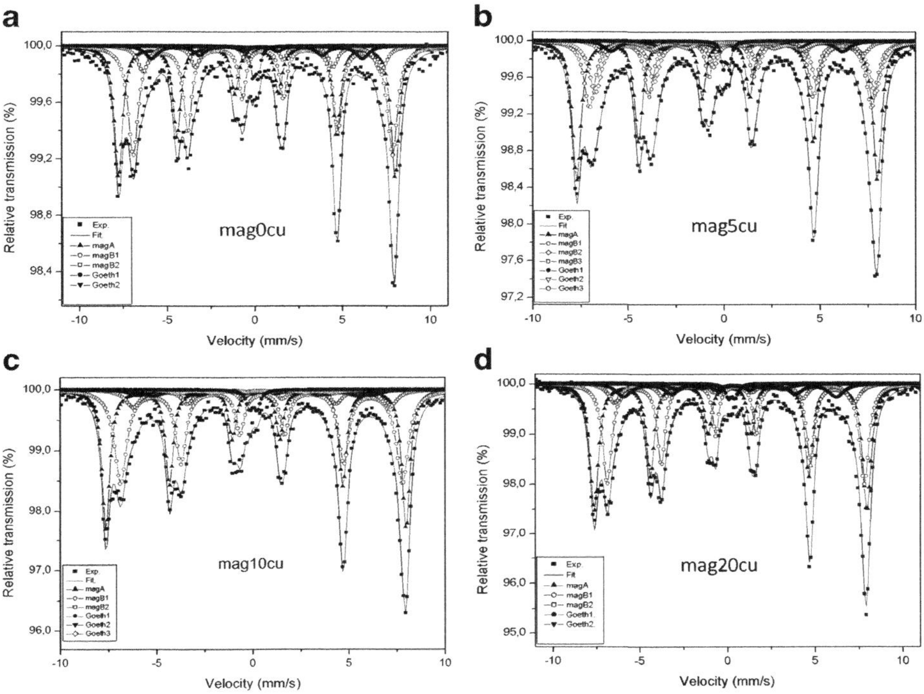

Fig. 3 Room temperature Mössbauer spectra of the samples

Table 4 Mössbauer parameters of the samples

Sample	B_{hfA} (± 0.2 T)	B_{hfB} (± 0.2 T)	IS_A (± 0.02 mm/s)	IS_B (± 0.02 mm/s)	QS_A (± 0.02 mm/s)	QS_B (± 0.02 mm/s)	$R=A(Fe^{2.5+})/A(Fe^{3+})$
mag0cu	48.8	45.8	0.23	0.57	0.00	0.00	1.31 ± 0.04
		42.1		0.60		−0.02	
mag5cu	48.7	45.7	0.24	0.47	−0.02	−0.06	1.25 ± 0.04
		45.4		0.70		−0.05	
		42.2		0.50		−0.03	
mag10cu	48.4	45.6	0.28	0.59	−0.01	−0.03	1.15 ± 0.03
		44.6		0.58		−0.04	
		41.4		0.56		−0.03	
mag20cu	48.4	45.5	0.25	0.58	0.01	−0.04	1.08 ± 0.03
		43.2		0.55		−0.02	
		42.4		0.47		−0.03	

Subscripts A and B represent tetrahedral and octahedral sites of magnetite, respectively

B_{hf} hyperfine magnetic field, *IS* isomer shift relative to α-Fe, *QS* quadrupole splitting, *R* area ratios between octahedral sub spectra and tetrahedral sub spectra

sites of a stoichiometric magnetite [2], which indicates that our samples also keep crystalline sites with properties similar to those of stoichiometric magnetite. A low field component, close to 42 T appears in all spectra, indicating that this contribution does not depend of the Cu^{2+} content used in the doping process, but it is due to the presence of vacancies or excess of Fe^{3+} in the samples. This component can be

Table 5 Mössbauer parameters of magnetic goethite impurity found in the samples

Sample	B_{hf} (±0.2 T)	IS (±0.02 mm/s)	QS (±0.02 mm/s)	W (±0.02 mm/s)	ΔW (±0.02 mm/s)	A (%)
mag0cu	37.8	0.36	−0.22	0.75	0.11	6.00 ± 0.12
mag5cu	37.5	0.36	−0.22	0.75	0.20	5.00 ± 0.10
mag10cu	37.8	0.36	−0.22	0.74	0.22	4.00 ± 0.12
mag20cu	37.7	0.36	−0.22	0.75	0.13	8.00 ± 0.16

Isomer shifts are relative to α-Fe

Table 6 Mössbauer parameters of the doublet attributed to superparamagnetic goethite impurity found in all samples

Sample	IS (±0.02 mm/s)	QS (±0.02 mm/s)	W (±0.02 mm/s)	A (%)
mag0cu	0.16	0.31	0.27	2.40 ± 0.05
mag5cu	0.23	0.32	0.26	1.90 ± 0.04
mag10cu	0.15	0.40	0.40	1.60 ± 0.03
mag20cu	0.36	0.38	0.40	1.40 ± 0.03

Isomer shifts are relative to α-Fe

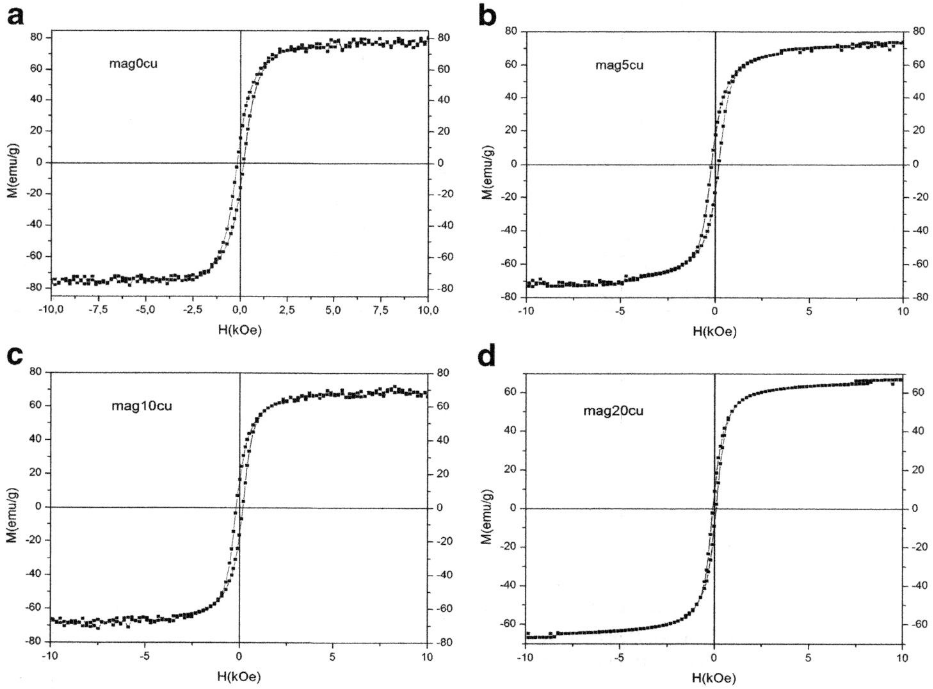

Fig. 4 Hysteresis loops of the samples

attributed to the fact that synthetic magnetite presents a high oxidation tendency, which means it can present an excess of Fe^{3+} at octahedral sites and vacancies possibly at both octahedral and tetrahedral sites. The unbalanced distribution of Fe^{2+}

Table 7 Hysteresis parameters of the samples

Sample	M_{sat}(emu/g)	M_{rem}(emu/g)	H_c(Oe)
mag0cu	77.5 ± 0.2	16.2 ± 0.2	179 ± 2
mag5cu	73.5 ± 0.1	14.4 ± 0.2	169 ± 2
mag10cu	69.7 ± 0.1	14.3 ± 0.2	164 ± 2
mag20cu	67.1 ± 0.1	7.0 ± 0.1	73 ± 2

and Fe^{3+} ions at octahedral sites leads to local atomic distributions where the Fe^{3+} ions are surrounded by different number of Fe^{2+}, Fe^{3+} ions and vacancies, producing different magnetic fields at these sites. Diamandescu [10] also reported three main peaks at octahedral sites of a magnetite with 0.342 atm.% Cu^{2+}.

All Mössbauer spectra include one sextet and one doublet, which were attributed to a small percentage of magnetic and superparamagnetic goethite, respectively. The spectral area of these two phases does not exceeded 10%. Goethite was detected in the infrared spectra of all samples by the presence of characteristic bands located at 800 and 900 cm^{-1}. The infrared spectra are not presented here for sake of brevity. This phase is poorly crystalline due to it could not be observed in the X-ray patterns. The hyperfine parameters of the goethite phase are presented in Tables 5 and 6.

3.5 Magnetization measurements

Magnetization analyses obtained by SQUID magnetometry are shown in Fig. 4 and Table 7. The samples presented hysteresis loops characteristic of multidomain systems [11]. The saturation magnetization (M_{sat}) decreases as the Cu^{2+} content increases, which is consistent with Fe^{2+} by Cu^{2+} substitution, because the magnetic moment of Cu^{2+} is 1 μ_B while that of Fe^{2+} is 4 μ_B.

4 Conclusions

We synthesized magnetite pure and magnetite in presence of Cu^{2+}. The magnetic and structural properties of the samples were changed by the presence of Cu^{2+} in the synthesis process. A weight percentage of copper, close to 70% of the value used in starting solutions was found in the final products. Both lattice parameter and particle size decreased as Cu^{2+} content increases, which is consistent with incorporation of Cu^{2+} within the structure of magnetite, vacancy generation at crystalline sites promoted by Cu^{2+} or both simultaneous effects. From our analyses we may not know what effect is the dominant one. Morphology of the particles was also degraded, starting from cubic shape in non doped and 5 atm.%-Cu^{2+} samples to undefined shape for 10 and 20 atm.%-Cu^{2+} doped samples. Magnetic hyperfine interactions measured by Mössbauer spectroscopy showed that Cu^{2+} has an appreciable effect on the magnetic hyperfine field of the octahedral sites, creating magnetic fields lower than those of a stoichiometric magnetite, as well as broad and asymmetric lines in the octahedral sub spectra, which is consistent with substitution of Fe^{2+} by Cu^{2+}, generation of vacancies or both simultaneous effects. Saturation magnetization and coercivity also were reduced due to the lower magnetic moment of Cu^{2+} with respect to Fe^{2+}. Our study may give important elements to the understanding of the changes in magnetic and structural properties of magnetite doped with metallic cations, which can be used to synthesize magnetite samples useful for industrial and technological applications.

References

1. Dorigueto, A.C., Fernandes, N.G., Persiano, A.L.C., Greneche, J.M.: Characterization of a natural magnetite. Phys. Chem. Miner. **30**, 250 (2003)
2. Vandenberghe, R.E.: Mössbauer spectroscopy and applications in geology. International training centre for post-graduate soil scientists, p. 25. State University Gent, Belgium (1990)
3. García, K.E., Morales, A.L., Barrero, C.A., Greneche, J.M. : New contributions to the understanding of rust layer formation in steels exposed to a total immersion test. Corrosion Sci. **48**, 2813–2830 (2006)
4. Ishikawa, T., Nakazaki, H., Yasukawa, A., Kandori, K., Seto, M.: Influences of Co^{2+}, Cu^{2+} and Cr^{3+} ions on the formation of magnetite. Corrosion Sci. **41**, 1670 (1999)
5. Pourroy, G., Valles-Minguez, A., Dintzer, T., Richard-Plouet, M.: Synthesis of Cu°/Cu(II) substituted magnetite nanocomposites in aqueous media. J. Alloys Compd. **327**, 267–269 (2001)
6. Cvejic, Z., Antic, B., Kremenovic, A., Rakic, S., Goya, G.F., Rechenberg, H.R., Jovalekic, C., Spasojevic, V.: Influence of heavy rare earth ions substitution on microstructure and magnetism of nanocrystalline magnetite. J. Alloy. Comp. **472**, 571–575 (2009)
7. Cornell, R.M., Schwertmann, U.: The iron oxides, pp. 130, 135. VCH mbH, Weinheim, Germany (1996)
8. Lutterotti, L.: Maud-Materials Analysis Using Diffraction-tutorial. http://www.ing.unitn.it/~maud/tutorial.html. Accessed 10 February 2010.
9. Vandenberghe, R., De Grave, E., De Bakker, P.M.A.: On the methodology of the analysis of Mössbauer spectra. Hyperfine Interact. **83**, 29–49 (1994)
10. Diamandescu, L., Tarabasanu-Mihaila, D., Teodorescu, V., Popescu-Pogrion, N.: Hydrothermal synthesis and structural characterization of some substituted magnetites. Mater. Lett. **37**, 340–348 (1998)
11. Khollam, Y.B., Dhage, S.R., Potdar, H.S., Deshpande, S.B., Bakare, P.P., Kulkarni, S.D., Date, S.K.: Microwave hydrothermal preparation of submicron-sized spherical magnetite (Fe_3O_4) powders. Mater. Lett. **56**, 571–577 (2002)

Hyperfine Interact (2011) 203:85–91
DOI 10.1007/s10751-011-0369-6

Synthesis of magnetic nanoparticles: effects of polyelectrolyte concentration and pH

J. P. Urquijo · Herley Casanova · Javier Garces · Alvaro L. Morales

Published online: 25 August 2011

Abstract This study refers to the effect of sodium polyacrylate concentration (1 to 5 mass %) and pH (10 to 12) on the synthesis of magnetic nanoparticles (magnetite–maghemite) and their characterization by Mössbauer spectroscopy. The magnetic particles were obtained by coprecipitation method using iron chloride (II) and iron chloride (III) as precursor reagents and sodium polyacrylate as stabilizing agent. All samples showed Mössbauer broad resonance lines in typical doublet and sextets patterns of magnetite or maghemite with corresponding wide particle size distributions. The stability of magnetic particles was carried out by measuring particle sizes with dynamic light scattering (DLS). The z-average values for magnetic particles were in the range 24 to 590 nm and no significant change in size was observed on aging by leaving this material in air for 20 days. X-ray diffraction patterns showed characteristic peaks of the spinel structure and have an increase in their broadening as the pH decreases, effect that is dominated by the decrease in crystallite sizes. The nanoparticles showed to be magnetic at pH 12 and at room temperature.

Keywords Mössbauer spectroscopy · Magnetic particles · Superparamagnetic · Nanoparticles · Dynamic light scattering

1 Introduction

Magnetic nanoparticles have attracted the interest of worldwide research during the last decade due to their magnetic and physicochemical properties, which vary from those of the same material in volume, with potential use for surface catalysis, clinical therapy and elaboration of microelectronic devices [1–3].

The magnetite nanoparticles can be obtained from aqueous solutions of iron salts. The size, magnetic properties and surface morphology of obtained particles depend

J. P. Urquijo (✉) · H. Casanova · J. Garces · A. L. Morales
Universidad de Antioquia, St 67 53-108 Medellín, Colombia
e-mail: jurquijo@fisica.udea.edu.co

on synthesis conditions. Among the most important variables to consider in aqueous media synthesis is the pH of reaction that defines the final phase of the material. In the literature, synthetic methods have been reported concerning the control of pH and ionic strength of the aqueous medium. Andrade and coworkers [4], reported the effect of pH in the range between 8.0 and 10.6 in the synthesis of magnetite nanoparticles by a reduction-precipitation method from an aqueous Fe^{3+} solution. They found that at pH between 9.7 and 10.6 magnetite and maghemite are the main products. Vayssieres and coworkers [5] controlled the particle size of magnetite particles obtained by coprecipitation chemical method, adjusting the pH and the ionic strength in the reaction medium, observing that the higher the pH and the ionic strength are, the smaller particle size was obtained. The size and stability of recent born nanoparticles is mainly determined by particle surface charge density (i.e. zeta potential), which is influenced by the pH due to protonation-deprotonation equilibrium of the surface hydroxylated groups. In the magnetite case, for pH above 8, the particle surface is negatively charged. At a given pH, the surface charge increases with the ionic strength, increasing the electrostatic repulsion between the particles, preventing their aggregation [6].

Other aspect to consider in the synthesis of magnetic nanoparticles is their stability against aggregation processes that change their magnetic properties. During the synthesis, the magnetic nanoparticles tend to aggregate themselves due to dipole-dipole magnetic or van der Waals interactions, leading to the formation of clusters that inhibit the desirable superparamagnetic behavior for some applications such as magnetic fluids [7–9]. The presence of a polymer in the reaction medium during the particle formation could control the particles size. The polymer adsorbed on the surface of the particles prevents their aggregation due to osmostic effects that induce a steric stabilization mechanism on particles. Moreover, the use of a polyelectrolyte gives a net charge to the particle surface, causing electrostatic repulsion and preventing their aggregation. By varying the polymer concentration it is possible to obtain a material with different particle sizes allowing it to act as polymer template [8, 10, 11].

Mössbauer spectroscopy allows studying the effect of pH and polymer presence in the magnetic and structural properties of magnetite nanoparticles. Andrade and coworkers [4] used room temperature Mössbauer spectroscopy to analyze the magnetite nanoparticles obtained by reduction of Fe^{3+}. They found that the spectra showed broad lines and the resonant lines from octahedral and tetrahedral sites of magnetite are not differentiable. This is probably due to superparamagnetic relaxation effects and collective magnetic excitation in the small particles.

To study the effects of polymers and surfactants in magnetic nanoparticles properties Kim and coworkers [12] obtained superparamagnetic iron oxides nanoparticles by coprecipitation chemical method in the presence of sodium oleate, starch and methoxypoly(ethylenglicol) (MEPG). The Mössbauer spectra of starch-magnetite samples showed superparamagnetic behavior, on the other hand, sodium oleate and MPEG showed broad doublet and sextet lines, indicating the presence of ferrimagnetic and superparamagnetic components. Similar results were reported by Pardoe and coworkers [13] for the synthesis of magnetite/maghemite nanoparticles in the presence of dextran and polyvinyl alcohol. The Mössbauer spectra of the samples presented two components, a sextet, indicating a magnetic component and a doublet, indicating a superparamagnetic component. The magnetic component in the sample iron oxide-dextran is larger than other samples, while the superparamagnetic characteristic is more important in the iron oxide-PVA sample. Other important factor in the magnetic properties of nanoparticles is their separation due the polymer presence. This effect was evaluated by Novacova and coworkers [14]. They found that the oxide concentration greatly influences the magnetic properties of magnetite-PVA composites. This can be seen in the Mössbauer spectra of the samples where a higher oxide volume concentration results in collapsed sextets. This is due to the magnetic interactions between the particles, indicating a higher magnetic component in samples where the oxide concentration is higher than polymer concentration.

In this work we study the preparation of magnetite nanoparticles obtained by a coprecipitation method and we determine the combined effect of pH and concentration of polyacrylate on some physical properties, which has not been considered by other researchers.

2 Experimental

Magnetite particles were prepared by chemical coprecipitation of Fe^{3+} and Fe^{2+} solutions in the presence of sodium polyacrylate following the procedure proposed by Lin and coworkers [8]. A volume of 80 mL of an aqueous mix solution of 0.06 M iron (II) chloride tetrahydrated and 0.12 M iron (III) chloride hexahydrated was drop-added to 100.0 mL of a sodium polyacrylate solution. Solutions were de-aired by purging them with $N_{2(g)}$ for 1 hour to prevent any oxidation of soluble Fe^{2+} during precipitation or post-synthesis oxidation of magnetite particles. Three concentration (1, 3, 5 mass%) and two pH (10 and 12) of the sodium polyacrylate (SPA) solution were evaluated for the synthesis of magnetic particles. The pH was adjusted by successive additions of 1.0 M NaOH solution to the reaction media. The obtained product was washed, dried and dry-stored under a $N_{2(g)}$ atmosphere. All chemicals used for the synthesis were analytical grade, with the exception of sodium poliacrylate which was of commercial grade.

Samples were characterized by Mössbauer spectroscopy, X-Ray diffraction and dynamic Light scattering. A Wissel Mössbauer spectrometer with a MR-260 transducer and a $^{57}Co(Rh)$ source was used. X-Ray spectra were obtained with a Rigaku 2200 spectrometer, using Cu–K_{α} radiation. Pass counting time was 1 s with an angular step of 0.02° 2θ. For dynamic light scattering measurements a Nanozeta S (Malvern) was employed. Mössbauer spectra were fitted with Dist3E software [15]. The PowderX [16] software was used to refine sample diffractograms.

3 Results and discussions

Figure 1 shows room temperature Mössbauer spectra of samples obtained by coprecipitation of iron salts at pH 10 (upper part of the figure) and pH 12(lower part of the figure). The spectrum of sample with 1 mass% SPA and pH 12 (Fig. 1d) is formed by broad sextets and a small doublet, evidencing strong relaxation effects. This might indicate the existence of small particles in the sample. As the concentration of the polymer increases, the relaxation effects increase and sextets collapsed into superparamagnetic patterns. The doublet also increases in area, indicating an increase in the superparamagnetic component in this sample. The system prepared at pH 10

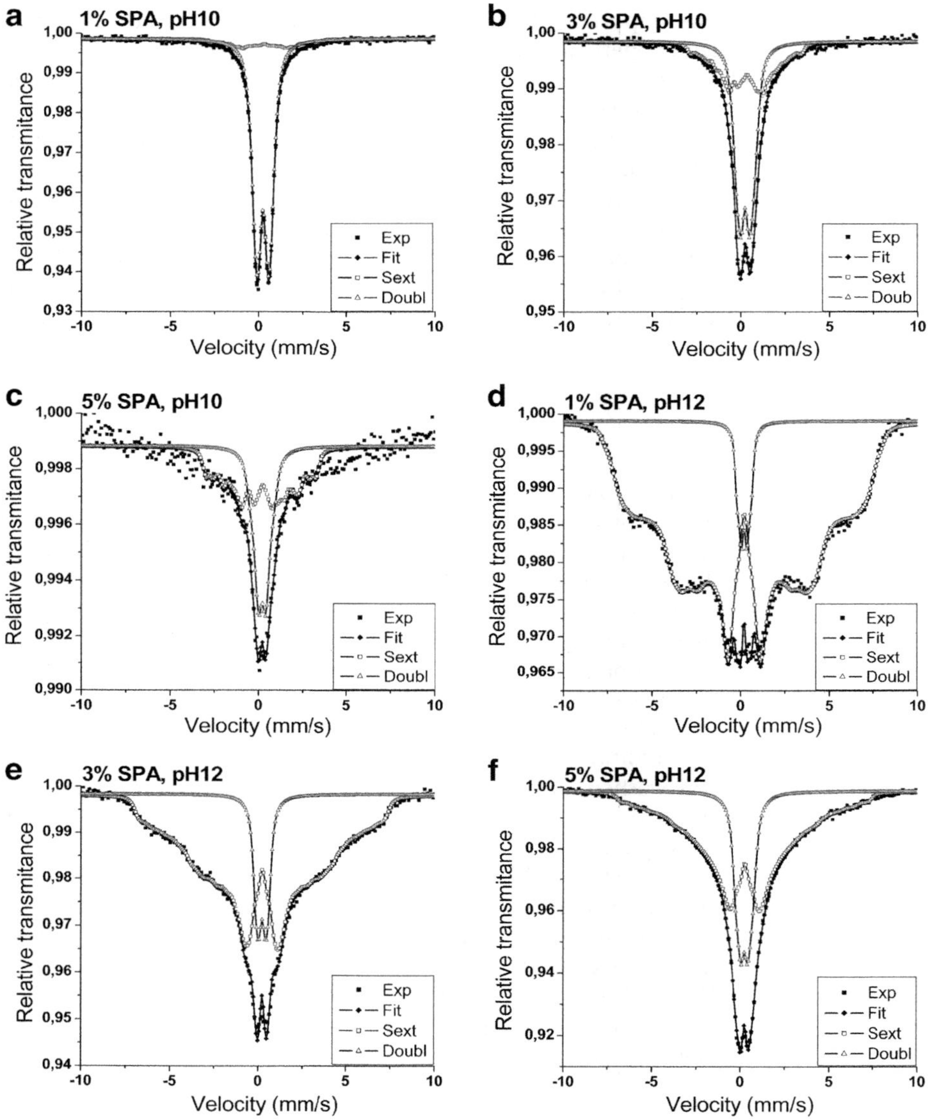

Fig. 1 Room temperature Mössbauer spectra for the samples obtained at; **a** pH = 10, 1% SPA; **b** pH = 10, 3% SPA; **c** pH = 10, 5% SPA; **d** pH = 12, 1% SPA; **e** pH = 12, 3% SPA; **f** pH = 12, 5% SPA

showed a further decrease in the magnetic spectral component probably associated to the presence of nanoparticles. However, the sample prepared with 5 mass% SPA at pH 10, there is a presence of a magnetic component probably due to aggregation of nanoparticles associated to the inefficient agitation during synthesis.

Figure 2 shows the X-ray patterns for samples obtained at pH 12. Reflection peaks present in all samples correspond to the spinel structure as reported in the literature

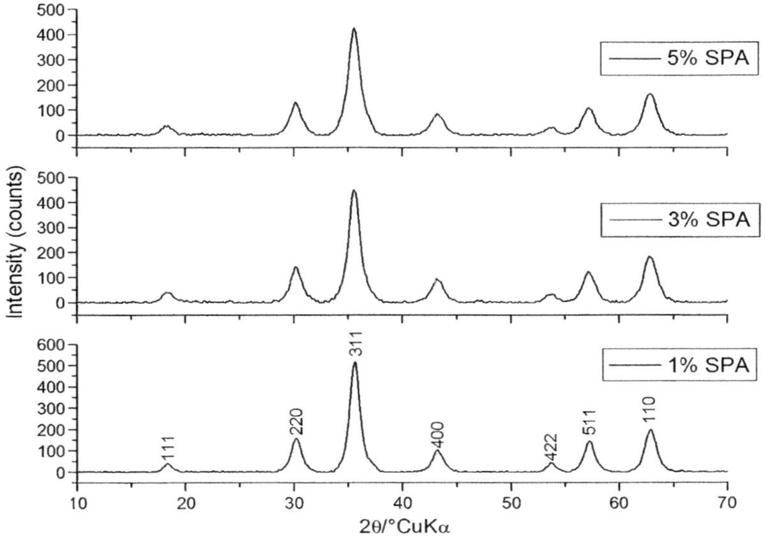

Fig. 2 X-ray patterns of the samples obtained at pH 12

Table 1 Average size of crystallite obtained by Scherrer equation for pH 12 samples

SPA concentration in sample (mass %)	Crystallite average size (nm)
1	10
3	9
5	8

[17]. It is expected that these samples are formed by a mixture of magnetite and maghemite and is not directly assessable the separation of contributions from these two cubic phases from powder XRD data. As the polymer concentration increases, the broadening of reflection peaks increases, due to a decrease of crystallite sizes.

The average crystallite sizes were estimated using Scherrer equation [18]. To find the average size of crystallite, the most intense peak was chosen. For our samples such peak was found at approximately $2\theta = 35.4°$, which correspond to the (311) plane of the cubic structure. Estimated crystallite sizes for samples are shown in Table 1.

Figure 3 shows results obtained by dynamic light scattering. The intensity particle size distribution of samples shows systems with high polydispersity at pH 12 (5% SPA), pH 10 (1% SPA) and pH 10 (3% SPA). Samples at pH 12 show a decrease of the z-average value with increasing concentration from 1 to 3 mass% SPA (Table 2). The sample obtained with 5%mass SPA and pH 12 showed a higher z-average than samples at lower SPA concentration, which is probably associated to the increase in viscosity that reduces the SPA adsorption velocity on new born particles, therefore particle aggregation is promoted. The samples obtained at pH 10 show a different tendency. For systems at 1 and 3 mass% SPA a broad particle size distribution was observed that indicates the presence of aggregated particles, which could be

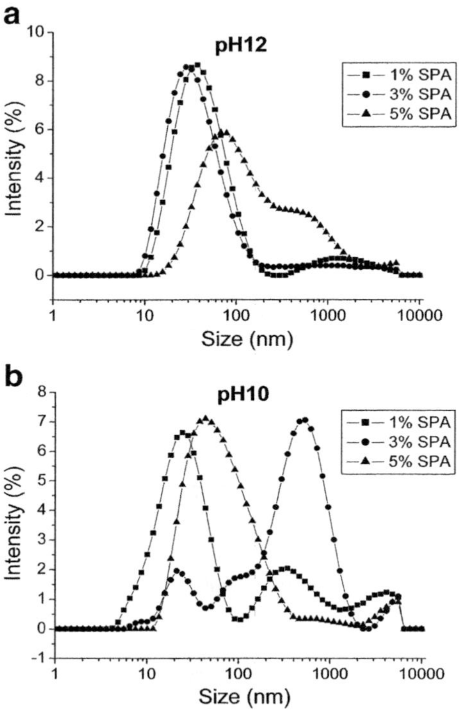

Fig. 3 Particle size distribution of samples at **a** pH 12; **b** pH 10

Table 2 Particle size as z-average for magnetic particles

SPA concentration (%mass)	pH 10	pH 12
1	24	37
3	590	29
5	43	72

explained in terms of stabilizer molecule sharing between particles producing bridging flocculation. At 5 mass% SPA there is enough stabilizing molecules to avoid the aggregation process and a system with a z-average value of 43 nm was produced.

4 Conclusions

In this study we obtained magnetite-maghemite nanoparticles using sodium polyacrylate in the reaction medium as stabilizing molecule. An increase in SPA concentration reduced the size of the crystallite at a constant pH. Samples obtained at pH 12 have a magnetically order structure at room temperature as determined by Mössbauer spectroscopy. Particles obtained at pH 10 have superparamagnetic characteristics probably due to the stabilizing effect of the SPA that produced smaller particles than at pH 12. Particle size analysis by dynamic light scattering showed that at pH 12, the hydrodynamic radius of the particles tends to decrease with increasing SPA concentration, except for the sample at pH 12 and 5 mass% SPA due to an increase in the system viscosity. At pH 10, the particles seem to be aggregated by bridging floculation because there is not enough polymer available to stabilize the newly born particles. Only when polymer was at high concentration (i.e. 5 mass%) particles did not aggregate to produce a nanoparticle system with a z-average size of 43 nm.

References

1. Boyer, C., Whittaker, M., Bulmus, V., Liu, J., Davis, T.: NPG Asia Mater. **2**, 23–30 (2010)
2. Teja, A.S., Koh, P.: Prog. Cryst. Growth Charact. Mater. **55**, 22–45 (2009)
3. López-López, M.T., Durán, J., Delgado, A., Gonzáles-Caballero F.: J. Colloid Interface Sci. **291**, 144–151 (2005)
4. Andrade, Â.L., Sousa, D., Pereira, M., Fabris, J., Dominguez, R.: Quim. Nova. **33**(3), 524–527 (2010)
5. Vayssierès, L., Chanéac, C., Tronc, E., Jolivet, J.P.: J. Colloid Interface Sci. **205**, 205–212 (1998)
6. Jolivet, J.P.: Metal Oxide Chemistry and Synthesis, p. 321. Wiley, New York (2000)
7. Di Marco, M., Guielbert, I., Port, M., Robic, C., Couvreur, P., Dubernet, C.: Int. J. Pharm. **331**, 197–203 (2007)
8. Lin, C, Lee, C., Chiu, W.: J. Colloid Interface Sci. **291**, 411–420 (2005)
9. Sun, Y., Duan, L., Guo, Z., DuanMu, Y., Ma, M., Xu, L., Zhang, Y., Gu, N.: J. Magn. Magn. Mater. **285**, 65–70 (2005)
10. Liu, S., Wei, X., Chu, M., Peng, J., Xu, Y.: Colloids Surf., B Biointerfaces **51**, 101–106 (2006)
11. Feng, B., Hong, R.Y., Wu, Y.J., Liu, G.H., Zhong L.H., Zheng, Y., Ding, J.M., Wei, O.G.: J. Alloys Compd. **473**(1–2), 356–362 (2009)
12. Kim, D.K., Mikhaylova, M., Zhang, Y., Muhammed, M.: Chem. Mater. **15**, 1617–1627 (2003)
13. Pardoe, H.A., Smirnov, E., Gendler T.: J. Magn. Magn. Mater. **300**, 41–46 (2006)
14. Novacova, A.A., Lanchinskaya, V.Y., Volkov, A.V., Gendler, T.S., Kiseleva, T.Y., Moskivina, M.A., Zesin, S.B.: J. Magn. Magn. Mater. **258–259**, 354–357 (2003)
15. Vandenberghe, R., De Grave, E., De Bakker, P.M.: Hyperfine Interact. **83**, 29–49 (1994)
16. Dong, C.: Institute of Physics, Chinese Academy of Sciences. CAD, Beijing. http://www.ccp14.ac.uk
17. JCPDS: X-Ray Diffraction Data Cards of the Joint Committee on Powder Diffraction Standards 196 (2001)
18. Klug, H.P., Alexander, L.E.: X ray Diffraction Procedures for Polycrystalline and Amorphous Materials, 2nd edn, p. 656. Wiley, New York (1974)

Hyperfine Interact (2011) 203:93–97
DOI 10.1007/s10751-011-0350-4

Mössbauer and XRD study of the $Fe_{65}Si_{35}$ alloy obtained by mechanical alloying

G. Y. Vélez · R. R. Rodríguez · C. A. Melo · G. A. Pérez Alcázar · Ligia E. Zamora · J. A. Tabares

Published online: 30 August 2011

Abstract A study was made on the alloy $Fe_{65}Si_{35}$ using x-ray diffraction and Mössbauer spectrometry. The alloy was obtained by mechanical alloying in a high energy planetary mill, with milling times of 15, 30, 50, 75 and 100 h. The results show that in the alloys two structural phases are present, a Fe-Si BCC disordered phase and ferromagnetic, and a Fe-Si SC phase, whose nature is paramagnetic and which decreases with milling time. In the temporal evolution of the milling two stages are differentiated: one between 15 and 75 h of milling, in which silicon atoms diffuse into the bcc matrix of iron and its effect is to reduce the hyperfine magnetic field; the other, after 75 h of milling, where the alloy is consolidated, the effect of the milling is only to increase the disorder of the system, increasing the magnetic order.

Keywords Fe-Si alloys · Mechanical alloying · X-rays diffraction · Mössbauer spectrometry

1 Introduction

Melted Fe-Si alloys have magnetic character, with a combination of both low magnetostriction and high saturation magnetization, features that make them very suitable for the manufacture of materials. However, melted alloys with high Si content are very brittle. Mechanical alloying (MA) allows obtaining nanocrystalline materials in a wide range of compositions [1] with the advantage that brittleness plays no

G. Y. Vélez (✉) · C. A. Melo · G. A. Pérez Alcázar · L. E. Zamora · J. A. Tabares
Departamento de Física, Universidad del Valle, A.A. 25360, Cali, Colombia
e-mail: giovelez29@gmail.com

R. R. Rodríguez
Corporación Universidad Autónoma de Occidente, Cali, Colombia

Fig. 1 XRD patterns of the $Fe_{65}Si_{35}$ alloy obtained at different milling times

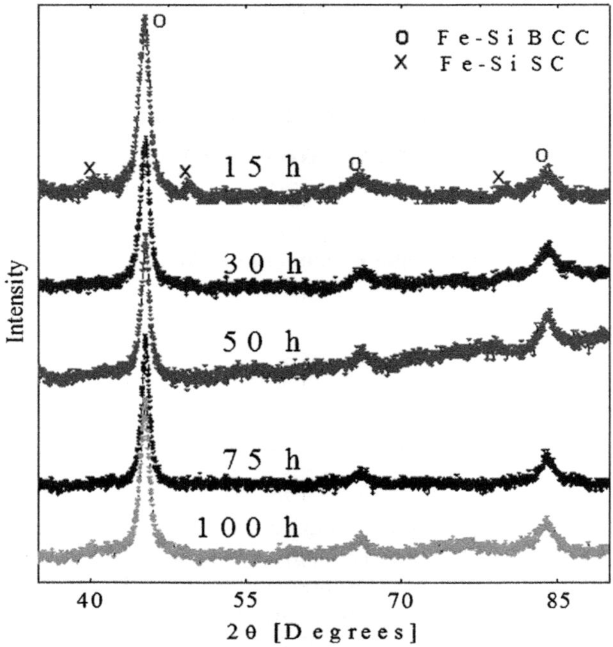

role whatever the silicon content is. Ball-milling techniques have become so popular because of their comparatively low cost, great flexibility in the selection of the processing parameters and large variety of attainable products. MA of intermetallic phases is a field of interest for applied and fundamental research. Disordering and grain refinement, typical modifications brought about in the materials during milling, are potentially interesting for overcoming the notorious brittleness of many intermetallic ordered phases [2]. The present work aims at the characterization of the structural and magnetic properties of $Fe_{65}Si_{35}$ (at.%) alloy powders produced by MA to characterize the relation between milling time and order-disorder effect. That study has been carried out by means of x-ray diffraction (XRD) and Mössbauer spectrometry (MS).

2 Experimental procedure

$Fe_{65}Si_{35}$ alloy was obtained by MA starting from elemental powders of iron and silicon of high purity (≥99.90%). The alloying was carried out in a high energy planetary ball mill *Fritsch-Pulverisette 5* with milling time of 15, 30, 50, 75 and 100 h. All samples were measured by XRD and MS at room temperature. The x-ray study was carried out in a X'Pert Pro diffractometer with Cu K-α radiation, and the patterns were refined with the MAUD program [3]. Mössbauer measurements were performed in a conventional spectrometer with a $^{57}Co/Rh$ source of 25 mCi and the spectra were fitted with the MOSFIT program [4].

Table 1 Structural parameters obtained from the refinement of the XRD patterns of the different samples

Milling time [hours]	Phase	$a\pm$ 0.005 [Å]	$\Phi\pm$ 30 [Å]	Wt.%
15	Fe-Si BCC	2.824	110	78
	Fe-Si SC	4.466	270	22
30	Fe-Si BCC	2.825	130	96
	Fe-Si SC	4.488	210	4
50	Fe-Si BCC	2.819	150	100
75	Fe-Si BCC	2.823	160	100
100	Fe-Si BCC	2.826	120	100

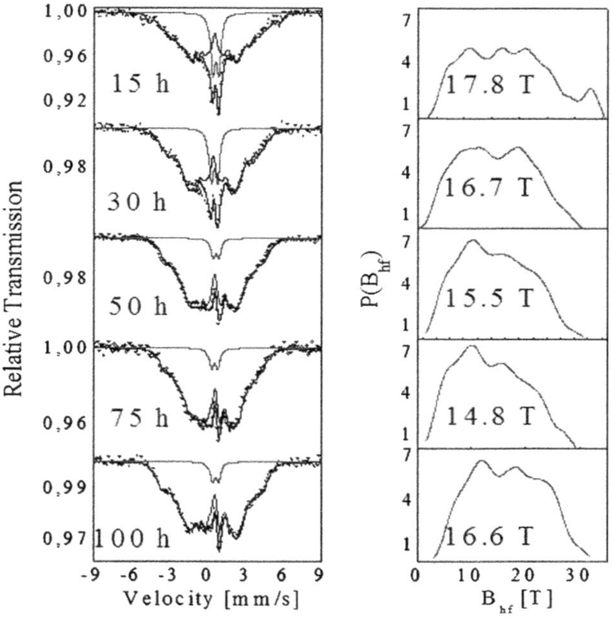

Fig. 2 Mössbauer spectra and their respective hyperfine magnetic field distributions for the $Fe_{65}Si_{35}$ alloy obtained at different milling times

3 Results and discussion

The XRD patterns of the $Fe_{65}Si_{35}$ alloy obtained for different milling times are shown in Fig. 1, the structural parameters obtained from the refinement are listed in Table 1. The result show that for 15 h of milling the Fe-Si alloy consists of two phases, one due to diffusion of Si atoms into the BCC iron matrix forming the BCC Fe-Si phase, and the other by the joint of iron and silicon atoms to form the SC Fe-Si phase. MA induces further transformations into the BCC and SC phases: from 15 to 30 h of milling the percentage of the SC phase decreases, and in contrast, the percentage of the BCC phase increases; after 50 h of milling the alloy has only the BCC Fe-Si phase. The lattice parameter *a* of the BCC phase does not change significantly with the milling time (2.825 Å). This value is less than the lattice parameter of the pure iron and confirms that the BCC phase is doped with silicon atoms [5, 6]. For all milling times, the crystallite sizes of the two phases show that the alloy is nanostructured (see Table 1), the crystallite size Φ of the SC phase decreases with the milling time while

Table 2 Mössbauer parameters obtained from the fit of the different spectra

Milling time	Distribution (Fe-Si BCC)		Doublet (Fe-Si SC)	
[hours]	$<B_{hf}> \pm 0.2$ [T]	$\delta \pm 0.01$ [mm/s]	$\Delta \pm 0.02$ [mm/s]	%
15	17.8	0.27	0.43	17
30	16.7	0.24	0.43	11
50	15.5	0.26	0.41	4
75	14.8	0.25	0.42	4
100	16.6	0.30	0.45	4

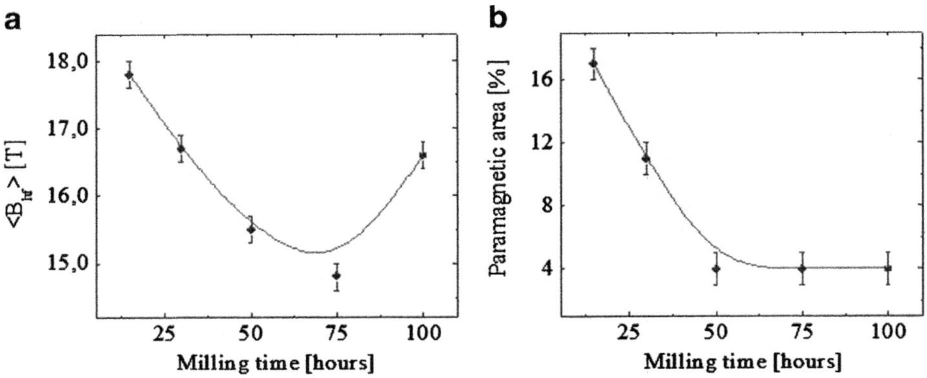

Fig. 3 **a** Mean hyperfine magnetic field and **b** percentage of paramagnetic area, both as a function of the milling time

the crystallite size of the BCC phase remains nearly constant, that means that the SC phase is more fragile than the BCC phase.

Figure 2 shows the Mössbauer spectra and their corresponding hyperfine magnetic field distributions [$P(B_{hf})$] of the $Fe_{65}Si_{35}$ alloy obtained for different milling times. The fit of all spectra was performed with two components, a $P(B_{hf})$ and a doublet. The hyperfine parameters obtained from the fit of the spectra can be seen in Table 2. The $P(B_{hf})$ is the majority component of the spectra and is associated with the Fe-Si BCC ferromagnetic phase [7]. From the $P(B_{hf})$ of the sample alloyed during 15 h is noted that it is disordered, and that some peaks, which correspond to dominant magnetic fields, including that associated with pure iron (33 T), are evident. After 30 h of milling the peaks of the $P(B_{hf})$ are broadened due to the disorder induced by the MA; it is also noted that the peaks corresponding to large magnetic fields decrease significantly, and this is mainly due to the diffusion of silicon atoms in the BCC Fe-Si matrix. The hyperfine parameters of the doublet do not change with milling time (see Table 2), these values are in agreement with those reported by Li Tie et al. in the study of the disordered FeSi alloy [8]. This doublet is due to the SC Fe-Si paramagnetic phase [9], however after 30 h milling the SC phase disappears and only the BCC remains. Then the doublet must correspond to Fe sites of the BCC phase rich in Si which have the same Mössbauer parameters as the SC phase.

By means of the $P(B_{hf})$ it was possible to establish that the MA induces both the diffusion of silicon atoms in the BCC Fe-Si matrix, and the increase of the atomic disorder of this structure. The effect of the diffusion is to reduce the magnetic order.

In contrast, the increase of the atomic disorder in binary alloys of the ferro-diluter type favors the ferromagnetism [2, 10]. Given these appreciations it is possible to interpret the behavior of the mean hyperfine magnetic field as a function of milling time shown in Fig. 3a. For the first hours of milling, diffusion and the increase of disorder are present, but the effects of diffusion are higher and mean hyperfine magnetic field decreases. The diffusion is triggered by the transformation of the SC Fe-Si phase to the BCC Fe-Si phase. In according with the XRD results this transformation ends after 30 h milling, but the diffusion of silicon atoms continue with the increase of the milling time and mean hyperfine magnetic field continues to decline. After 75 h of milling the diffusion ends and the atomic disorder is the only effect present, so the magnetic order increases.

4 Conclusions

In this work was carried out a structural and hyperfine study of a $Fe_{65}Si_{35}$ alloy obtained by mechanical alloying for different milling times. It was determined that the alloy has two structural phases, the BCC Fe-Si ferromagnetic and disordered phase, and the SC Fe-Si paramagnetic phase which decreases with the milling time. The ferromagnetism of the alloy is affected by the diffusion of silicon atoms in the BCC Fe-Si structure as well as by the increase in the atomic disorder of this structure. The diffusion of silicon atoms in the BCC matrix dominates between 0 and 75 h of milling and the mean hyperfine magnetic field decreases from 17.8 to 14.8 T. After 75 h of milling the effect of the mechanical alloying is only to increase the atomic disorder of the BCC Fe-Si phase and the mean hyperfine magnetic field increases from 14.8 to 16.6 T.

Acknowledgements This work was supported by COLCIENCIAS (Colombian Agency), the Excellence Center of New Materials (CENM) and the Universidad del Valle.

References

1. García, E., Adeva, P., Cristina, M.C., Martín, A., Carmona, F., Cebollada, F., Martín, V.E., Leonato, M., González, J.M.: Mater. Sci. Eng. **A134**, 1394 (1991)
2. Gialanella, S., Amils, X., Baro, M.D., Delcroix, P., Le Caër, G., Lutterotti, L., Suriñach, S.: Acta Mater. **46**(9), 3305 (1998)
3. Lutteroti, L., Scardi, P.J.: J. Appl. Crystallogr. **23**, 246 (1990)
4. Varret, F., Teillet, J.: Unpublished MOSFIT Program
5. Abdellaoui, M., Barradi, T., Gaffet, E.: J. Alloys Compd. **198**, 155–164 (1993)
6. Abdellaoui, M., Djega-Mariadassou, C., Gaffet, E.: J. Alloys Compd. **259**, 241–248 (1997)
7. Ding, J., Ling, Y., Chen, L.F., Deng, C.R. Shi, Y., Chow, Y.S., Gang, T.B.: J. Alloys Comp. **314**, 262–267 (2001)
8. Li, T., Li, Y.-Z., Zhang Y.-h.: J. Phys Condens. Matter **9**, 1381–1388 (1997)
9. García Escorial, A., Adeva, P., Cristina, M.C., Martín, A., Carmona, F., Martín, V.E., Leonato, M., González, J.M.: Mater. Sci. Eng. **134**, 1394–1397 (2001)
10. Zamora, L.E., Pérez Alcázar, G.A., Vélez, G.Y., Betencur, J.D., Marco, J.F., Romero, J.J., Martínez, A., Palomares, F.J., González, J.M.: Phys. Rev. B **79** 094418 (2009)

Hyperfine Interact (2011) 203:99–104
DOI 10.1007/s10751-011-0370-0

Mössbauer and magnetic studies of $Mn_{0.1}Sr_{0.2}Co_{0.7}Fe_2O_4$ nanoferrite

H. M. I. Abdallah · J. Z. Msomi · T. Moyo · J. J. Dolo · A. Lančok

Published online: 25 August 2011

Abstract The magnetic properties of $Mn_{0.1}Sr_{0.2}Co_{0.7}Fe_2O_4$ nanoferrite with particle size of about 8 nm were investigated using magnetization and Mössbauer spectroscopy measurements. The sample shows a large increase in coercive field from 0.045 kOe at room temperature to about 3.00 kOe at 4 K. Room temperature coercive fields increased with increase in the annealing temperature between 300°C and 800°C. Our results show evidence of transformation from single domain to multi-domain structure with thermal annealing.

Keywords Nanoparticles · Coercive fields · Mössbauer spectroscopy · Spin freezing

1 Introduction

Ferrites are important materials in many magneto-electric and magneto-mechanical applications. Many previous studies have focused on the $CoFe_2O_4$ and similar compounds with Mn and Mg substitutions. The strength of superexchange interactions in these materials depends on the type of magnetic ions on tetrahedral (A) and octahedral (B) sites. This has remarkable effects on their physical (or magnetic) properties.

H. M. I. Abdallah · T. Moyo
School of Physics, University of KwaZulu-Natal, P/Bag X54000, Durban 4000, South Africa

H. M. I. Abdallah
e-mail: Abdallah@ukzn.ac.za

J. Z. Msomi (✉) · J. J. Dolo
Department of Physics, University of the Free State, P/Bag X13, Phuthaditjhaba 9866, South Africa
e-mail: msomijz@qwa.uovs.ac.za

A. Lančok
Institute of Inorganic Chemistry, AS CR, v.v.i.250 68 Husinec, Řež 1001, Czech Republic

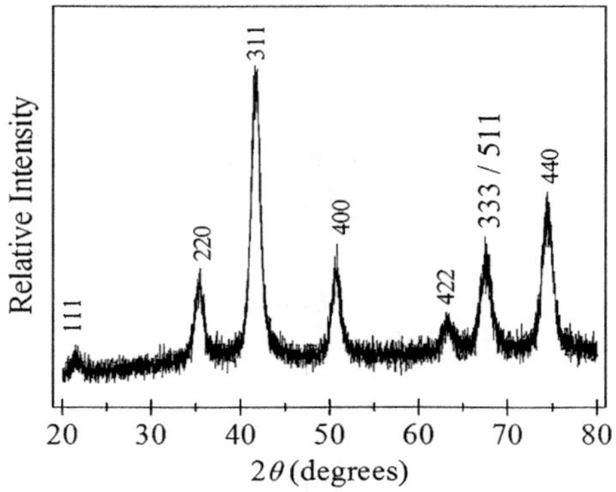

Fig. 1 XRD spectrum for raw $Mn_{0.1}Sr_{0.2}Co_{0.7}Fe_2O_4$ oxide

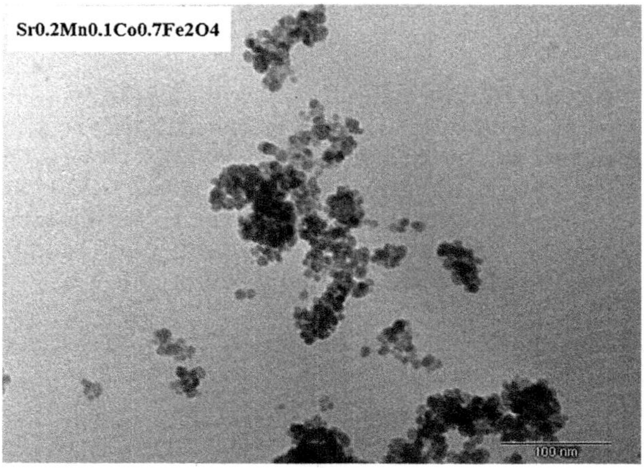

Fig. 2 TEM image of raw $Mn_{0.1}Sr_{0.2}Co_{0.7}Fe_2O_4$ oxide

The properties also depend on the synthesis methods and the nanosize nature of the compounds. Mn–Co ferrites have application as stress sensors due to their large magneto-mechanical effect and high sensitivity to stress [1]. Efforts continue to find the best substitutions to improve the properties. We have been interested in the study of Co, Mn and Mg ions in ferrites [2, 3]. In the present work, we extend these studies to include substitution by larger Sr ions.

2 Experimental

We have produced $Mn_{0.1}Sr_{0.2}Co_{0.7}Fe_2O_4$ fine powder by glycol-thermal process following procedures discussed in reference [3] and investigated the magnetic properties. X-ray diffraction (XRD) measurements were performed on a Phillips diffractometer (type: PW 1710) using CoK_α radiation ($\lambda = 1.7903$ Å).

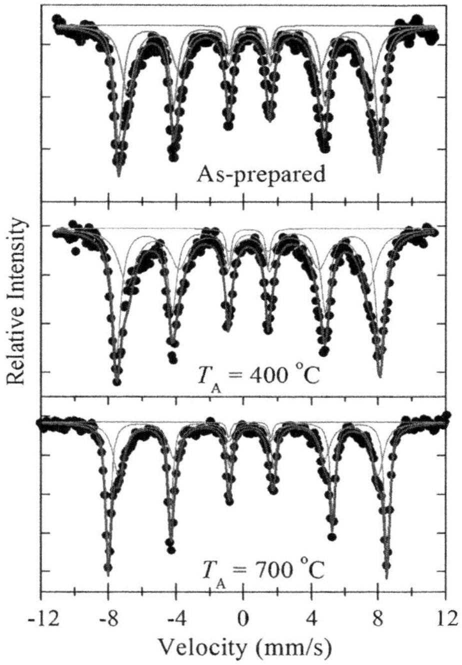

Fig. 3 Mössbauer spectra recorded at room temperature for annealed samples at 400 and 700°C

Table 1 Values of isomer shifts (δ), hyperfine fields (B_{hf}), line widths (Γ) and Fe fraction (f) at A and B sites for $Mn_{0.1}Sr_{0.2}Co_{0.7}Fe_2O_4$ oxides

Sample	δ_A (mm/s) ±0.02	δ_B (mm/s) ±0.01	B_{hf}^A (kOe) ±3	B_{hf}^B (kOe) ±1	Γ^A (mm/s) ±0.05	Γ^B (mm/s) ±0.02	f^A (%)	f^B (%)
As-prepared	0.29	0.32	445	481	0.49	0.29	46.2	50.4
400°C	0.29	0.31	444	485	0.60	0.32	44.0	52.3
700°C	0.32	0.38	477	512	0.41	0.18	45.9	51.9

3 Results and discussion

In Fig. 1 we show the XRD spectra for the as-prepared $Mn_{0.1}Sr_{0.2}Co_{0.7}Fe_2O_4$ compound. The peaks are characteristic of pure cubic spinel structure confirming single-phase formation of the synthesized compound. The sample has a cubic inverse spinel structure with a lattice parameter of 8.359 ± 0.001 Å, which similar to that of similar compounds [1]. The grain size of 8.06 ± 0.01 nm was calculated from the most intense (311) XRD peak using the Debye–Scherer formula. Transmission electron microscopy (TEM) measurements in Fig. 2, show spherical particles with an average particle size of 8.47 ± 0.04 nm, which is close to the value estimated from the XRD data.

The Mössbauer spectra were determined using a conventional spectrometer with a ^{57}Co source sealed in Rh matrix and vibrated at constant acceleration. Figure 3 shows the Mössbauer spectra of the as-prepared and annealed samples. The spectra

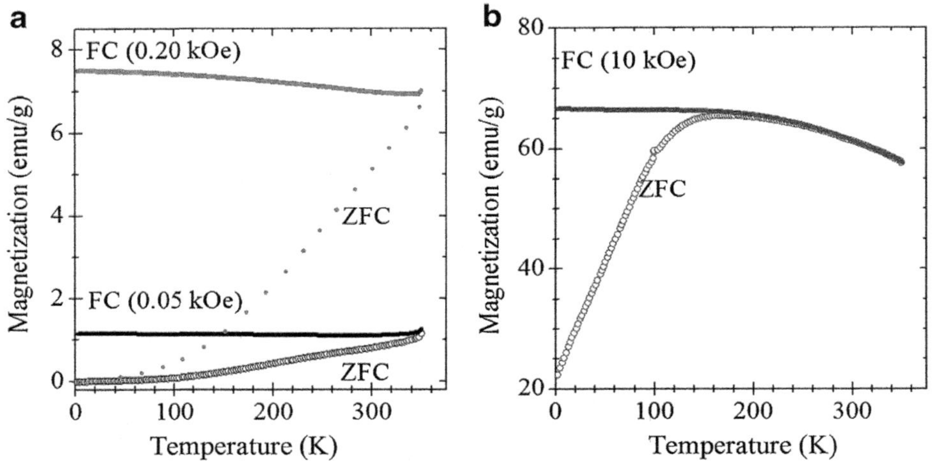

Fig. 4 Variation of FC and ZFC Magnetization curves as a function of measuring temperature for as-prepared sample of $Mn_{0.1}Sr_{0.2}Co_{0.7}Fe_2O_4$

Fig. 5 Magnetization curves as a function of measuring temperature for the as-prepared $Mn_{0.1}Sr_{0.2}Co_{0.7}Fe_2O_4$ sample

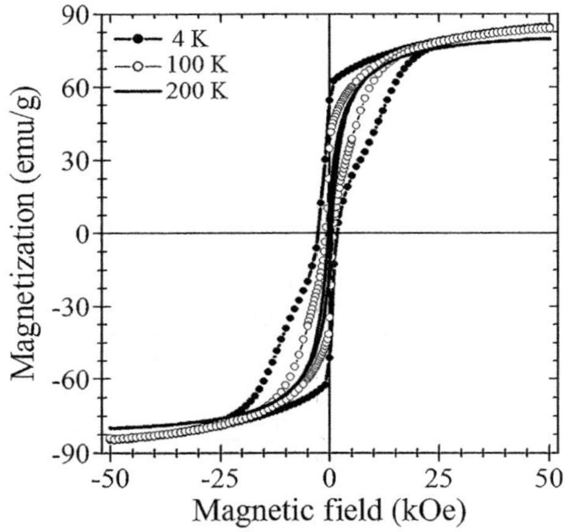

show well resolved magnetic splitting. There is no appreciable change in Mössbauer spectra with increasing annealing temperature up to about 700°C. The spectra are be fitted with two sextets and one doublet. Sextets are associated with trivalent Fe ions on tetrahedral (A) and octahedral (B) sites. A doublet indicates existence of some spins in a paramagnetic state. The sextets were allocated to A or B sites based on the fitted values of isomer shifts and hyperfine fields. Table 1 gives the best-fit parameters to the Mössbauer data.

The hyperfine fields appear to increase with increasing annealing temperature. This trend is associated with increase in particle sizes. Lower hyperfine fields correlate well with smaller grains because of increased spin fluctuations. Beyond the hyperfine field correlation, the spin fluctuation may also give rise to broadened spectra as observed in the as-prepared and 400°C annealed spectra.

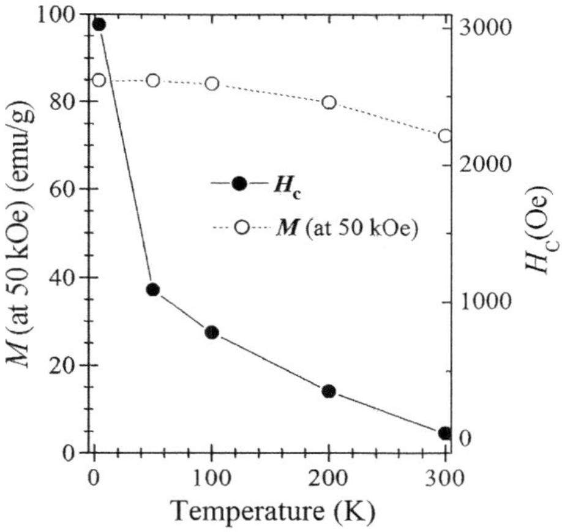

Fig. 6 Variation of coercive field as a function of measuring temperature for the as-prepared $Mn_{0.1}Sr_{0.2}Co_{0.7}Fe_2O_4$ sample

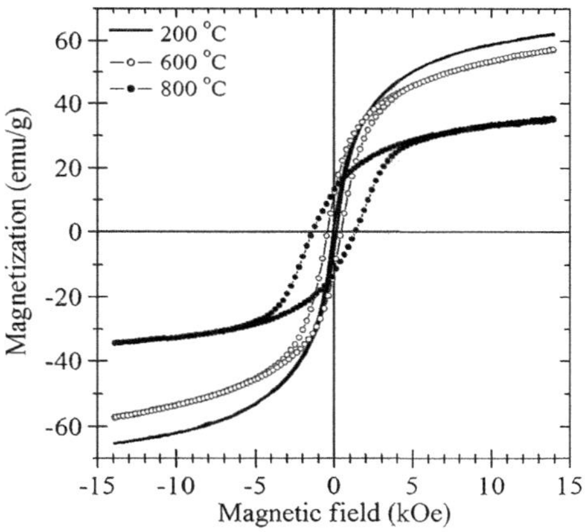

Fig. 7 Hysteresis loops at different annealing temperature for the $Mn_{0.1}Sr_{0.2}Co_{0.7}Fe_2O_4$ sample

The variation of the magnetization as a function of applied magnetic field (up to 5 T) and temperature (4–350 K) was determined using a SQUID MPMS produced by Quantum Design. Figures 4a and b show the field cooled (FC) and zero field (ZFC) magnetization curves for as-prepared sample of $Mn_{0.1}Sr_{0.2}Co_{0.7}Fe_2O_4$. The FC magnetization curves obtained with increasing temperature from 4 K to about 350 K in external magnetic fields of 0.05, 0.20 and 10.00 kOe. The upper and lower curves are for FC and ZFC measurements. Magnetization increases with increasing applied external magnetic field as expected. These effects are associated with spin–spin freezing and thermal relaxation effects typical for nanosized ferrites.

Figure 5 shows the hysteresis loops of the as-prepared $Mn_{0.1}Sr_{0.2}Co_{0.7}Fe_2O_4$ oxide as function of measuring temperature. There is no saturation even in the fields of up 50 kOe. This may be due to the presence of canted spin structure. The magnetization curves recorded at 50, 100 and 200 K appear distorted. This may be due to the freezing of disordered spins [4]. In Fig. 6, we show the magnetic parameters derived from the hysteresis loops as a function of measuring temperature. We observe a sharp increase in coercive field with decreasing measuring temperature. At 4 K the

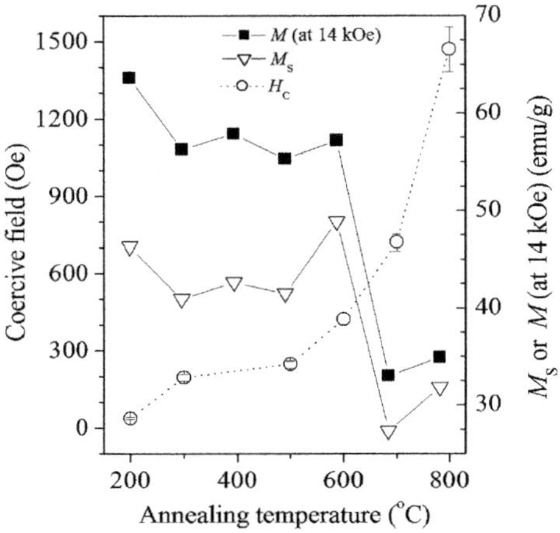

Fig. 8 Coercive field and saturation magnetization as a function of annealing temperature for the $Mn_{0.1}Sr_{0.2}Co_{0.7}Fe_2O_4$ sample

coercive field increased from about 0.045 to 3.024 kOe. The compound becomes magnetically hard at lower temperatures due to the freezing of the spins. This makes magnetization reversal harder to achieve. The saturation magnetization increases more slowly with decreasing temperature.

We have also investigated the effect of annealing temperature on the hysteresis loops in Fig. 7. The as-prepared oxide has a smaller coercive field, which increases by annealing at higher temperature. Figure 8 shows the variation of coercive fields and magnetizations as a function of annealing temperature.

4 Conclusions

In conclusion, we have produced single-phase cubic spinel $Mn_{0.1}Sr_{0.2}Co_{0.7}Fe_2O_4$ oxide by glycol-thermal method. The sample produced has single domain structure and evolves to multi-domain structure with increased annealing temperature. The spin freezing at low temperature is associated with increased magnetic hardness. The magnetic properties derived from magnetization and Mössbauer spectroscopy are sensitive to the microstructure of the sample.

References

1. Kambale, R.C., Shaikh, P.A., Harare, N.S., Bilur, V.A., Kolekar, Y.D., Bhosale, C.H., Rajpure, K.Y.: J. Alloys Compd. **490**, 568 (2010)
2. Abdallah, H.M.I., Moyo, T., Msomi, J.Z.: J. Supercond. Nov. Magn. **24**, 669–673 (2011)
3. Abdallah, H.M.I., Moyo, T., Msomi, J.: J. Phys.: Conf. Ser. **217**, 012141 (2010)
4. Upadhyay, C., Verma, H.C., Sathe, V., Pimpale, A.V.: J. Magn. Magn. Mater. **312**, 271 (2007)

Hyperfine Interact (2011) 203:105–112
DOI 10.1007/s10751-011-0351-3

Improving the empirical model for plasma nitrided AISI 316L corrosion resistance based on Mössbauer spectroscopy

M. Campos · S. D. de Souza · S. de Souza · M. Olzon-Dionysio

Published online: 30 August 2011

Abstract Traditional plasma nitriding treatments using temperatures ranging from approximately 650 to 730 K can improve wear, corrosion resistance and surface hardness on stainless steels. The nitrided layer consists of some iron nitrides: the cubic γ' phase (Fe_4N), the hexagonal phase ε ($Fe_{2-3}N$) and a nitrogen supersatured solid phase γ_N. An empirical model is proposed to explain the corrosion resistance of AISI 316L and ASTM F138 nitrided samples based on Mössbauer Spectroscopy results: the larger the ratio between ε and γ' phase fractions of the sample, the better its resistance corrosion is. In this work, this model is examined using some new results of AISI 316L samples, nitrided under the same previous conditions of gas composition and temperature, but at different pressure, for 3, 4 and 5 h. The sample nitrided for 4 h, whose value for ε/γ' is maximum (= 0.73), shows a slightly better response than the other two samples, nitrided for 5 and 3 h (ε/γ' = 0.72 and 0.59, respectively). Moreover, these samples show very similar behavior. Therefore, this set of samples was not suitable to test the empirical model. However, the comparison between the present results of potentiodynamic polarization curves and those obtained previously at 4 and 4.5 torr, could indicated that the corrosion resistance of the sample which only presents the γ_N phase was the worst of them. Moreover, the empirical model seems not to be ready to explain the response to corrosion and it should be improved including the γ_N phase.

Keywords AISI 316L · Mossbauer spectroscopy · Corrosion resistance · Empirical model

M. Campos · S. D. de Souza · M. Olzon-Dionysio (✉)
Departamento de Física, Universidade Federal de São Carlos, 13565-950, São Carlos, SP, Brazil
e-mail: dmod@df.ufscar.br

S. de Souza
Centro de Ciência e Tecnologia de Materiais, Instituto de Pesquisas Energéticas e Nucleares, 05508-000, São Paulo, SP, Brazil

1 Introduction

In spite of their good corrosion resistance in many aggressive environments, stainless steels show the effects of pitting corrosion in chloride ion solutions. Moreover, their industrial applications are limited because of the low hardness and low wear resistance. Traditional plasma nitriding treatments using treatment temperatures from approximately 650 to 730 K can improve the wear and corrosion resistance, as well as the surface hardness. The nitrided layer consists of some iron nitrides (f.c.c. γ' (Fe_4N) and hexagonal ε ($Fe_{2-3}N$)) and the metastable phase. This is known as expanded austenite γ_N, which is mainly responsible for improving nitrided AISI 316L properties [1] according to some authors. However, most of them use X-Ray Diffraction (XRD) in their studies. The difractograms present some difficulties, such as the enlargement of the γ_N peaks. Once γ'and ε peak positions are close to the γ_N peak positions, these phases can not be reliably identified. In recent studies on AISI 316L and ASTM F138 nitrided samples, we used not only XRD, but also Conversion Electron Mössbauer Spectroscopy (CEMS), as well as observing their corrosion resistance [2, 3]. As the phases show distinct hyperfine parameters, the Mössbauer spectra can provide better information. The results have demonstrated that the ε/γ' fraction ratio seemed to play an important role in the corrosion resistance as the best sample showed the maximum value for this ratio. Therefore, we proposed an empirical model. The larger this ratio value of the sample is, the better its corrosion resistance is. Some new results of AISI 316L samples, nitrided under the same conditions of gas composition and temperature, but at different pressures, are used in this work to examine the empirical model.

2 Materials and methods

AISI 316L stainless steel was produced by Villares Metals. It was annealed in the industry at 1323 K for 60 min and then cooled in water. Its chemical composition (in wt. %) is: 0.03% C, 17.0% Cr, 10. 0% Ni, 2.15% Mo and Fe balance. The samples (diameter, 19.70 mm, thick, 3.85 mm) were polished, cleaned ultrasonically in an acetone bath and finally air-dried. Afterwards, they were plasma nitrided in equipment with a DC power supply, which consists of a hermetically closed vessel containing two electrodes: an anode and a cathode, where the metallic work piece is placed. A nitriding atmosphere, consisting of a mixture of 20% N_2-80% H_2, at a pressure of 6 Torr was used. The voltage applied between the electrodes was adjusted at ≈400 V (and current density ≈485 mA) to maintain the temperature at 400°C. This was measured using an alumel–chromel (type K) long rod thermocouple placed underneath the sample. The samples were nitrided at different nitriding time intervals (t_N) of 3, 4, and 5 h.

The Mössbauer spectra of untreated and nitrided samples were obtained in the backscattering geometry using a conventional constant acceleration Mössbauer spectrometer. A self-built detection chamber was used with a 95% He + 5% CH_4 gas mixture flux. A ^{57}Co source in Rh matrix with a nominal activity of 100 mCi was used. The CEMS measurements were performed at room temperature and the isomer shifts were given relative to α-Fe.

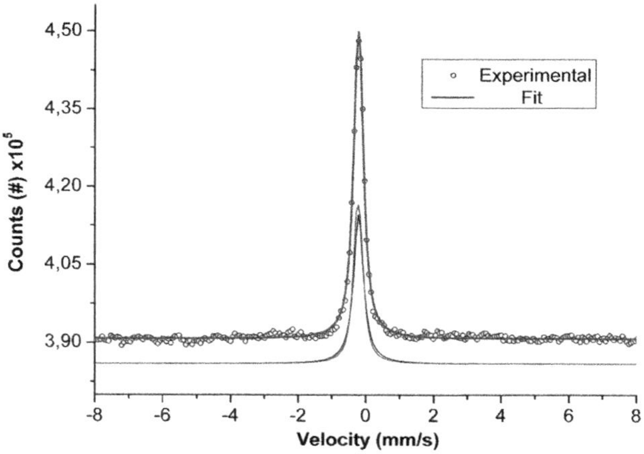

Fig. 1 CEMS spectra for the AISI 316L sample before nitriding

The corrosion performance of the samples was investigated in 3% NaCl aerated electrolytic solution using an Autolab PGSTAT30 Potentiostat/Galvanostat. Potentiodynamic polarization curves were obtained using the EG&G-Pal M352 Software. The scanning potential was in the range of -1.0 $V_{(SCE)}$ (cathodic potential) to $+0.8$ $V_{(SCE)}$ (anodic potential), and the scan rate was 1 mV s^{-1}. Electrochemical experiments were performed in a conventional Pyrex cell using untreated and nitrided AISI 316L samples as working electrodes and a platinum sheet as a counter electrode. The potentials were related to the saturated calomel electrode (SCE) in a KCl solution and all the electrochemical experiments were done at room temperature.

3 Results and discussion

The depth of the nitrided layer, measured in a Scanning Electron Microscopy, was in the range of 3.0 to 4.0 μm. Figure 1 shows the CEMS spectrum for the untreated sample. It shows that the surface of this sample is formed by 51% of an austenitic f.c.c. phase represented by one single-line (isomer shift IS = $-0.09(1)$ mm/s). The remaining 49% was represented by one double-line component (IS = $-0.09(1)$ mm/s and quadrupole splitting EQ = 0.13(1) mm/s), which means that the Fe atom neighbors affect the FCC cubic symmetry.

Figure 2a, b and c show the CEMS data for the nitrided samples. The spectra were fitted as a superposition of the sub-spectra of different phases: a nitrogen supersatured solid phase γ_N, a hexagonal ε phase and the γ' phase. Table 1 presents the Mössbauer parameters used for these different phases, which are in agreement with previous results [4–11].

Figure 3 shows the phase fractions obtained from the Mössbauer fittings as a function of t_N, where $\gamma_N = \gamma_N(p) + \gamma_N(m)$. The γ_N fraction stays constant at 40%. When t_N increases from 3 to 4 h, the ε phase concentration increases as well and the γ' phase concentration decreases. This is because the largest amount of nitrogen that is introduced should be used to transform the γ' phase into the ε phase, which

Table 1 Hyperfine parameters used to fit the different CEMS spectra for nitriding time of 3, 4, and 5 h. H is the magnetic hyperfine field, EQ is the quadruple splitting, IS is the isomer shift

Phase	H (T)	EQ (mm/s)	IS (mm/s)	Reference
γ_N(p)	–	–	-0.09 ± 0.05^b	[4]
γ_N(m)	15.5	–	0.25	[5]
γ_N(m)	16.8	–	0.18	[6]
$\gamma'-Fe_4N$	21.8	–0.21	0.16	[7]
$\gamma'-Fe_4N$	34.2	–	0.35	[8]
$\varepsilon-Fe_{2,47}N$	–	0.86	0.31	[9]
$\varepsilon-Fe_{2,5}N$	25.6	–	0.12	[10]
$\varepsilon-Fe_{2,6}N$	–	0.68 ± 0.02^a	0.39 ± 0.09^c	[9]
$\varepsilon-Fe_3N$	28.8	–	0.09	[11]

[a] 0.66 and 0.70 for 3 and 4 h

[b] −0.14, −0.14 and −0.04 for 3, 4 and 5 h

[c] 0.47and 0.30 for 3 and 4 h

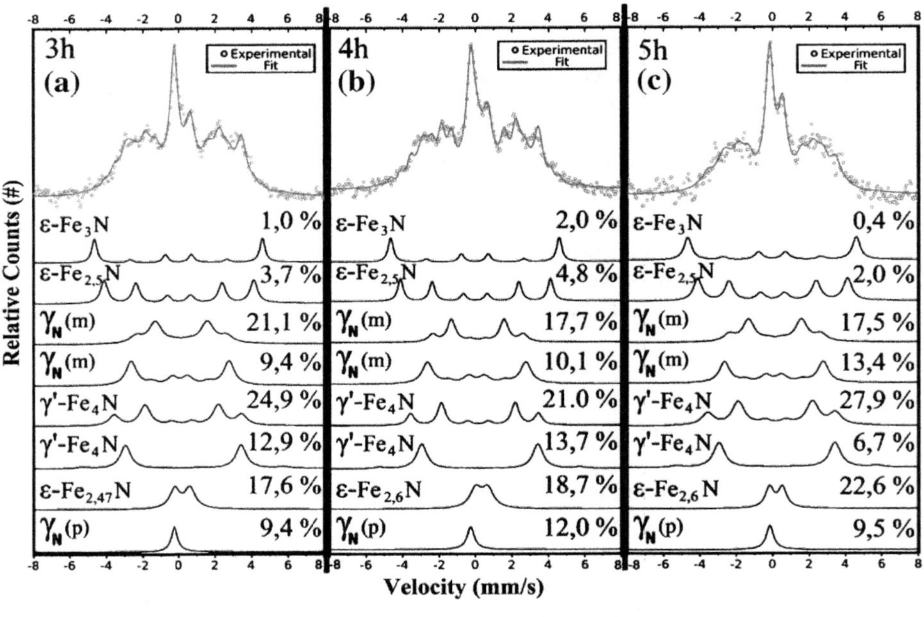

Fig. 2 **a–c** CEMS spectra for the AISI 316L samples nitrided at 3, 4 and 5 h

is richer in nitrogen. After 4 h, the system seems to come into equilibrium as the concentrations remain constant. The broad nature of the resonance signal is due to variations in the local N, Cr, Ni environments around Fe [12].

Figure 3 also shows the data from the reference [2], where we studied the same sample nitrided at the same temperature conditions, but at a pressure of 4.5 Torr. We can compare the time of 4 h, which was used in both studies. The concentration values of ε and γ' phases are very close in both studies. However, the value of the γ_N phase of this study is about 10% above that obtained previously, where we also

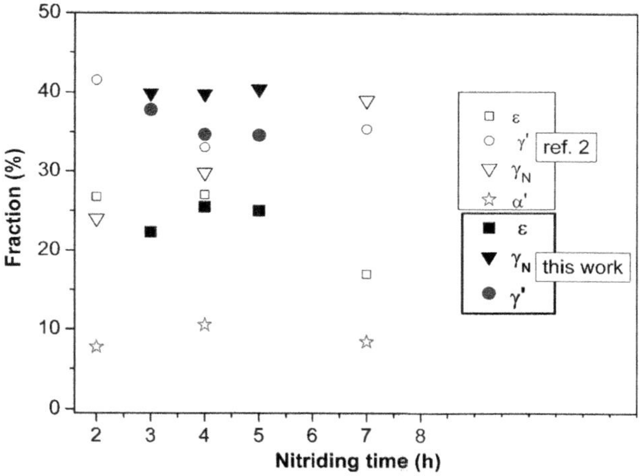

Fig. 3 Fractions of the adjusted phases as a function of different time intervals used during nitriding at 6 and 4.5 Torr [2]

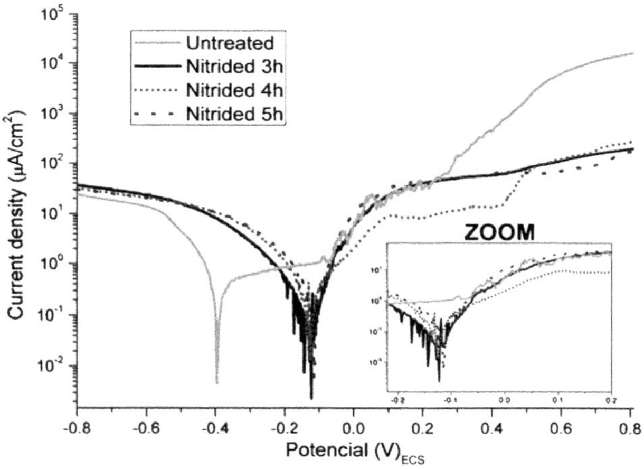

Fig. 4 Polarization curves in 3% NaCl aerated electrolytic solution of the ASIS 316L samples untreated and nitrided for 3, 4 and 5 h

observed the presence of about 10% of the martensite α' phase, formed on the near surface during mechanical polishing of the sample [13].

Figure 4 shows the potentiodynamic polarization curves for untreated and nitrided samples.

Although the sample nitrided for 4 h, whose value for ε/γ' is a maximum $(= 0.73)$, has a slightly better response than the other two samples, nitrided for 5 and 3 h ($\varepsilon/\gamma' = 0.72$ and 0.59, respectively), these samples show very similar behavior. Therefore, this set of samples is not sufficient to test the empirical model.

It is interesting to compare the present corrosion results with some preceding ones [2, 14], where the samples were nitrided at 400°C for 4 h at 4.0 and 4.5 Torr. Figure 5(a) and (b) show the potentiodynamic polarization curves for untreated and nitrided samples for both studies.

All the nitrided samples show a higher corrosion potential (E_{corr}) compared to the untreated sample. The samples nitrided at 4.5 and 6 Torr, which have ε and

Fig. 5 Polarization curves in 3% NaCl aerated electrolytic solution of samples nitrited at **a** 4 Torr, **b** 4.5 Torr

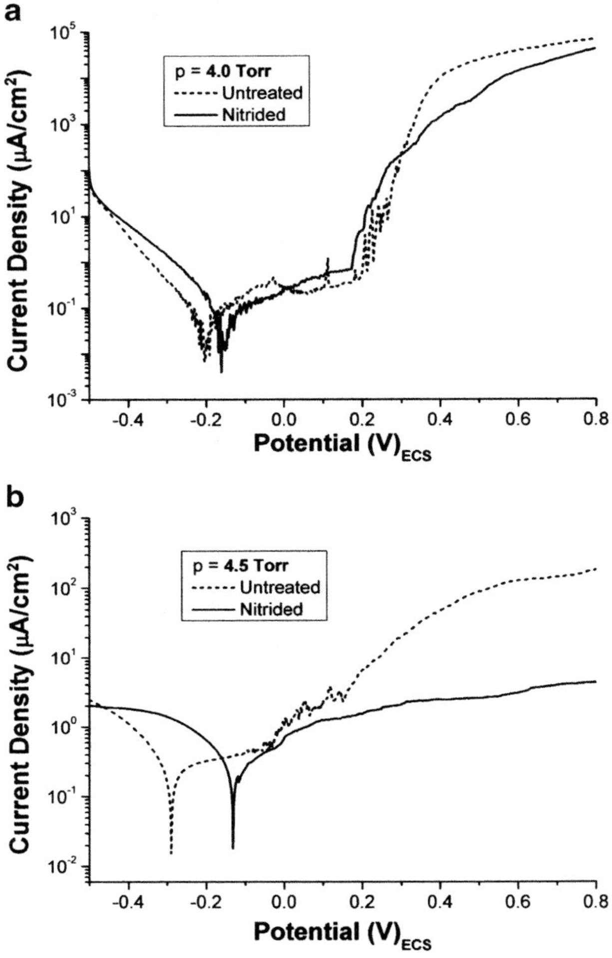

γ' phases besides γ_N, showed a greater difference than the one nitrided at 4 Torr, which only has a γ_N phase. Indeed, an examination of the curve behavior shows that the sample nitrided at 4 Torr shows a rapid increase in current density, indicating localized corrosion, commonly known as pitting corrosion.

As is known, the lower the current density of the nitrided sample (j_{nitr}), in relation to the current of the untreated sample (j_{untr}), the better the corrosion resistance is. Therefore, to compare the nitrided sample response, it is appropriate to compare the ratio j_{nitr}/j_{untr} along the curve for the three studies. These values are given in Table 2.

Some authors [1] have regarded the presence of the γ_N phase as being responsible for better corrosion resistance. However, this comparison shows that the sample which presents only this phase was the worst. We now compare the values in the Table 2 for the sample nitrided at 4.5 Torr in [2], which shows the highest value for $\varepsilon/\gamma'(=0.84)$, with the values for the three samples in this work. It can be observed that [2] sample is more resistant to corrosion than the samples nitrided at $t_N = 3$

Table 2 Ratio between current density of nitrided sample (j_{nitr}), in relation to the current of untreated sample (j_{untr}) along the polarization curve of the samples nitrided at different pressures [2, 14] and times

Sample	ε/γ'	j_{nit}/j_{untr} $E = 0.0V$	j_{nit}/j_{untr} $E = 0.1V$	j_{nit}/j_{untr} $E = 0.2V$	j_{nit}/j_{untr} $E = 0.3V$	j_{nit}/j_{untr} $E = 0.4V$	j_{nit}/j_{untr} $E = 0.5V$	j_{nit}/j_{untr} $E = 0.6V$
[14]	–(Only γ_N phase)	2.55	5.12	10.47	1.23	0.098	0.061	0.064
[2]	0.84	0.57	0.58	0.23	0.10	0.051	0.028	0.024
This work								
3h	0.59	0.87	0.99	1.27	0.38	0.13	0.039	0.018
4h	0.73	0.43	0.46	0.28	0.12	0.031	0.036	0.022
5h	0.72	1.58	1.38	1.32	0.38	0.13	0.032	0.011

and 5h. On the other hand, if we compare the values of second (ref. [2] sample) with fourth ($t_N = 4h$) rows in Table 2, it is difficult to choose the best one. However, Fig. 4 shows that the passive region (in the range between 0.14 and 0.43 V) is better defined for sample nitrided at 4h, than for sample nitrided at 4.5 Torr (Fig. 5b). In this region, the second row shows values which are slightly lower than those in fourth row. It indicates that the sample in [2] seems to be a little more resistant than the sample nitrided at 4h. Thus, it can be concluded that this empirical model is not yet ready to explain the response to corrosion. It is very likely that the γ_N phase also has an important role in the corrosion process and this model should be improved also including the γ_N phase. We are currently developing a systematic study using a higher number of samples with this aim.

4 Conclusions

The AISI 316L samples nitrided for 3, 4 and 5 h, at 6 Torr showed the presence of three phases: ε, γ' and γ_N. The proportions of these phases were used to examine the empirical model which correlates the maximum ratio ε/γ' with the best corrosion resistance. The sample nitrided for 4 h, whose value for ε/γ' is maximum (= 0.73), shows a slightly better response than the other two samples, nitrided for 5 and 3 h ($\varepsilon/\gamma' = 0.72$ and 0.59, respectively). Moreover, these samples show very similar behavior. Therefore, this set of samples was not sufficient to test the empirical model.

However, the comparison between the present results of potentiodynamic polarization curves and those obtained previously at 4 and 4.5 torr, could indicated that: 1) the corrosion resistance of the sample which only presents the γ_N phase was the worst of them, 2) the empirical model is not yet ready to explain the response to corrosion and it should be improved including the γ_N phase.

Acknowledgements The Brazilian agencies FAPESP (Fundação de Amparo e Pesquisa do Estado de São Paulo) and CAPES(Coordenação de Aperfeiçoamento de Pessoal de Nível Superior) gave financial support for this investigation.

References

1. Fossati, A., Borgioli, F., Galvanetto, E., Bacci, T.: Corrosion resistance properties of glow-discharge nitrided AISI 316L austenitic stainless steel in NaCl solutions. Corros. Sci. **48**, 1513–1527 (2006)
2. Olzon-Dionysio, M., de Souza, S.D., Basso, R.L.O., de Souza, S.: Application of Mössbauer spectroscopy to the study of corrosion resistance in NaCl solution of plasma nitrided AISI 316L stainless steel. Surf. Coat. Technol. **202**, 3607–3614 (2008)
3. de Souza, S.D., Olzon-Dionysio, M., Basso, R.L.O., de Souza, S.: Mössbauer spectroscopy study on the corrosion resitance of plasma nitrided ASTM F138 stainless steel in chloride solution. Mater. Charact. **61**, 992–999 (2010)
4. Jiràsková, Y., Havlíček, S., Schneeweiss, O., Peřina, V., Blawert, C.: Characterization of iron nitrides prepared by spark erosion, plasma nitriding, and plasma immersion ion implantation. J. Magn. Magn. Mater. **234**, 477–488 (2001)
5. Ozturk, O., Williamson, D.L.: Phase and composition depth distribution analyses of low energy, high flux N implanted stainless steel. J. Appl. Phys. **77**, 3839–3850 (1995)
6. Gontijo, L.C., Machado, R., Miola, E.J., Casteletti, L.C., Alcântara, N.G., Nascente, P.A.P.: Study of the S phase formed on plasma-nitrided AISI 316L stainless steel. Mater. Sci. Eng. A **431**, 315–321 (2006)
7. Kuhnen, C.A., de Figueiredo, R. S. , Drago, V., da Silva, E. Z. : Mössbauer studies and electronic structure of γ'-Fe4N. J. Magn. Magn. Mat. **111**, 95–104 (1992)
8. Nozik, A.J., Wood Jr., J.C., Haacke, G.: High resolution Mössbauer spectrum of Fe_4N. Sol. St. Comm. **7**, 1677–1679 (1969)
9. Firrao, D., Rosso, M., Principi, G., Frattini, R.: The Influence of carbon on nitrogen Substitution in iron ε-phases. J. Math. Sci. **17**, 1773–1788 (1982)
10. Mekata, M., Yoshimura, H., Takaki, H.: Magnetic study on hexagonal nitrides of 3d transition metals. J. Phys. Soc. Jpn. **33**, 62–69 (1972)
11. DeCristofaro, N., Kaplow, R.: Mössbauer spectroscopy of hexagonal iron-nitrogen alloys. Metall. Mater. Trans. A **8**, 425–430 (1977)
12. Ozturk, O., Williamson, D.L.: Thermal stability of the high-N solid-solution layer on stainless steel. Surf. Coat. Technol. **158–159**, 288–294 (2002)
13. Wang, L., Xu, X., Xu, B., Yu, Z., Hei, Z.: Low temperature nitriding and carburizing of AISI304 stainless steel by a low pressure plasma arc source. Surf. Coat. Technol. **131**, 563–567 (2000)
14. Olzon-Dionysio, M., Campos, M., Kapp, M., de Souza, S., de Souza, S.D.: Influences of plasma nitriding edge effect on properties of 316L stainless steel. Surf. Coat. Technol. **204**, 3623–3628 (2010)

Hyperfine Interact (2011) 203:113–118
DOI 10.1007/s10751-011-0352-2

Kinetic study of the thermal transformation of limonite to hematite by X-ray diffraction, μ-Raman and Mössbauer spectroscopy

P. R. Palacios · Angel Bustamante · P. Romero-Gómez · J. C. González

Published online: 25 August 2011

Abstract A kinetic study about the phase limonite ($FeO(OH)$-nH_2O) was performed through X-ray diffraction, μ-Raman spectroscopy and Mössbauer spectroscopy. The oxide powder sample was extracted from Taraco district, Huancané province of Puno (Peru). X-ray diffraction identified the phase goethite as the main mineralogical component, and then the sample was subjected to *in-situ* heat treatment in the temperature range: 100 to 500°C in oxidizing (air) and inert (nitrogen) atmospheres. The goethite phase remains stable in this range: room temperature to 200°C. Between 200°C to 250°C there is a phase transition: α-$Fe^{3+}O(OH) \rightarrow \alpha$-$Fe_2O_3$, i.e., from goethite to hematite phase, taking as evidence the evolution of the diffraction profiles. At 200°C spectra shows the start of broadened magnetic component and it was adjusted through of a magnetic distribution giving a mean field of 38.6T and a relative area of 52.9%, which is a characteristic of goethite. Also, it is noticed the presence of a small amount of hematite with a mean field of 49.0T linked with a superparamagnetic broadened doublet of relative area of 47.1% where the domains of the particles have sizes smaller than 100 Å and it is evidence the superparamagnetic limit; i.e., the superparamagnetic effect tends toward a distribution of magnetic fields. Moreover, the Raman spectra of the *in-situ* thermal treatment, support the transition at 290°C through the transformation

P. R. Palacios (✉) · A. Bustamante
Laboratorio de Cerámicos y Nanomateriales, Facultad de Ciencias Físicas,
Universidad Nacional Mayor de San Marcos, Ap. Postal 14-0149, Lima, Peru
e-mail: roggerpr@gmail.com

P. Romero-Gómez · J. C. González
Grupo de Investigación de Superficies, Intercaras y Láminas Delgadas,
Instituto de Ciencia de Materiales de Sevilla – CSIC – Univ. Sevilla,
Calle Américo Vespucio 49, Isla de la Cartuja, 41092 Seville, Spain

of characteristic bands of goethite to hematite phase at the frequency range from 200 to 1,800 cm^{-1}.

Keywords Limonite · X-ray diffraction · μ-Raman spectroscopy · Mössbauer spectroscopy

1 Introduction

It is known that goethite phase transforms to hematite phase easily in an oxidant atmosphere, the former phase is an orthorhombic structure [1] and the second one is a rhombohedral or hexagonal structure [2]. In the reference [3] claims that the transformation of goethite phase comes from hematite compound at temperatures among 370 to 600°C. It plays a fundamental key about the origin of the pigment's tonality of the more stable iron ions (Fe^{2+},Fe^{3+}) [4].

The south zone of Peru has important commercial quantities of iron minerals, so it is necessary to have a deep knowledge of the chemical composition and physical properties of this mineral. A right assessment just about of the quality of the raw material, it will allow us to use these iron oxides like as high quality pigments and their subsequent application in paints and/or ceramic glasses [5].

2 Experimental

The mineral in powder, Limonite (LPT), was extracted from Taraco district in Huancané Province of the Puno region (Perú). It was divided in three equal parts for its study and analysis. The in-situ X ray diffraction (XRD) analysis was performed in a Panalitycal X'Pert Pro MPD diffractometer, in Bragg-Brentano geometry, using the Cu-Kα radiation at the 2θ range from 10° to 80°, and step of $\Delta 2\theta = 0{,}016°$. The in-situ thermal treatments were carried out in a high temperature furnace HTK 1200 (Anton Paar) with a temperature rate of 10°C /min, the range of work was from room temperature (30°C) to 500°C, at two kinds of atmospheres: air and nitrogen flux, and obtaining XRD patterns each 50°C.

The in-situ Mössbauer measurements were obtained by using a high temperature Wissel furnace model MBF-1100 with a ^{57}Co source. The μ-Raman spectra were carried out by using a YAG laser ($\lambda = 785$ nm) and the RH800 Horiba Jovin-Yvon Raman spectrometer with CCD detector cooling by air. The laser power was kept low as possible (0,5mW) to avoid the sample degradation. Moreover, Olympus BX-41 metallographic microscope is attached to the spectrometer. The μ-Raman spectra were collected in backscattering configuration. The in-situ μ-Raman measurements were determined by employing a Linkam furnace CCR1000 (10°C /min step), attached to the microscope, in oxidant atmosphere.

3 Results and discussion

The Fig. 1 illustrates the in-situ XRD patterns of the sample LPT for selected temperatures; we can clearly observe the amorphous character, related to the background of the spectrum at room temperature (RT). It was disappearing but the crystalline

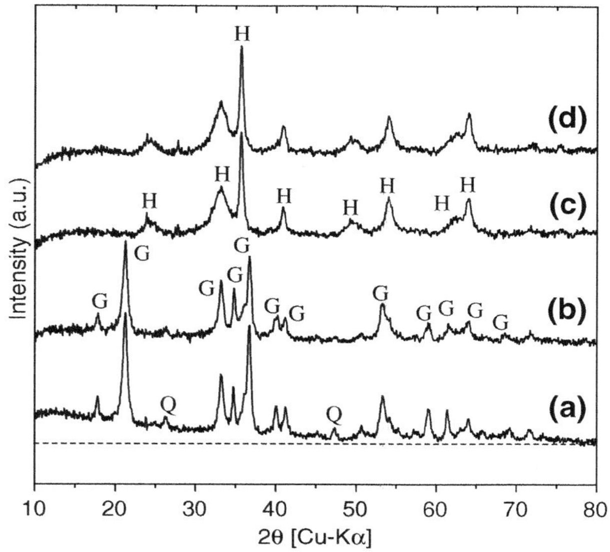

Fig. 1 XRD patterns of the sample LPT at the temperatures: **a** RT, **b** 200°C, **c** 250°C and **d** 300°C. Note the crystalline and amorphous characteristics at room temperature, also, the change of the XRD profiles of the goethite (G) to hematite (H) at 250°C

character increases when the temperature is raised. The crystalline phase until 200°C is identified as goethite (α-FeOOH, Powder Diffraction File 02-0273) and the second phase is detected as quartz (SiO_2, PDF 85-0462) at $2\theta = 27{,}69°$. When the mineral was heated to 250°C, there was a phase transition: $\alpha\text{-}Fe^{3+}O(OH) \rightarrow \alpha\text{-}Fe_2O_3$, where we identified the presence of hematite phase as main component ($\alpha\text{-}Fe_2O_3$, PDF 06-0502), and quartz as a second phase, for more details of XRD patterns you can see reference [6].

The Fig. 2 illustrates the Mössbauer spectra (MS) at RT shows a sextet with broadened lines which are characteristics of a magnetic distribution, that collapse at 150°C. Showing a main doublet D_1 ($IS_1 = 0.287$ mm/s and $\Delta_1 = 0.53$ mm/s) with relative area of 79.3% and a secondary broadened doublet D_2 ($IS_2 = 0.276$ mm/s and $\Delta_2 = 1.32$ mm/s) with relative area of 20.7% as a result of the loss of H_2O which is inherent to limonite and thermal effects. At 200°C spectra shows the start of a broadened magnetic component and it was adjusted through of a magnetic distribution. As a result, we can obtain a mean field of 38.6T with relative area of 52.9%, which is characteristic of goethite. Also, it is noticed the presence of a small amount of hematite with a mean field of 49.0T linked with a superparamagnetic broadened doublet with relative area of 47.1% in which the domains of the particles have sizes smaller than 100 Å. It is evidence the superparamagnetic limit; i.e., the superparamagnetic effect tends toward a distribution of magnetic fields. At temperature 220, 240 and 260°C, the spectra show well-defined distribution fields: 40.4, 39.1 and 38.7T respectively (with hyperfine parameters of $IS_{220°C} = 0.215$ mm/s, $\Delta_{220°C} = -0.17$ mm/s, $IS_{240°C} = 0.197$ mm/s, $\Delta_{240°C} = -0.20$ mm/s and $IS_{260°C} = 0.194$ mm/s, $\Delta_{260°C} = -0.19$ mm/s). These spectra correspond to hematite, which was confirmed by XRD patterns. The Mössbauer spectra at 240 and 260°C are not shown in this job because there is not a change of the evolution even 280°C. At 350°C shows that phase is totally hematite.

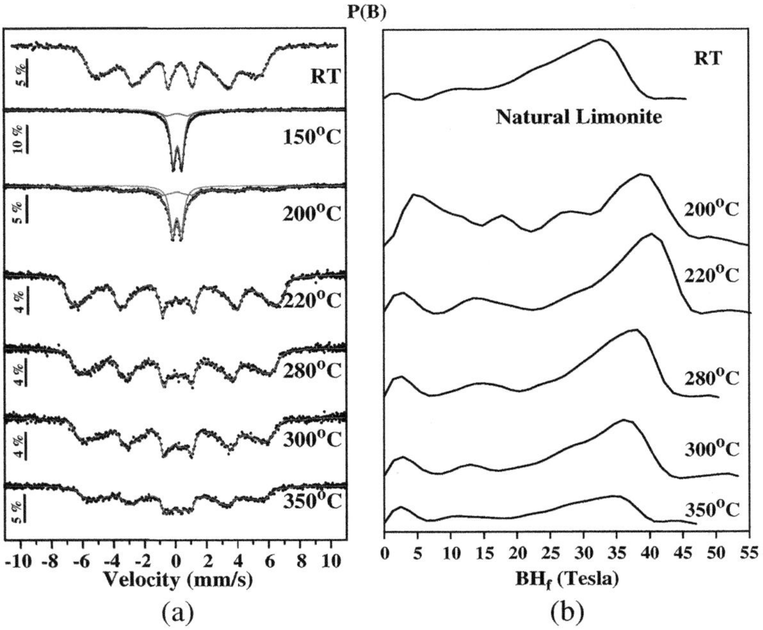

Fig. 2 **a** Mössbauer spectra at several temperatures. All the spectra were fitted using hyperfine magnetic field distributions, and one small or residual superparamagnetic doublet. **b** The most probable fields are from RT to 350°C: 38.6, 40,4, 38.2, 36.0 and 34.3 T respectively

Fig. 3 μ-Raman spectrum of LPT at room temperature. Inset: natural single crystals of hematite and goethite

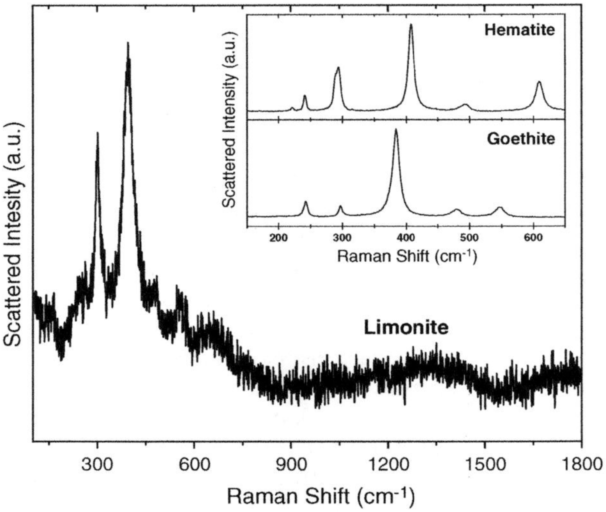

The third part of the LTP sample was thermally in-situ treated with laser radiation in oxidant atmosphere, and then Raman scattered radiation was collected. In Fig. 3 shows the amorphous and crystalline components of the spectrum, the former

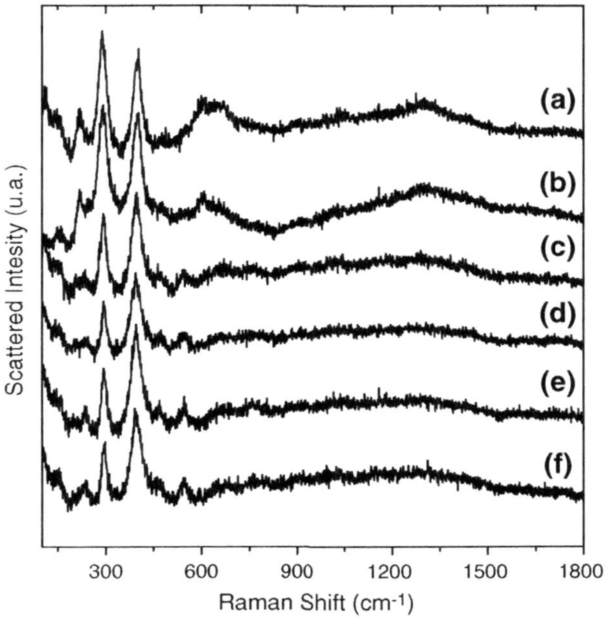

Fig. 4 μ-Raman Spectra of the sample LPT at selected temperatures: **a** 300°C, **b** 290°C, **c** 280°C, **d** 270°C, **e** 260°C and **f** 250°C. Note the phase transformation from goethite to hematite by the evolution of the bands at 610 and 1,322 cm^{-1}

described by the background and the last one specified the Raman bands in the frequency regions characteristic of crystalline grains. Below 280°C, the characteristic bands of goethite are observed (see Fig. 4), also we can observe that the effect of temperature limits to expansion and change in the position of the bands; as a result there is the appearance of new profiles. We emphasize that hematite is not in its stable phase (above 500°C), because the compound hematite is in an ordering process of its crystal structure at temperatures from 300°C to 500°C, this fact is due to the absence of two features Raman bands: 246 y 498 cm^{-1}. If we compare the Raman bands of Fig. 4 with the goethite and hematite spectra of Fig. 3, these bands present wide profiles and not smooth background. This situation is characteristic of materials that undergo an atomic rearrangement process in their structure such as a phase transformation. Therefore, the appearance of harmonic and anharmonic contributions, to a lesser extent, in the Raman spectra is expected to happen.

4 Conclusions

The crystalline and amorphousness phase transformations of the mineral limonite (hydrated goethite) can be monitored by XRD, μ-Raman and Mössbauer spectroscopy. The XRD patterns and Mössbauer spectra show that the main mineralogical component was goethite until 200°C and hematite phase over 250°C. The Raman spectra support the transition at 290°C through the transformation of characteristic bands of goethite to hematite phase. Therefore, the phase transformation of LTP sample was located in the range of 250 to 300°C, i.e., the transition from goethite to hematite phase was analyzed by the three techniques.

Acknowledgements The authors thank to the Academic Vice Rectorate and Faculty of Physical Sciences at San Marcos University in Peru for its economical support to the work. To Materials Science Institute of Sevilla – CSIC – Univ. Sevilla by the in-situ measurments of XRD and μ-Raman. Finally, JC González thanks to the Spanish Scientific Research Council (CSIC) for his JAE-Doc contract (2009-2012) at the ICMSE-CSIC-US. (Spain).

References

1. Hurlbut, C.S., Cornelis, K.: Manual of Mineralogy, 20th edn. Wiley, ISBN 0-471-80580-7 (1985)
2. Palache, C., Berman, H., Frondel, C.: Dana's system of mineralogy, 7th edn., vol. I, pp. 527–534 (1944)
3. Bustamante, A., Cabrera, J., Garcia, V., Urday, E., Abdu, Y.A., Scorzelli, R.B.: Hyperfine Interact. **166**, 593–597 (2005)
4. Bosio, V., Pogliani, C., Donati, E.: VI southern hemisphere on minerals technology. Río de Janeiro (Brasil) **2**, 405–409 (2001)
5. Bustamante, A., Cabrera, J.: Hyperfine Interact. **166**, 593–597 (2005)
6. Palacios, P.R., Bustamante, A.B., Gonzáles, J.C.: Rev. Investig. Fís. **13**, 1–6. ISBN 1605–7744 (2010)

Hyperfine Interact (2011) 203:119–124
DOI 10.1007/s10751-011-0371-z

Aging effect in CaLaBa$\{Cu_{1-x}Fe_x\}_3O_{7-\delta}$ with $0 \leq x \leq 0.07$ studied by Mössbauer spectroscopy

Angel Bustamante · Luis De Los Santos Valladares · Jesús Flores · Crispin H. W. Barnes · Yutaka Majima

Published online: 25 August 2011

Abstract In this work, we study the long-term aging effect caused by Fe atoms in the superconductor CaLaBa$\{Cu_{1-x}Fe_x\}_3O_{7-\delta}$ with $0 \leq x \leq 0.07$. XRD confirms that this system has a YBCO-like structure. The critical temperature (T_c) is strongly affected by aging and depends on the amount of Fe in the structure. Room temperature Mössbauer spectroscopy reveals the presence of the typical species **A**, **B**–**B′**, **C** and new species **E′** and **F**. Interestingly; **A**, which corresponds to the Fe^{3+} atom located in the Cu(1) of the chains with spin $S_z = 3/2$, shows a drastic reduction which means migration to the species **B**, **B′** and **C**. Species **B** and **B′** correspond to the Fe^{3+} in the Cu(2) site forming planar quasi-octahedral and planar square pyramidal, while the **C** specie is a square pyramidal with O(5) respectively (spin $S_z = 3/2$ in all these cases). Aging causes loss of superconductivity in the samples with 5 and 7% of iron content.

Keywords Mössbauer spectroscopy · Cuprates · Aging effect · Spatial group P4/mmm

A. Bustamante (✉) · J. Flores
Laboratorio de Cerámicos y Nanomateriales, Facultad de Ciencias Físicas,
Universidad Nacional Mayor de San Marcos, Apartado Postal 14–0149, Lima 14, Perú
e-mail: angelbd1@gmail.com

L. De Los Santos Valladares · C. H. W. Barnes
Cavendish Laboratory, University of Cambridge, J.J. Thomson Av. Cambridge CB3 0HE, UK

L. De Los Santos Valladares
e-mail: ld301@cam.ac.uk

L. De Los Santos Valladares · Y. Majima
Materials and Structures Laboratory, Tokyo Institute of Technology,
4259 Nagatsuta-cho, Yokohama 226-8503, Japan

Y. Majima
CREST, Japan Science and Technology Agency (JST), 4259 Nagatsuta-cho,
Yokohama 226-8503, Japan

1 Introduction

The superconductor $CaLaBaCu_3O_{7-\delta}$ (La1113) has been intensively studied because it has similar properties to the well known $YBa_2Cu_3O_{7-\delta}$ (YBCO) and its promising application in nanotechnology [1, 2]. The crystalline structure of La1113 is tetragonal (P4/mmm) and its critical temperature (T_C) is around 80 K. The literature reports indicate that this superconductor does not show significant structural variations after single atom substitution [3–5]. However, La1113 becomes orthorhombic (Pmmm) when polyhedrons of oxyanions are located in the Cu(1) site [6–8]. The ^{57}Fe substitution in the Cu-site in both structures has been of great interest since it is possible to analyse local distortions and charge transfer by Mössbauer spectroscopy [9–16]. Short-term room temperature aging effects in the ^{57}Fe hyperfine structure of the YBCO structure have been previously reported by P. Boolchand et al [15]. They found differences between the Mössbauer spectra of fresh and 3-day aged YBCO and suggested that it is caused by desorption of oxygen from the Cu-O chains. In this work we study the long-term room temperature aging effects in the La1113 structure caused by ^{57}Fe substitution. The pristine samples were prepared in 1998 with 1%, 3%, 5% and 7% of ^{57}Fe concentration. After 12 years, the T_C in the lower concentration samples (1% and 3%) decreased, whereas the superconductivity in the higher concentration ones (5% and 7%) was destroyed.

2 Experimental

The preparation of the samples is previously reported in reference [16]. After initial characterization they were stored at room temperature for 12 years. These long-term aged samples were characterized by X-ray diffraction (XRD). The XRD data of the samples with concentrations of 1%, 3%, 5% and 7% were collected from 8° to 80° (0.02° step) using a powder universal diffractometer Bruker D8 Focus with Cu K_α radiation ($\lambda = 1.5406$ Å). The Mössbauer spectra (MS) were recorded at room temperature (RT) using a^{57}Co in Rh source with average activity 25 mCi and the collected data were fitted by using the program Normos. The superconducting states were studied by the DC-susceptibility measurements obtained by a magnetic property measurement system (MPMS-SQUID) from Quantum Design. During measurements, an external DC magnetic field of 50 Oe was applied from 10 K to 110 K in Zero Field Cooling (ZFC) mode.

3 Results and discussions

Figure 1 shows the XRD spectra of the aged samples with doping concentrations of 1%, 3%, 5% and 7% Fe. The reflection at 32.7° corresponds to the phase of interest in each diffractogram. Two impurities are also detected at 23.9° and 24.3° and indicated with asterisk (*). The first was identified as rhombohedra carbon (PDF 74–2328) and the second one is orthorhombic carbon (PDF 89–8493) [17]. The presence of impurity decreases with x and has an influence on the magnetic measurements as discussed below. Comparison of the spectra of interest to the pure La1113 (PDF2 No. 46–0580 [17]) suggests a tetragonal structure similar to that for YBCO with principal

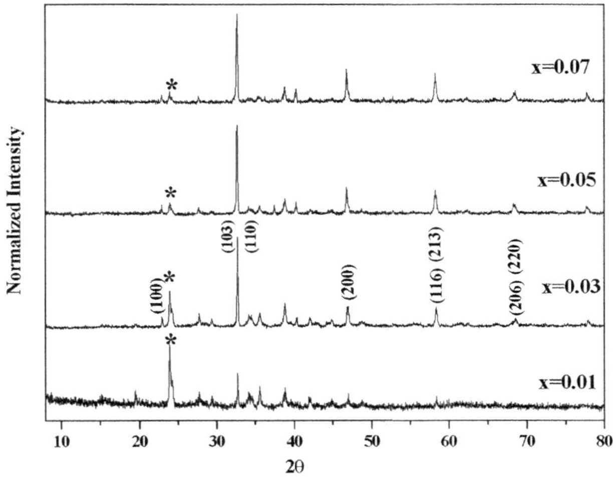

Fig. 1 XRD patterns of the aged CaLaBa$\{Cu_{1-x}Fe_x\}O_{7-\delta}$ samples with x = 1%, 3%, 5% and 7%

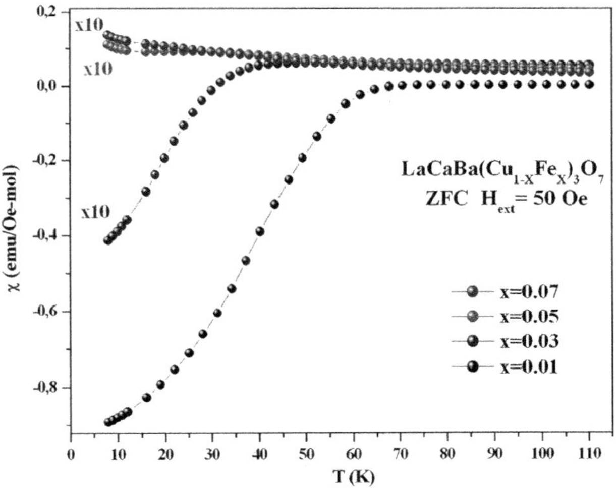

Fig. 2 Magnetic susceptibility of CaLaBa$\{Cu_{1-x}Fe_x\}O_{7-\delta}$ samples with x = 1%, 3%, 5% and 7%

Miller indexes (103), (110), (200), (116), (213), (206) and (220) as they are indicated in Fig. 1.

The T_Cs of the pristine samples were 74, 59, 40 and 39 K for 1%, 3%, 5% and 7% of Fe, respectively [16]. Similar to the pristine samples, the superconductivity states of the aged samples were determined from their Meissner effect. This is a common procedure to observe superconductivity in materials and to estimate their T_C. In brief, the Meissner effect can be analysed from the diamagnetic behaviour in the susceptibility measurements obtained by MPMS-SQUID. The material under study is classified as a superconductor if it presents negative values in the susceptibility loop below an onset temperature ($T_{C(onset)}$). Figure 2 shows the DC susceptibility responses for all the aged samples. The data corresponds to the phase of interest together with the impurity in each sample. Note that the magnetism of the impurity

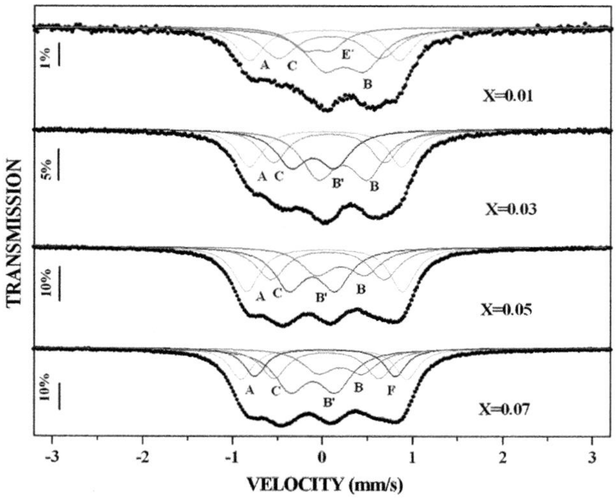

Fig. 3 RT MS of $CaLaBa\{Cu_{1-x}Fe_x\}O_{7-\delta}$ samples with x = 1%, 3%, 5% and 7%

Table 1 Summary of Fe-dopant sites observed after 12 years and compared to the fresh samples

Site	Δ mm/s	δ mm/s	Γ mm/s	A % aged	A % fresh	ΔA=aged-fresh %
A	1% 1.66	0.132	0.37	23	51	−32
	3% 1.69	0.147	0.35	22	43	−21
	5% 1.73	0.138	0.35	28	38	−10
	7% 1.84	0.131	0.31	18	27	−9
B	1% 0.45	0.351	0.50	35	09	+26
	3% 0.54	0.340	0.45	33	17	+16
	5% 0.53	0.319	0.43	20	08	+12
	7% 0.49	0.313	0.43	17		
B′	3% 0.48	0.010	0.41	24		
	5% 0.51	0.001	0.41	29		
	7% 0.50	0.010	0.44	31	23	+8
C	1% 1.14	0.189	0.49	28	12	+16
	3% 1.24	0.184	0.38	21	16	+5
	5% 1.26	0.169	0.40	23	16	+7
	7% 1.17	0.158	0.35	19	23	−4
E′	1% 0.30	0.053	0.38	14	New	+14
F	7% 1.56	0.144	0.29	15	New	+15

affects the zero level of the signals which is better observed from the mismatch of the loops above 60 K. However it does not affect the onset critical temperatures. The aged samples with 1% and 3% of iron content are still superconductors but with new onset T_Cs of 60 K and 38 K, respectively. The susceptibility loops of samples with 5% and 7% of iron content, meanwhile, show paramagnetic behaviour indicating that the long-term aging has destroyed the superconductivity.

RT Mössbauer spectra (MS) of the long-term aged samples are shown in Fig. 3. The spectra confirm that the iron occupies the Cu(1) and Cu(2) sites in the structure [9–15]. After fitting the spectra, many doublets or species, labelled as **A**, **B**, **B′** and **C** are detected as in the pristine case [16]. They correspond to ^{57}Fe in a different local environment and do not show any magnetic component at high velocity. However, the **G** specie of the pristine samples (with $\delta = -0.044$ mm/s and $\mathbf{\Delta} = 0.65$ mm/s for 1%, $\delta = -0.055$ mm/s and $\mathbf{\Delta} = 0.62$ mm/s for 3%, $\delta = -0.019$ mm/s and $\mathbf{\Delta} = 0.58$ mm/s for 5%, and $\delta = -0.033$ mm/s and $\mathbf{\Delta} = 0.57$ mm/s for 7% [16]) were not detected in the aged samples. The corresponding Mössbauer hyperfine parameters of the aged samples are listed in Table 1. Several reports identify **A** as originating from Fe^{3+} located in the square planar chain ($S_z = 3/2$), **B** represents the Fe^{3+} located in the plane-site of a quasi-octahedron ($S_z = 5/2$), **B′** represents the Fe^{3+}located in the plane-site of a square pyramid ($S_z = 5/2$) and **C** represents the Fe^{3+} located in the chain-site or square pyramid in which O(5) sits ($S_z = 3/2$) [16]. The **A** species presents a notorious decrease in all the spectra corresponding to the aged samples (see Table 1). The decrease in the area of **A** for the four iron concentrations reveals the migration of Fe from the Cu(1) site to Cu(2) in the CuO_2 plane causing **B** and **B′** to increase. The presence of the Fe in the CuO_2 planes destroys Cooper pairs, causing the critical temperature to decrease (for 1% and 3%) or the superconductivity to be destroyed (for 5% and 7%). This effect is also accompanied with an increase in specie **C**. Furthermore, two new species assigned as **E′** in 1% and **F** in 7% appear. They could be originated from the specie **G** of the pristine samples and thus represent a new coordination of the iron.

4 Conclusions

Long-term aging in the system $CaLaBa\{Cu_{1-x}Fe_x\}O_{7-\delta}$ causes the T_C to decrease. This effect depends on iron content, at 1% the T_C decreases by 14 K, whereas at 3% the T_C decreases by 21 K. Higher iron content destroys the superconductivity (as detected for 5% and 7%) and the material displays paramagnetic behavior. The area of species **A** in the Mössbauer spectra also depends on the Fe content; a high concentration of Fe reduces the intensity of this doublet. It would be caused by the migration of Fe from the Cu(1) site in the chain to the Cu(2) site in the CuO_2 plane, destroying the Cooper pairs and thus causing the critical temperature to decrease or to destroy the superconductivity.

Acknowledgements This work was supported by the Vice-council of Research of the University of San Marcos (Contract No. 091301051). The work in Cambridge was supported by the European Union program ALBAN (Contract No. E06D101257PE). The work in Japan was supported by the Japan Society for the Promotion of Science (JSPS contract N° PE10027). The work in Japan was also partially supported by Grants-in-Aid for Scientific Research on Innovative Areas (No. 20108011, π-Space) from the Ministry of Education, Culture, Sports, Science and Technology (MEXT); a Collaborative Research Project of the Materials and Structures Laboratory (Tokyo Institute of Technology) and the World Class University Program of the Sunchon National University, Korea. The authors are indebted to Mirtha Pillaca for the XRD measurements and Miriam Mejía for the Mössbauer measurements.

References

1. De Los Santos Valladares, L., Bustamante Domínguez, A., Gonzalez, J.C., Feijoo Levano, J., Osorio Anaya, A., Mitrelias, T., Majima, Y., Barnes, C.H.W.: Open Supercond. J. **2**, 19–27 (2010)
2. De Los Santos Valladares, L., Bustamante Domínguez, A., Llandro, J., Suzuki, S., Mitrelias, T., Bellido Quispe, R., Barnes, C.H.W., Majima, Y.: Jpn. J. Appl. Phys. **49**, 093102 (2010)

3. Yamaya, K., Yagi, T., Okajima, Y.: Solid State Commun. **87**, 1113 (1993)
4. Awana, V.P.S., Narlokar, A.V.: Mod. Phys. Lett. B **15**(14), 415–453 (2001)
5. Singh, R., et al.: Phys. Rev. B. **55**(N 2), 1216–1222 (1997)
6. De Los Santos Valladares, L., Bustamante Domínguez, A., Flores Santibañez, J., González, J.C.: Physica C **408–410**, 44–45 (2004)
7. De Los Santos Valladares, L., Bustamante Domínguez, A., González, J.G., Flores Santibañez, J., Obradors, X.: J. Phys. Chem. Solids **67**, 605–609 (2006)
8. Bustamante Domínguez, A., Bellido Quispe, R., De Los Santos Valladares, L., González, J.C.: Physica C **408–410**, 884–885 (2004)
9. Baggio-Saitovitch, E., Souza Azevedo, I., Scorzelli, R.B., Saitovitch, H., Da Cunha, S.F., Guimarães, A.P., Silva, P.R., Takeuchi, A.Y.: Phys. Rev. B **37**, 7967–7970 (1988)
10. Baggio-Saitovitch, E., Scorzelli, R.B., Souza Azevedo, I., Micklitz, H.: Phys. Rev. B **41**, 2103–2108 (1990)
11. Boolchand, P., McDaniel, D.: Hyperfine Interact. **72**, 125–152 (1992)
12. Kimball, C.W., Dabrowski, B., Liang, Y., Peng, W.: Hyperfine Interact. **72**, 153–172 (1992)
13. Chechersky, V., Nath, A.: Hyperfine Interact. **72**, 173–189 (1992)
14. Boolchand, P., McDaniel, D., Blue, C., Wu, Y., Enzweiler, R., Elgaid, K., Burrows, R.: Hyperfine Interact. **68**, 15–26 (1991)
15. Boolchand, P., Blue, C., Elgaid, K., Huff, W., Kilinc, A., McDaniel, D., Biswas, P., Zhou, D., Oostens, J.: Hyperfine Interact. **62**, 73–88 (1990)
16. Bustamante Domínguez, A., Landinez Tellez, D.A., Albino Aguiar, J.: Hyperfine Interact. (C) **4**, 141–146 (1999)
17. PCPDFWIN, Version 2.1, JCPDS, The International Centre for Diffraction Data (2000)

Hyperfine Interact (2011) 203:125–132
DOI 10.1007/s10751-011-0372-y

Mössbauer characterization of joints of steel pieces in transient liquid phase bonding experiences

N. Di Luozzo · P. F. Martínez Stenger · J. P. Canal · M. R. Fontana · B. Arcondo

Published online: 25 August 2011

Abstract Joining of seamless, low carbon, steel tubes were performed by means of Transient Liquid Phase Bonding process employing a foil of Fe-Si-B metallic glass as filler material. The influence of the main parameters of the process was evaluated: temperature, holding time, pressure and post weld heat treatment. Powder samples were obtained from the joint of tubes and characterized employing Mössbauer Spectroscopy in transmission geometry. The sampling was performed both in tubes successfully welded and in those which show joint defects. The results obtained are correlated with the obtained microstructure and the diffusion of Si and B during the process.

Keywords Transient liquid phase bonding · Mössbauer spectroscopy · Steel tubes

1 Introduction

The extensive use of a variety of tubular products for drilling and workover operations, e.g. coiled tubing, is being prevented by the inability to girth-weld the present materials with an acceptable fatigue life [1].

Girth welded butt joint are often performed using manual gas tungsten arc welding (GTAW) procedures. It is well known that the present welding method has severe limitations in field application. Also there is a correlation between the quality and repeatability of the welds and the skill of the welding operator [2].

The transient liquid phase bonding (TLPB) process is one of the technologies selected to overcome the known problems involved in shop welding and field welding repairs of coiled tubing and their poor fatigue properties. The objective of the

N. Di Luozzo · P. F. Martínez Stenger · J. P. Canal · M. R. Fontana · B. Arcondo (✉)
Laboratorio de Sólidos Amorfos, Facultad de Ingeniería, INTECIN (UBA-CONICET),
Paseo Colón 850, 1063, Buenos Aires, Argentina
e-mail: barcond@fi.uba.ar

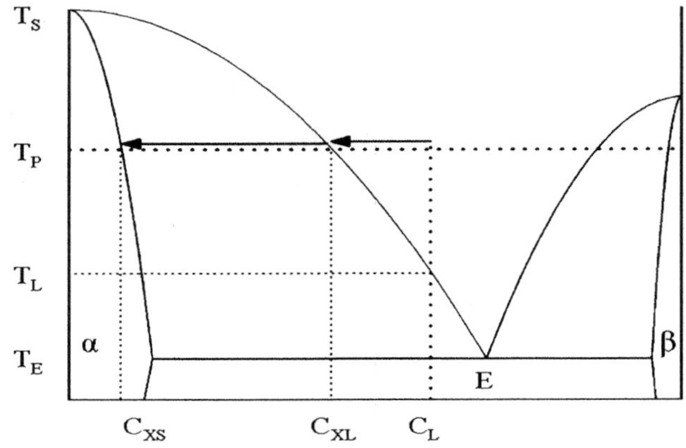

Fig. 1 Phase equilibrium diagram of α and β system. T_P is the isothermal welding temperature, C_L the initial composition of the transient liquid gap, and C_{XL} the composition of the liquid gap

technology is to join long lengths of seamless tubing prior to reeling using the TLPB process in conjunction with an amorphous foil as a filler material.

In recent years, the TLPB process has been widely studied [3–6] in order to improve the characteristics of the joints by means of composition and microstructure homogenization along the joined pieces.

The purpose of this work is to study the relationship between the microstructure and Mössbauer parameters of the as-received (AR) tubes and, for the welded tubes, the base metal (BM)—the metal that is not affected by the welding thermal cycle, the heat affected zone (HAZ) and the joint.

2 Physical description—transient liquid phase bonding (TLPB)—binary composition case

Let us assume that two tubes of a solid α are aligned with their butted surfaces in contact with an amorphous foil. The amorphous foil composition C_L is not far from a eutectic point (E) whose components are α and β (a schematic binary phase diagram is shown in Fig. 1). The melting temperature of the amorphous foil (T_L) is lower than the melting temperature of α (T_S). If the temperature at the joint is raised to T_P, being $T_L < T_P < T_S$, a transient liquid phase is formed. As the temperature T_P is held constant, the following process will take place:

- The dissolution of the Base Metal (with composition $C_S < C_{XS}$) into the liquid gap occurs, thus widening the gap (the concentration of the solid at the interface is C_{XS}).
- When the concentration in the liquid reaches C_{XL}, the solidification starts and the gap proceeds to close up to its total disappearance.
- The completion of the solid-state diffusion is performed [7].

The amorphous nature of the foil is not a precondition for the TLPB process. However, amorphous foils present many advantages when compared to the corresponding crystalline materials. Metallic glasses have a very large ductility therefore, in the form of thin layers; amorphous metals are easier to handle than crystals. In addition,

Table 1 Steel tubes composition (wt.%)

C	Mn	Si	S	P	Ni
0.10	1.26	0.28	0.001	0.011	0.41
Cr	Mo	V	Nb	Ti	Cu
0.09	0.07	0.002	0.023	0.022	0.25

Table 2 Metglas® 2605 SA1 amorphous foil composition (wt.%)

B	Si	Fe
1–5	5–10	Balance

amorphous layers improve the contact with the tubes when pressure is applied, whereas crystalline layers break.

From previous works [5, 8], we know that for low carbon steel tubes, $Fe_{96.2}B_{3.8}$ wt.% amorphous foils of 20 μm thickness as filler material, an applied pressure of 4 MPa and a holding time of 7 min at $T_P \approx 1,250°C$, the solidification process was completed. From now on, this pressure and thermal cycle are referred as standard procedures.

3 Experimental procedure

TLPB processes under a reducing atmosphere (Ar +10% H_2) have been performed to join seamless low carbon steel tubes (see composition in Table 1) with an external diameter of 25.4 mm, a length of 80 mm and a wall thickness of 2.3 mm.

Metglas® 2605 SA1 amorphous foil (see composition in Table 2) of 22 μm thickness has been employed as filler material.

The arrangement was placed horizontally into the coil of an induction furnace, and a pressure of 4 MPa was applied (Fig. 2).

In this work two pressure and thermal cycles were considered:

- Specimen A: a holding time t of 2 min at $T_P \approx 1,150°C$, with an applied pressure of 4 MPa. After that, the sample was subjected to a post weld heat treatment (PWHT) with a holding time t′ of 15 min at $T' \approx 1,350°C$, without applying pressure.
- Specimen B: a holding time t of 7 min at $T_P \approx 1,350°C$, with an applied pressure of 4 MPa.

It is worth noting the advantages of the pressure and thermal cycle of specimen A. The shorter holding time t at a lower T_P is enough to assure the proper start of the isothermal solidification process, with the least utilization of the welding station. Outside the welding station, the sample can be PWHT anywhere, resuming the diffusion of B and Si. This cycle is much more flexible compared with the cycle used in Specimen B.

With the aim to increase the diffusion of Si in Fe, we selected both T′ for specimen A and T_P for specimen B higher than temperature of the standard procedure. The influence of both cycles on the welded joint was evaluated.

Fig. 2 Experimental arrangement **a** diagram **b** current arrangement

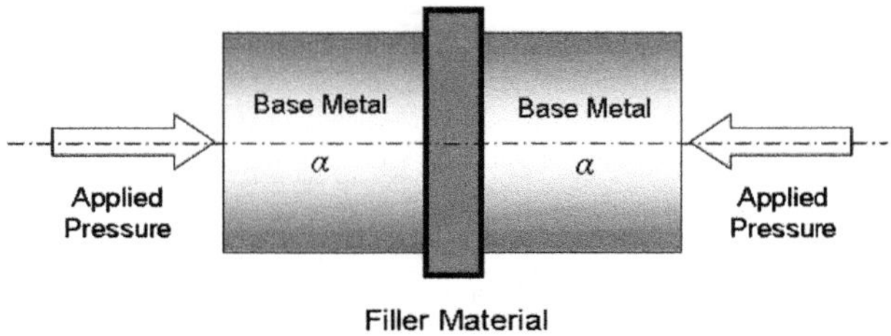

(a)

(b)

Moreover, for Mössbauer Spectroscopy, powder samples were obtained from different points along the joined tubes. The sampling was performed both in tubes which show a sound weld and in those which show joint defects.

All powdered samples were obtained from the tubes at the mid-thickness location. Three samples were prepared from each specimen:

- The joint.
- The heat affected zone (HAZ), at 3 mm from the joint.
- The base metal (BM), at 10 mm from the joint.

Also, a sample from the as-received tube was prepared as a reference material.

The samples, with a thickness of 19 mg/cm^2, were analyzed by transmission Mössbauer effect spectroscopy at room temperature with a conventional constant acceleration spectrometer, which was equipped with a proportional counter. The γ-radiation source was ^{57}Co (Rh) of $\sim$ 20 mCi. The absorption spectra were fitted by using the NORMOS program [9]. The isomer shift (δ) data are given relative to α-Fe.

4 Results and discussion

4.1 As-received steel

The SEM micrograph of the as-received steel is shown in Fig. 3. As a cold drawn tubular product, the structure is strongly banded. It consists of ferrite (dark), martensite-austenite (gray) and carbides (light).

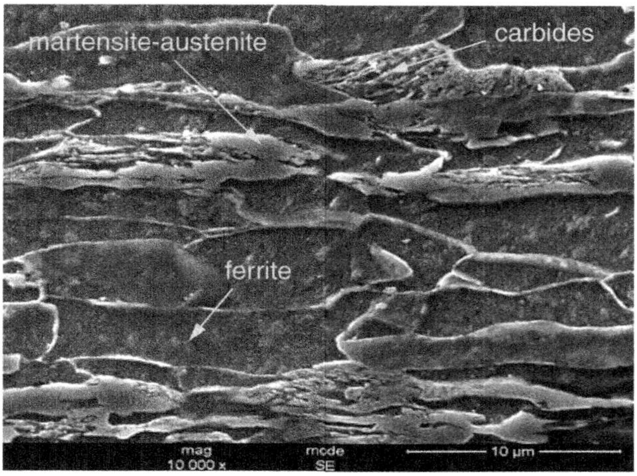

Fig. 3 Scanning electron microscopy image of the as received tube material

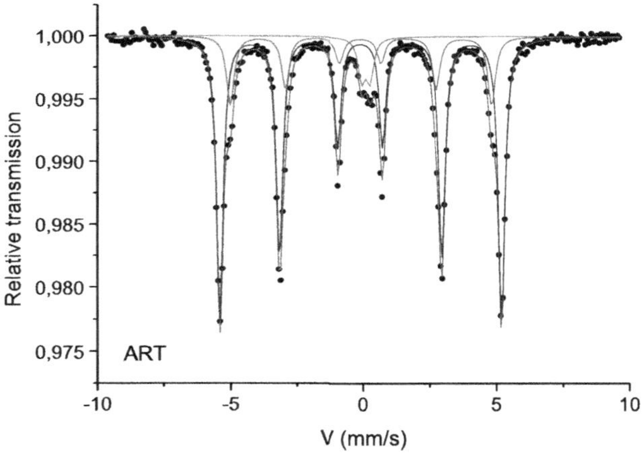

Fig. 4 Mössbauer spectrum—as-received steel

The Mössbauer transmission spectrum of this steel shows, apart of the sextets typical of low-carbon steels (named interactions I and II), a paramagnetic phase that consists of a doublet with a positive isomer shift (named interaction III) (Fig. 4). The hyperfine parameters are reported in Table 3.

Both sextets (interactions I and II), due to their hyperfine parameter, can be attributed either to ferrite where iron atoms can be surrounded by iron atoms or having as near neighbor substitutional atoms and to bainitic ferrite/ martensite where iron atoms can be surrounded by iron atoms or having carbon as interstitial atoms in its neighborhood [10].

On the other hand, the common spectrum of Fe-C austenite consists of a singlet, which corresponds to the iron atoms having no carbon atoms in the nearest neighborhood, and a doublet, which is from the iron atoms having one carbon atom in the nearest neighborhood [11].

Therefore, the doublet subspectrum (interaction III) is ascribed to the presence of retained austenite, with a high concentration of solutes, mainly C and Mn.

Table 3 Mössbauer fitting results

Interaction	Variable	JOINT-A	JOINT-B	HAZ-A	HAZ-B	AR
I	$\Gamma \pm 0.01$	0.32	0.31	0.32	0.32	0.30
	SA ± 2	69	72	75	75	74
	$\delta \pm 0.01$	0.00	0.00	0.00	0.00	0.00
	$\Delta \pm 0.01$	0.01	0.00	0.00	0.00	0.00
	BHf ± 0.1	33.1	33.2	32.5	33.2	32.9
II	$\Gamma \pm 0.01$	0.37	0.39	0.34	0.37	0.35
	SA ± 2	25	27	22	23	21
	$\delta \pm 0.01$	0.01	0.01	0.00	0.01	0.01
	$\Delta \pm 0.02$	−0.03	0.01	0.00	0.01	0.01
	BHf ± 0.1	30.6	30.8	30.2	30.8	30.5
III	$\Gamma \pm 0.01$	0.35	0.35	0.35	0.35	0.32
	SA ± 2	6	1	3	2	5
	$\delta \pm 0.02$	0.23	0.20	0.19	0.18	0.22
	$\Delta \pm 0.02$	0.29	0.25	0.21	0.25	0.31

Sample A (*A*); Sample B (*B*); As-Received (*AR*); Heat Affected Zone (*HAZ*); Linewidth (Γ) (mm/s); Spectral Area (*SA*) (%); Isomer shift (δ) relative to α-Fe (mm/s); Quadrupolar Splitting (Δ) (mm/s); Hyperfine Magnetic Field (*BHf*) (T)

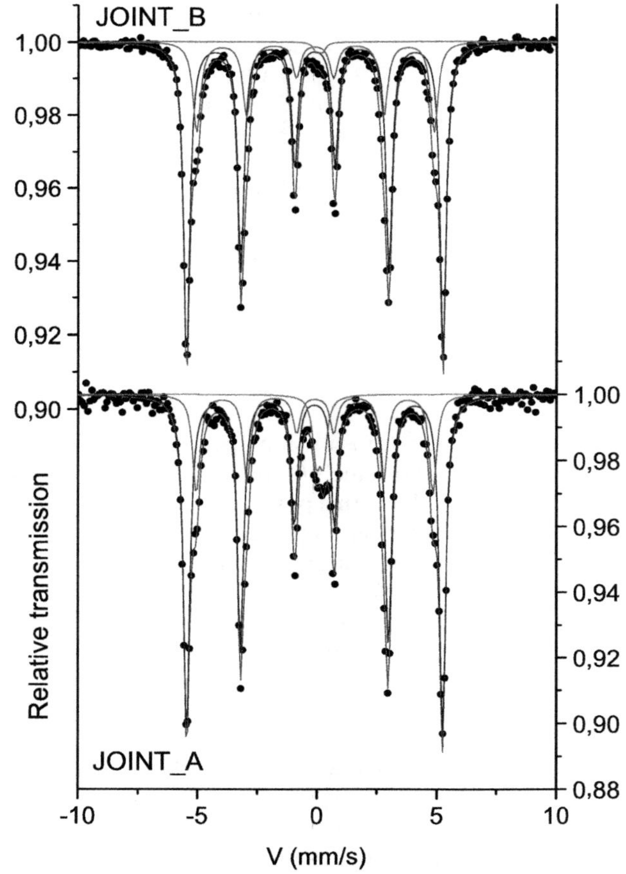

Fig. 5 Mössbauer spectra—Samples A and B at the Joint

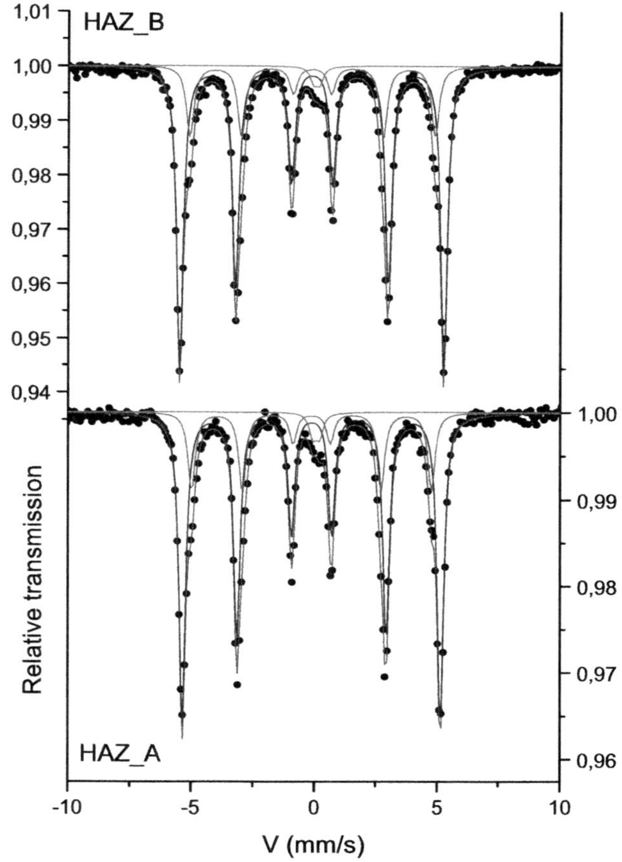

Fig. 6 Mössbauer spectra—Samples A and B at the HAZ

4.2 TLP bonding

4.2.1 Samples of specimen A

At the joint, the Mössbauer transmission spectrum was fitted with two sextets (interactions I and II) typical of low-carbon steels plus a paramagnetic phase in (Fig. 5). The hyperfine parameters are very similar to that of the as-received steel (see Table 3), except for the larger doublet relative area (interaction III). On the other hand, at the HAZ, the paramagnetic phase contribution (interaction III) is weaker (Fig. 6).

4.2.2 Samples of specimen B

In contrast with sample A, Mössbauer transmission spectrum shows a weak paramagnetic phase (interaction III) contribution both at the joint and at the HAZ (Figs. 5 and 6).

In both samples A and B, silicon and carbon atoms in the neighborhoods of iron atoms strongly affect the ferrite sextet. The increasing area (SA) of interaction II

(Table 3) may be attributed to the increase of the concentration of substitutional atoms (Si) in substituted ferrite. As mentioned before both sextets (interactions I and II) can be attributed to either substituted ferrite, interstitial carbon atoms in bainitic ferrite/martensite or both.

4.2.3 Effect of the diffusion of Si and B

Because of the chemical composition along the welded tubes, it is therefore tempting to relate the different Mössbauer parameters to changes in the concentration of the diffusion elements. Since silicon is a ferrite former, the presence of important amounts of retained austenite can only be related to the presence of boron.

In the case of boron added steels, boron stabilizes the austenite grain boundary by the segregation at high temperature, and thus delays the transformation [12].

The Joint—in contact with the filler material—is the zone that is able to attain the higher boron concentration. But the joint at the sample A was the double of time at $\approx 1,350°C$, compared with the joint at sample B (there is more time for diffusion and subsequent segregation of boron at austenite grain boundaries). Therefore, there are higher probabilities to stabilize austenite in specimen A. This is in accordance with the obtained Mössbauer spectra.

5 Conclusions

From these results it can be concluded that Mössbauer spectroscopy is an interesting tool to evaluate the characteristics of a TLPB joint, taking advantage of the austenite former behavior of boron. Also, the presence of silicon is clearly visible through the increasing of the relative area of the low hyperfine magnetic field sextet. However, Mössbauer spectroscopy is not the best characterization tool to resolve whether the ferromagnetic phase is ferrite or martensite, or both.

Further experiences are planned in backscattering geometry evaluating systematically the role of the different process parameters.

Acknowledgement The authors acknowledge support of this work by Consejo Nacional de Investigaciones Científicas y Técnicas and Universidad de Buenos Aires (I039), Argentina.

References

1. Cameron, G., Ikeda, A., Tochikawa, T., Higuchi, K.: SPE/IADC drilling conference (1997). doi:10.2118/37645-MS
2. Luft, H.B.: SPE/ICoTA coiled tubing roundtable (1999). doi:10.2118/54481-MS
3. Khan, T.I., Saha, R.K.: J. Mater. Sci. **42**, 9187–9193 (2007)
4. Shirzadi, A.A., Zhub, Y., Bhadeshia, H.K.D.H.: Mater. Sci. Eng. A **496**, 501–506 (2008)
5. Di Luozzo, N., Fontana, M., Arcondo, B.: J. Mater. Sci. **43**(14), 4938–4944 (2008)
6. Yuan, X., Kim, M.B., Kang, C.Y.: Metall. Trans. A (2010). doi:10.1007/s11661-010-0534-6
7. Tuah-Poku, I., Dollar, M., Massalski, T.B.: Metall. Trans. A **19**A, 675–686 (1988)
8. Di Luozzo, N., Fontana, M., Arcondo, B.: J. Mater. Sci. **42**(11), 4044–4050 (2007)
9. Brand, R.A.: NORMOS programs (1990)
10. Bruna, P., Pradell, T., Crespo, D., García-Mateo, C., Bhadeshia, H.K.D.H.: In industrial applications of the Mössbauer effect. AIP Conf. Proc. **765**, 338–343 (2004)
11. Gavriljuk, V.G., Shanina, B.D., Berns, H.: Acta Mater. **48**, 3879–3893 (2000)
12. Moon, J., Kim, S., Lee, J., Hwang, B., Lee, C.G., Lee, C.: Mater. Sci. Eng. A **497**, 153–159 (2008)

Hyperfine Interact (2011) 203:133–141
DOI 10.1007/s10751-011-0375-8

Mineralogy of the clay fraction of soils from the moray cusco archaeological site: a study by energy dispersive X-ray fluorescence, X-ray diffractometry and Mössbauer spectroscopy

María L. Cerón Loayza · Jorge A. Bravo Cabrejos · Mirian E. Mejía Santillán

Published online: 13 October 2011

Abstract The purpose of this work is to report the advances in the elemental and structural characterization of the clay fraction of soils from the terraces of the Moray Archaeological site, located 38 km north of the city of Cusco, Cusco Region. One sample was collected from each of the twelve terraces of this site and its clay fraction was separated by sedimentation. Previously the pH of the raw samples was measured resulting that all of the samples were from alkaline to strongly alkaline. Energy dispersive X-ray fluorescence (EDXRF) was used for the elemental characterization, and X-ray diffractometry (XRD) and transmission Mössbauer spectroscopy (TMS), using the γ14.4 keV nuclear resonance transition in ^{57}Fe, were used for the structural characterization of the clays and clay minerals present in each sample. The EDXRF analyses of all the samples show the presence of relatively high concentrations of sulfur in some of the samples and relatively high concentrations of calcium in all of the samples, which may be related to the high alkalinity of the samples. By XRD it is observed the presence of quartz, calcite, gypsum, cronstedtite, 2:1 phyllosilicates, and iron oxides. The mineralogical analysis of Fe by TMS shows that it is present in the form of hematite and occupying Fe^{2+} and Fe^{3+} sites in phyllosilicates, cronstedtite, and other minerals not yet identified.

Keywords Energy dispersive X-ray fluorescence · Mössbauer spectroscopy · X-ray diffractometry · Moray · Cusco

1 Introduction

The Moray terraces are located 7 km from Maras, in the Inka's Sacred Valley, 38 km northeast from the city of Cusco, Cusco Region at 3.500 masl [1]. Moray is

M. L. Cerón Loayza (✉) · J. A. Bravo Cabrejos · M. E. Mejía Santillán
Laboratorio de Análisis de Suelos, Laboratorio de Espectroscopia Mössbauer, Facultad de Ciencias Físicas, Universidad Nacional Mayor de San Marcos, Apartado, 14–0149, Lima 14, Perú
e-mail: malucelo@hotmail.com

Fig. 1 Site of this study: Moray terraces near the city of Cusco

Table 1 pH values of the Moray-Cusco samples

Samples	pH	Classification
S1-Arc	8.7	Strongly
S2-Arc	8.1	Alkaline
S3-Arc	7.9	Alkaline
S4-Arc	7.4	Weakly alkaline
S5-Arc	8.3	Alkaline
S6-Arc	8.5	Alkaline
S7-Arc	9.0	Strongly alkaline
S8-Arc	9.3	Strongly alkaline
S9-Arc	8.6	Strongly alkaline
S10-Arc	8.3	Alkaline
S11-Arc	8.7	Strongly alkaline
S12-Arc	9.2	Strongly alkaline

an outstanding archaeological complex, conformed by large terraces that superpose concentrically. Their configuration produces a gradient in microclimates with a temperature higher at the base of the concentric terraces that decreases gradually at the higher outer terraces. These beautiful terraces may have constituted a large agriculture laboratory and contributed important advances in agricultural technology in ancient times, when agriculture was the main labor activity and the foundation of the economic development. The samples were obtained *in situ* from the soil of all the terraces as it is shown in Fig. 1, with the purpose of characterizing each sample with the analytical techniques available at the Faculty of Physical Sciences of San Marcos.

2 Materials and methods

The physical-chemical analysis such as: pH, granulometric analysis, textural classification, and water content, were carried out at the Laboratory of Soil Analysis (LSA) of the Faculty of Physical Sciences following standard procedures [2]. Table 1 shows the pH values for all the raw samples. The texture (% mass of sand, silt and clay), of all the raw samples was obtained by sedimentation. Moreover, all the samples were analyzed by the analytical techniques available at this Faculty,

Table 2 Results of the quantitative analysis (mass %) by EDXRF

Moray samples (mass %)												
Elements	S1-Arc	S2-Arc	S3-Arc	S4-Arc	S5-Arc	S6-Arc	S7-Arc	S8-Arc	S9-Arc	S10-Arc	S11-Arc	S12-Arc
Al	6.1(4)	6.7(4)	6.7(4)	7.5(4)	4.2(3)	9.3(5)	6.7(4)	6.7(4)	6.7(4)	5.3(4)	6.7(4)	6.76(5)
Si	16.0(3)	22.1(4)	21.9(5)	18.2(4)	28.3(6)	21.7(4)	15.3(3)	15.3(3)	22.1(4)	9.7(3)	22.1(4)	19.7(6)
S	6.9(1)	0.62(2)	0.56(2)	3.2(1)	0.28(1)	0.59(2)	7.3(1)	2.04(4)	2.32(4)	14.7(3)	0.037(4)	0.19(1)
Cl	0.22(1)	0.26(1)	0.29(1)	0.24(1)	0.075(5)	0.11(1)	0.24(1)	0.24(1)	0.24(1)	0.31(1)	0.12(1)	0.22(1)
K	1.68(4)	2.1(1)	2.02(4)	2.15(3)	2.13(3)	2.51(5)	1.84(3)	1.91(4)	2.21(4)	1.96(4)	1.92(4)	1.65(3)
Ca	7.26(2)	3.7(1)	4.47(7)	8.9(2)	4.66(7)	4.21(6)	10.8(2)	7.4(1)	7.8(1)	19.5(4)	4.00(7)	3.98(6)
Ti	0.22(1)	0.40(1)	0.40(1)	0.40(1)	0.40(1)	0.48(1)	0.20(1)	0.30(1)	0.37(1)	0.14(1)	0.37(1)	0.34(1)
V	0.02(1)	0.015(1)	0.015(1)	0.023(1)	0.006(1)	0.008(1)	0.015(1)	0.015(1)	0.015(1)		0.015(1)	0.021(1)
Cr		0.008(1)	0.008(1)		0.010(2)	0.006(1)						
Mn	0.040(1)	0.088(1)	0.088(3)	0.080(2)	0.095(2)	0.088(2)	0.080(2)	0.080(2)	0.078(3)	0.065(2)	0.078(2)	0.10(1)
Fe	2.07(4)	3.10(1)	3.00(5)	2.70(4)	2.98(6)	3.14(5)	2.06(4)	3.23(6)	2.69 (4)	1.76(3)	3.57(7)	3.10(4)
Ni	0.012(1)	0.014(1)	0.010(1)	0.014(1)	0.09(1)	0.010(1)	0.014(1)	0.014(1)	0.014(1)	0.024(1)	0.014(1)	0.013(1)
Cu	0.014(1)	0.022(1)	0.011(1)	0.016(1)	0.014(1)	0.0070(2)	0.016(1)	0.016(1)	0.014(1)	0.012(1)	0.014(1)	0.013(1)
Zn	0.030(1)	0.028(1)	0.022(1)	0.036(1)	0.020(1)	0.014(1)	0.038(1)	0.036(1)	0.036(1)	0.043(1)	0.036(1)	0.026(1)
Rb	0.014(1)	0.013(1)	0.013(1)	0.019(1)	0.015(1)	0.014(1)	0.013(1)	0.013(1)	0.013(1)		0.013(1)	0.020(1)
Sr	0.075(2)	0.038(1)	0.024(1)	0.041(1)	0.026(1)	0.019(1)	0.066(1)	0.061(1)	0.061(1)	0.095(2)	0.046(1)	0.035(1)
Zr	0.042(1)	0.062(2)	0.036(1)	0.053(1)	0.046(1)	0.049(1)	0.025(1)	0.025(1)	0.055(1)	0.025(1)	0.046(1)	0.021(1)

The estimated uncertainty in the measurement of the mass concentrations is of order of 10%

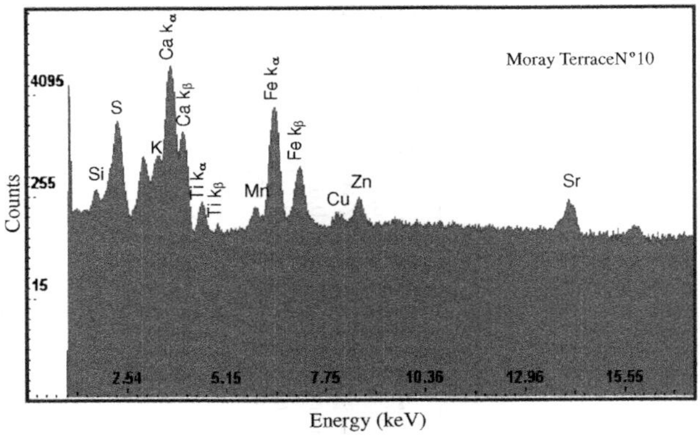

Fig. 2 EDXRF spectrum of the S10-Arc sample

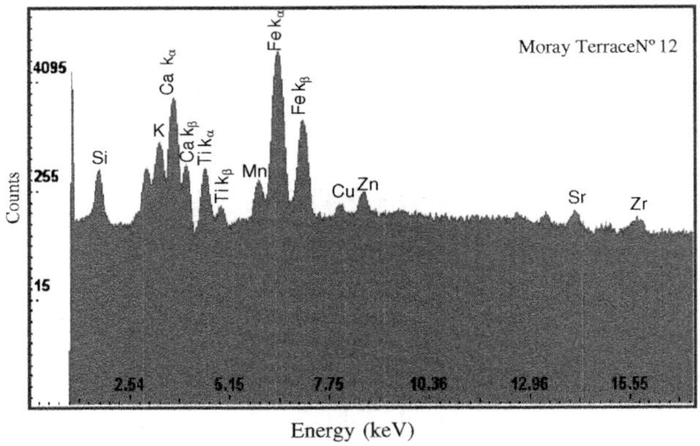

Fig. 3 EDXRF spectrum of the S12-Arc sample

such as: energy dispersive X-ray fluorescence (FRXED), X-ray diffractometry, and transmission Mössbauer spectroscopy (EMT). The analysis by XRF was done using a portable equipment that allows the detection of elements with atomic number higher than 12. The sample was prepared in the form of a tablet. In order to obtain more detailed information about the presence of compounds containing Fe the technique of TMS was used. The Mössbauer spectra of all the samples were taken at room temperature (RT) at the Laboratory of Archaeometry of San Marcos University; a ^{57}Co source in a Rh matrix was used and the spectra were fitted using the Normos program by Brand [3] in its version for crystalline sites (Normos Site).

3 Discussion and results

Table 2 gives the results of the quantitative analysis by FRXED for all of the clay fractions of the samples from the Moray terraces. However, here only the most relevant contrasting results in their concentrations are given. It is observed that in the terraces S1-Arc (lowest) to S10-Arc the presence of the following elements: Si, S, K, Ca, Ti, Mn, Fe, Cu, Zn, Sr y Zr, with relatively high concentrations of Ca in all of them and relatively high concentrations of sOulfur in samples S1-, S4-, S7-,

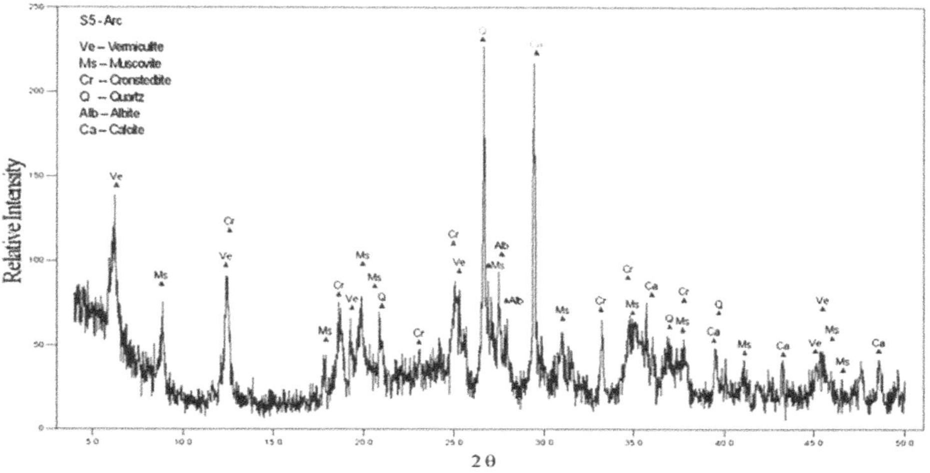

Fig. 4 X-ray diffraction pattern of the S5-Arc sample

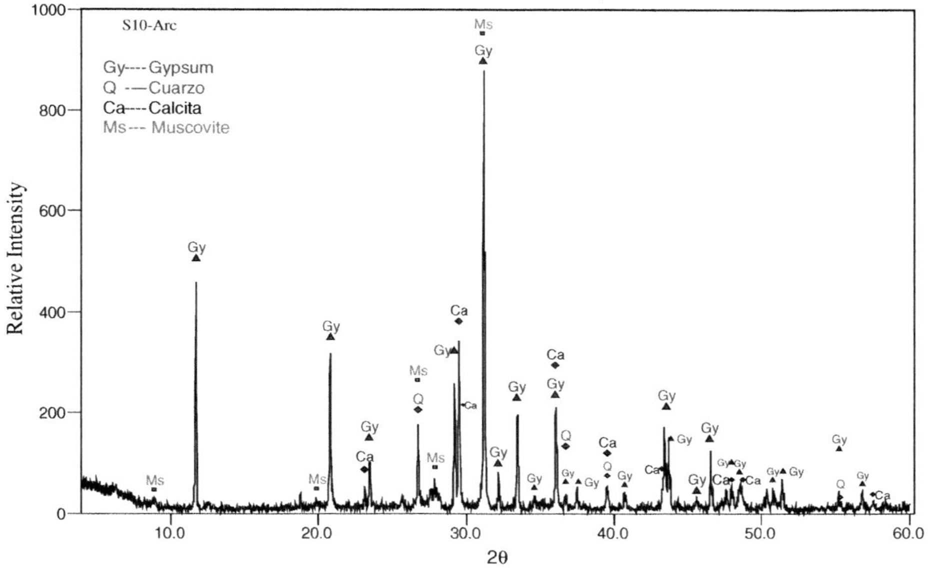

Fig. 5 X-ray diffraction pattern of the S10-Arc sample

S8-, S9- and S10-Arc; samples S10-Arc and S11- Arc show the highest and lowest concentrations respectively as it can be observed in Figs. 2 and 3. Likewise, it can be seen that the concentration of Fe is slightly larger in S12- Arc compared to S10-Arc, whereas the opposite result occurs in the case of Ca.

The results by XRD analysis show that all the terraces contain quartz and calcite, and one or more phases that may correspond to the phylosillicates such as vermiculite, montmorillonite, muscovite and/or illite. In samples S5-Arc and S6-Arc it is

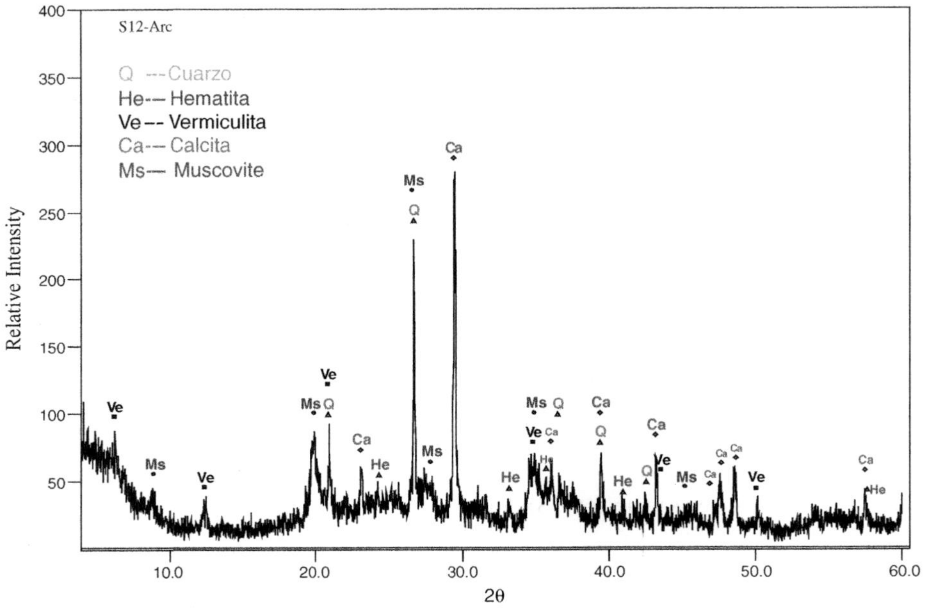

Fig. 6 X-ray diffraction pattern of the S12-Arc sample

Fig. 7 Mössbauer spectrum of the S5-Arc sample

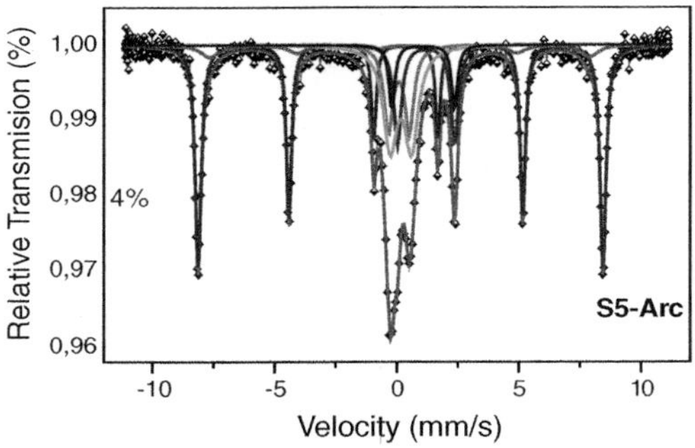

observed the presence of the mineral cronstedtite $Fe(II)_2Fe(III)(SiFe(III)O_5(OH)_4$ (see Fig. 4). For the case of S4-Arc, S7-Arc and S10-Arc it is observed also the presence of gypsum (see Fig. 5). It is worth mentioning that in the case of the S10-Arc the presence of the mineral gypsum ($CaSO_4$–$2H_2O$), identified by XRD, corroborates the finding of sulfur as observed by EDXRF; a similar comment can be made in the case of the minerals calcite ($CaCO_3$) and vermiculite ($Ca_{0.72}$ ($Si_{5.69}Al_{2.25}Fe$ $(III)_{0.06}$) ($Mg_{5.50}Ti_{0.11}$ $Mn_{0.01}Fe(II)_{0.04}Fe(III)_{0.45})O_{22}(OH)_2/8H_2O$ and in the case of S12-Arc it is observed the presence of the iron oxide hematite (see Fig. 6).

The results by TMS show the presence in all of the samples of two magnetic sextets assigned to hematite and seven paramagnetic doublets. By grouping the hyperfine parameters and comparing their values with known parameters for clay minerals observed by XRD, it has been possible to establish the following assignments: doublets

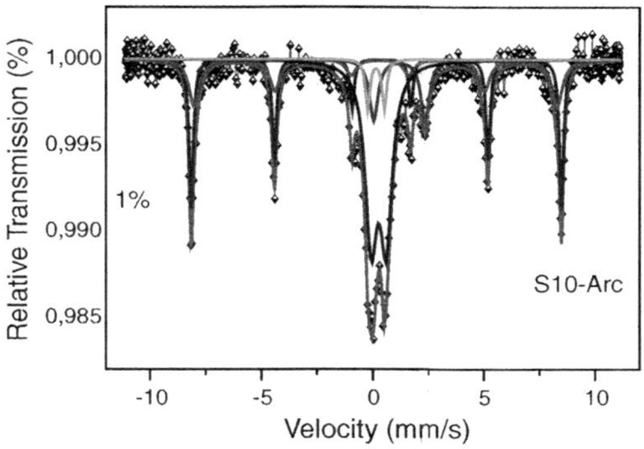

Fig. 8 Mössbauer spectrum of the S10-Arc sample

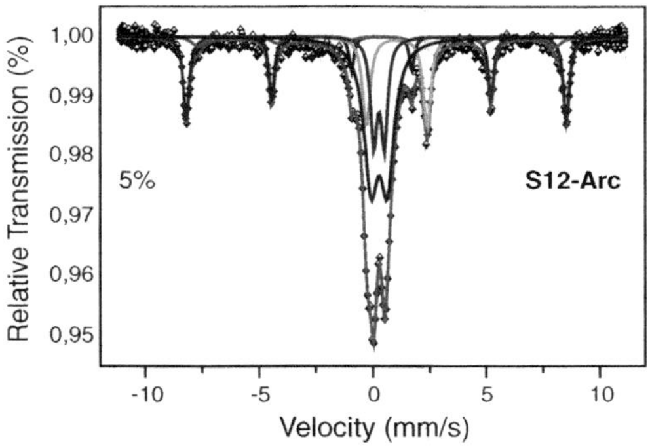

Fig. 9 Mössbauer spectrum of the S12-Arc sample

D3 and D4 are assigned to sites occupied by Fe^{2+} and Fe^{+3} respectively in montmorillonite; the presence of this clay mineral is corroborated by XRD in S7-Arc; however, it is not observed in the difractograms of S5-Arc and S10-Arc probably due to a low abundance of this mineral to be observed by XRD; in S3-Arc and S10-Arc, it is observed the presence of the doublet D5 which is assigned to muscovite, which is also observed by XRD; the doublet D6 is assigned to an Fe^{3+} site in illite in S4-, S6-, S9- and S11-Arc; and finally, the doublet D7 is assigned to an iron Fe^{+3} site in crosntedtite [4, 5], which is observed in S3-, S5- and S6-Arc; this mineral is also observed by XRD. In Fig. 7 it is shown the Mössbauer spectrum of S5-Arc, where it is observed the presence of cronstedtite.

In the case of S12-Arc, due to its higher concentration of Fe, it is observed a higher nuclear resonance absorption area compared to that of S10-Arc (see Figs. 8 and 9); this agrees with the results pertaining the elemental composition obtained by EDXRF. Table 3 lists the hyperfine parameters for all the analyzed clay fractions. For the preparation of this Table paramagnetic sites, with similar ISO and QUA values, have been group together in order to identify possible connection with observed mineral structures.

Table 3 Mössbauer hyperfine parameters of the clay fractions from the terraces of Moray-Cusco

Samples													
Sites	Parameter	S1-Arc	S2-Arc	S3-Arc	S4-Arc	S5-Arc	S6-Arc	S7-Arc	S8-Arc	S9-Arc	S10-Arc	S11-Arc	S12-Arc
Hematite (I)	δ (mm/s)	0.30(1)	0.27(1)	0.26(1)	0.27(1)	0.27(1)	0.26(1)	0.27(1)	0.27(1)	0.27(1)	0.27(1)	0.28(1)	0.28(1)
	2ε (mm/s)	−0.26(1)	−0.21(1)	−0.22(1)	−0.22(1)	−0.21(1)	−0.22(1)	−0.22(1)	−0.22(1)	−0.22(1)	−0.21(1)	−0.23(1)	−0.22(1)
	B (T)	51.36(5)	52.32(2)	52.49(2)	52.49(2)	51.39(1)	51.14(2)	51.38(2)	51.77(2)	51.96(1)	51.54(2)	51.89(3)	51.83(3)
Hematite (II)	δ (mm/s)			0.28(1)	0.26(5)	0.23(6)	0.26(1)	0.28(2)	0.20(6)	0.28(9)	0.22(2)	0.28(4)	0.27(5)
	2ε (mm/s)			−0.19(9)	−0.17(9)	−0.22(1)	−0.19(3)	−0.20(4)	−0.15(11)	−0.25(3)	−0.17(1)	−0.19(7)	−0.06(9)
	B (T)			49.9(4)	49.1(4)	47.7(7)	50.2(2)	49.0(2)	46.3(4)	50.7(3)	50.45(2)	50.3(4)	48.9(4)
Fe^{2+} D1	δ (mm/s)	1.00(1)	1.00(1)	1.05(1)	1.05(2)	1.04(1)	1.17(1)			1.02(1)			1.05(1)
(Vermiculite)	QUA (mm/s)	2.66(2)	2.72(2)	2.71(2)	2.69(3)	2.66(2)	2.34(1)			2.71(1)			2.66(1)
Fe^{3+} D2	δ (mm/s)	0.21(1)	0.22(1)	0.26(1)	0.27(1)	0.23(2)				0.24(1)			0.28(1)
(Vermiculite)	QUA (mm/s)	0.65(2)	0.65(2)	0.68(1)	0.55(1)	0.65(6)				0.68(2)			0.70(1)
Fe^{2+} D3 (Montmorillonite)	δ (mm/s)	1.25(1)	1.30(2)			1.17(3)		1.22(1)	1.19(1)		1.20(2)	1.25(1)	
	QUA (mm/s)	2.32(2)	2.48(5)			2.36(5)		2.28(1)	2.37(2)		2.28(4)	2.29(1)	
Fe^{3+} D4 (Montmorillonite)	δ (mm/s)							0.20(1)	0.08(1)		0.12(2)		
	QUA (mm/s)							0.74(1)	0.82(3)		0.72(2)		
D5 Muscovite	δ (mm/s)								0.25(2)		0.24(1)	0.21(1)	0.28(1)
	QUA (mm/s)								0.73(2)		0.69(2)	0.65(1)	0.46(1)
Fe^{3+} D6 (Illite)	δ (mm/s)				0.32(4)		0.24(3)			0.77(7)		0.23(1)	
	QUA (mm/s)				1.27(7)		1.09(6)			1.19(1)		1.27(4)	
Fe^{3+} D7 (Cronstedtite)	δ (mm/s)			0.03		0.09(3)	0.08(1)						
	QUA (mm/s)			0.80		0.89(9)	0.79(3)						

The uncertainties are: 0.04 T in B, 0.01 mm/s in δ; and 0.02 mm/s in 2ε

δ = isomershift relative to α-Fe in mm/s. QUA = quadrupole split in mm/s. B = hyperfine magnetic field in teslas

4 Conclusions

There is a general and complementary agreement among the results obtained for the elemental analysis by EDXRF with the results obtained for the structural analysis of the clay fractions of the soil samples by XRD and TMS.

The analyses by EDXRF allowed the determination of the elemental composition of the clay fractions of the soil samples. It is observed that the concentration of S is very low in the two upper S11-Arc and S12-Arc compared with the S10-Arc, as seen in Table 2, which gives the results of the quantitative analysis. Moreover, the concentration of Ca is higher than that of iron in all the samples, this difference being largest in S10-Arc and smallest in S11-Arc. This may be explained in terms of a hypothesis that the soils used to build the terraces were brought from quarries in the neighborhood of Moray with different elemental composition and probably with different agricultural applications. At the present time a commercial fertilizer rich in S and Ca is sold in the market.

The results by XRD show that quartz and calcite are present in all the samples. The mineral cronstedtite has been detected in S5-Arc and S6-Arc by XRD. The mineral gypsum, which contains sulfur in its structure, is observed in all the samples with high concentration of sulfur. The minerals vermiculite, montmorillonite, muscovite and illite may be present in most of the samples; however, due to a large overlap of the reflection peaks of these phases, it will require further work in sample preparation using chemical and temperature treatments to obtain a more precise identification of these phases.

The analysis by TMS at RT shows the presence of magnetic phases which have been assigned to hematite, which is usually found in soils, and of seven paramagnetic doublets related to sites occupied by $Fe^{2+}$and Fe^{3+} which have been assigned to clay minerals observed by XRD such as: vermiculite, montmorillonite, muscovite, illite, and cronstedtite. In the case of S5-Arc one of the sites occupied by Fe^{3+} with $\delta =$ 0.06 mm/s and QUA $=$ 0.78 mm/s, and in the case of S6-Arc with $\delta = 0.06$ mm/s and QUA $= 0.78$ mm/s are assigned to the mineral cronstedtite, which is observed by XRD.

Due to the archaeological relevance of Moray, this study will continue with the purpose of identifying the quarries from which the soils were collected.

References

1. Informe de Actividades IPGH Perú: Instituto Panamericano de Geografía e Historia, OEA (2009)
2. Manual de Ensayos de Suelos. Laboratorio de Análisis de Suelos. Facultad de Ciencias Físicas-UNMSM (2004)
3. Brand, R.A.: NORMOS: Mössbauer Fitting Program (1995)
4. Stevens, J. (ed.): Mössbauer Minerals Handbook. Mössbauer Effect Data Center, Ashville, NC (1998)
5. Cerón Loayza, M.L.: Estudio Mineralógico de Suelos Agrícolas. Tesis de Licenciatura. Facultad de Ciencias Físicas. Universidad Nacional Mayor de San Marcos (2001)

Hyperfine Interact (2011) 203:143–147
DOI 10.1007/s10751-011-0357-x

Aggregates of iron in ytterbium films

W. T. Herrera · C. Rojas-Ayala · I. S. Dinóla · E. C. Passamani · M. Kraken · J. Engelke · E. Baggio-Saitovitch · F. J. Litterst

Published online: 25 August 2011

Abstract Films of ytterbium doped with iron (0.3–1.6 at.% Fe) have been prepared by co-deposition of atomic beams of the elements onto Kapton substrates under high vacuum. Iron is immiscible in ytterbium and various iron species are expected to occur. XRD reveals a mixture of fcc and hcp ytterbium. The Mössbauer spectra are interpreted with contributions from monomeric iron on interstitial sites, two types of neighboring substitutional iron aggregates and two types of clusters formed upon diffusion. The present data supplement earlier data obtained for 0.5 and 5 at.% Fe. There is no systematic correlation of cluster formation with increasing iron concentration. There are indications that clusters are formed more easily for thin films. We conclude that cluster formation occurs mainly on the film surface and is limited by the slow diffusion from inside films.

Keywords Mössbauer spectroscopy · Interstitial site · Aggregates · Clusters · Ytterbium film · Vapor deposition · Diffusion

Presented at LACAME 2010, Lima.

Work supported by a Brazilian-German research program financed by CNPq and DFG 444BRA-113/55/0–1.

W. T. Herrera · C. Rojas-Ayala · I. S. Dinóla · E. Baggio-Saitovitch · F. J. Litterst
Centro Brasileiro de Pesquisas Físicas, 22290-180, Rio de Janeiro, Brazil

E. C. Passamani
Departamento de Física, Universidade Federal do Espírito Santo, 29075-910, Vitória, ES, Brazil

M. Kraken · J. Engelke · F. J. Litterst (✉)
Institut für Physik der Kondensierten Materie, Technische Universität Braunschweig, 38106 Braunschweig, Germany
e-mail: j.litterst@tu-bs.de

Table 1 Hyperfine parameters for the room temperature Mössbauer spectra of the Yb films (data for 0.5 and 5 at.% are from [1])

Iron conc. (at %)	Thickness (μm)		D1	D2	D3	D4	S
5.0	0.2	δ (mm/s)	–	–	−0.05 (2)	0.22 (2)	
		Δ (mm/s)	–	–	0.39 (1)	0.89(2)	
		Area (%)	–	–	56 (2)	44(2)	
1.5	6.6	δ (mm/s)	0.44(5)	0.18(5)	−0.05	–	−0.32(5)
		Δ (mm/s)	0.97(5)	1.07(5)	0.45(5)	–	–
		Area (%)	21(2)	16(2)	56 (2)	–	7(2)
0.6	2.0	δ (mm/s)	–	–	−0.06	–	−0.32(5)
		Δ (mm/s)	–	–	0.46 (5)	–	–
		Area (%)	–	–	93 (2)	–	7(2)
0.5	0.9	δ (mm/s)	–	–	−0.03	–	−0.26(1)
		Δ (mm/s)	–	–	0.47 (1)	–	–
		Area (%)	–	–	96 (2)	–	4(2)
0.3	7.0	δ (mm/s)	0.43(5)	0.18(5)	−0.05	–	−0.30(5)
		Δ (mm/s)	0.97(5)	1.06(5)	0.45 (5)	–	–
		Area (%)	53(2)	35(2)	5 (2)	–	7(2)

δ: isomer shifts rel. α-iron. Δ: quadrupole splitting

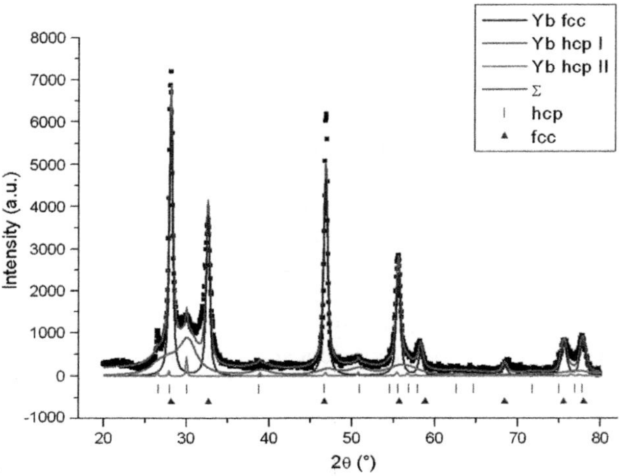

Fig. 1 X-ray diffractogram of a 0.6 at.% Fe in Yb film revealing a mixture of fcc and hcp Yb

1 Introduction

Formation of magnetic moments at iron positioned at substitutional and interstitial sites of ytterbium has been the subject of earlier studies [1, 2]. For both types of positions magnetic iron moments could be confirmed. The reported ^{57}Fe Mössbauer spectra, however, are complex revealing a superposition of several sub-spectra. Aim of the present study was to accomplish better systematics allowing a new tentative assignment of sub-spectra to possible iron species. We therefore also include data from the earlier study [1].

Iron is immiscible in ytterbium. From [1] it is known that stable conditions of the different iron species found in films deposited at low temperatures can only

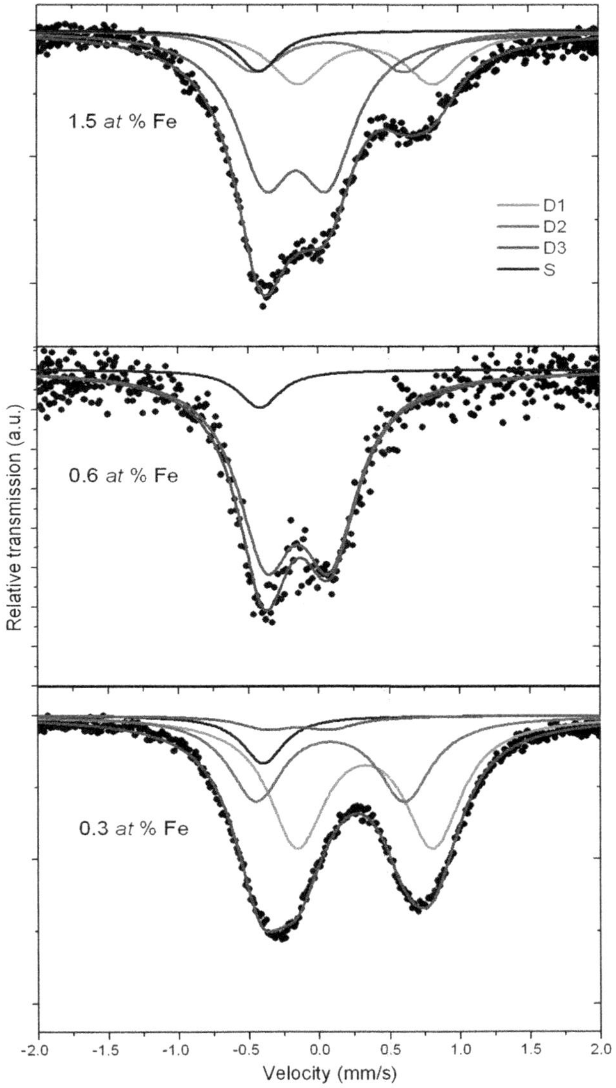

Fig. 2 ^{57}Fe Mössbauer absorption spectra of iron in ytterbium films with 0.3, 0.6 and 1.5 at.%

be reached when samples are kept at room temperature for several days. Here we therefore restrict ourselves to the discussion of spectra from in this way annealed films or obtained by deposition of substrates kept at room temperature.

2 Experimental

Samples with 0.3, 0.6 and 1.5 at.% Fe were prepared by vapor co-deposition from atomic beams of Yb (4n) and Fe (90% enriched in ^{57}Fe) under vacua of $\sim 10^{-9}$ mbar onto Kapton foils kept at room temperature. The films reported in [1] contain 0.5

and 5 at.% Fe, resp., and were prepared in the same installation yet with a substrate temperature of 20 K during deposition. Spectra of these samples were taken after warming to room temperature.

Film thicknesses are on the order of μm (see Table 1). The films are sufficiently stable against oxidation for allowing short-time handling at air.

^{57}Fe absorption spectra were recorded in a standard Mössbauer spectrometer using a ^{57}Co**Rh** source.

3 Results and discussion

A typical X-ray diffractogram is shown in Fig. 1. The Yb films have a mixed structure of typically about 60% fcc and 40% of distorted hcp phase (indicated as hcp I) with grain sizes of ca. 20 nm. The relative intensities of diffraction peaks indicate considerable texture of the films. As seen in Fig. 1 a small part (hcp II) of the hcp phase reveals sharp peaks that may be related to larger grains at the surface.

Mössbauer absorption spectra of Yb films containing 0.3, 0.6, and 1.5 at.% Fe obtained at room temperature are shown in Fig. 2. For those of 0.5 and 5 at.% Fe we refer to [1]. Visual inspection of the spectra clearly indicates strong differences. All samples reveal a superposition of several sub-spectra.

Overall four quadrupole doublets and one single line are identified. For all samples studied, these four species have the same hyperfine parameters, yet different relative spectral weights (see Table 1, note, that the labeling of sub-spectra is different from that used in [1]). As discussed in [1] one may give tentative assignments to the configuration of the various species. Iron responsible for the single line S is obviously positioned in a cubic surrounding. Its negative isomer indicates a site with enhanced s-electron density as expected for an interstitial site.

This is also in agreement with data from implantation experiments [2]. We associate the two doublets with high quadrupole splitting (D1 and D2) with pairs of neighboring iron on substitutional (yet structurally probably relaxed) sites as indicated by the higher isomer shift. Reason for the different pair configurations can be sought in the mixture of fcc and hcp Yb found in these films. Doublet D3 is found for all studied samples. It has smaller quadrupole splitting and intermediate isomer shift and is attributed to iron clusters formed either already during film deposition or upon annealing. Whereas the spectra of the 5 at.% sample [1] do not show D1 and D2, they reveal instead another doublet D4, which due to the high iron concentration should also be attributed to another kind of clusters.

4 Conclusion

Contrary to our expectation there cannot be found a systematic variation of the relative contributions from the various iron species with concentration. The number of iron species has increased compared to the earlier found [1, 2]. We could confirm spectrum S and its assignment as an interstitial monomer.

As well the "dimer" aggregate species D1 and D2 and the "cluster" spectra D3 and D4 need further consideration. It has to be corroborated if the formation of the aggregates D1 and D2 is related to the observed mixture of fcc and hcp Yb phases.

From Table 1 one may conclude that formation of clusters is easier achieved for the thinner films, which would suggest that clusters are mainly formed at the film surface limited by diffusion from the inner part of the films to the surface. Also in this context the phase mixture may play a role causing dislocations that should influence diffusion. More systematics concerning the thickness dependence of clustering is still needed.

In order to learn more on the magnetism of the various iron species, Mössbauer experiments at low temperatures are presently performed. As urgent supplementary experiments for gaining more information on the local structure, EXAFS will be studied.

References

1. Passamani, E.C., Morales, M.T., Baggio-Saitovitch, E.: Phys. Rev. B **60**, 1111 (1999)
2. Kapoor, J., Riegel, D., Li, Y., Polaczyk, C., Andres, J., Mezei, F., Sielemann, R., Yoshida, Y., Brewer, W.D., de Mello, L.A., Frota-Pessoa, S.: Phys. Rev. Lett. **78**, 1279 (1997)

Hyperfine Interact (2011) 203:149–153
DOI 10.1007/s10751-011-0373-x

Magnetic dynamics of dilute iron nano-clusters in silver films from Mössbauer spectroscopy and muon spin rotation

W. T. Herrera · I. Dinola · E. Baggio-Saitovitch · M. Kraken · F. J. Litterst

Published online: 30 August 2011

Abstract A silver film containing nanometer size clusters of iron (nominal conc. 1 *at*%) has been studied by Mössbauer spectroscopy and Low-Energy Muon Spin Rotation. Below about 20 K spin glass freezing due to interparticle interactions is found from both methods. Whereas Mössbauer spectra are insensitive to the fast fluctuations of cluster moments above spin glass freezing temperature, muon spin rotation in magnetic fields applied perpendicular to the polarized muon spins allows tracing the fluctuations of superparamagnetic moments. The temperature dependence of the damping of the muon spin rotation signal shows Arrhenius behavior between 10 to 100 K. Depending on the assumed shape of damping the activation energy of superparamagnetic fluctuations of cluster moments ranges between about 20 K $\cdot k_B$ and 40 K $\cdot k_B$. Above about 120 K muon spin depolarization indicates diffusion and trapping of muons.

Keywords Mössbauer spectroscopy · Muon spin rotation · Correlation time · Superparamagnetism · Nano-clusters

Presented at LACAME 2010, Lima.

Work supported by a Brazilian–German research program financed by CNPq and DFG 444BRA-113/55/0-1.

W. T. Herrera · I. Dinola · E. Baggio-Saitovitch
Centro Brasileiro de Pesquisas Físicas, Rua Dr. Xavier Sigaud 150, 22290-180, Rio de Janeiro, Brazil

W. T. Herrera
e-mail: william@cbpf.br

M. Kraken · F. J. Litterst (✉)
Inst. Physik der Kondensierten Materie, Technische Universität Braunschweig, Mendelssohnstrasse 3, 38106 Braunschweig, Germany
e-mail: j.litterst@tu-bs.de

1 Introduction

Very dilute (< 10 ppm) iron in silver reveals Kondo behaviour as shown from early Mössbauer experiments on ^{57}Co in Ag [1]. Since iron is practically immiscible in silver, higher concentrations will lead to formation of clusters. Their magnetic and electronic behaviour is complex due to counteracting magnetic coupling and Kondo effect [2]. Silver films (typical thickness 1 μm) containing iron, have been prepared by vapor codeposition from atomic beams [2, 3]. Depending on substrate temperature and iron concentration, nanometer sized clusters of iron are formed. Samples were characterized via XRD, $EXAFS$, resistivity, magnetization, susceptibility and Mössbauer spectroscopic measurements [2, 3]. Up to about 2 at% Fe concentration films contain dominantly two types of clusters with only up to about 20 iron atoms/cluster. Relative contributions of these clusters vary systematically with concentration. The magnetic behavior can be described by an ensemble of ferromagnetic particles with 30–40 μ_B. Depending on concentration magnetic hyperfine interaction at ^{57}Fe appears below about 5–15 K. Anomalies in magnetic data are found in the same temperature range where hyperfine magnetic splitting has its onset, indicating the freezing of interacting cluster moments. Interaction of the magnetic moments with the conduction electron sea is reflected in resistivity data revealing a shallow minimum. It is suppressed by an applied magnetic field and indicates Kondo effect. We therefore conclude that magnetic freezing is mainly related to $RKKY$ interaction, i.e., spin polarization of the conduction electron sea.

Mössbauer spectroscopy and Muon Spin Rotation offer a complementary view on the magnetic cluster dynamics due to the different positions of the probe atoms / particles: the iron nuclei sense the dynamics from inside clusters; the positive muons implanted into the film are positioned at interstitial sites of silver. Here we report on a preliminay series of comparative Mössbauer and Muon Spin Rotation results performed on the same sample.

In contrast to an earlier Muon Spin Rotation study [4] of monodisperse superparamagnetic iron particles implanted into silver, the nano-clusters studied here are about 50 times smaller. The shape of the particles cannot yet be resolved by TEM due to their smallness. EXAFS data are consistent with clusters comprising only few iron atoms in agreement with the results from magnetic data.

2 Experimental

The chosen film nominally contained 1 at% Fe with a cluster concentration of ca. 0.05%. Preparation was performed under high-vacuum conditions as described in [2, 3]. The Kapton substrate was kept at room temperature. Mössbauer absorption spectra between 5 K and 300 K were taken in a standard spectrometer. Muon Spin Rotation (μSR) experiments were performed at the Low-Energy μSR (LEM) facility of Paul Scherrer Institute, Switzerland using positive muons with an energy of 14 keV to guarantee stopping of muons within the film thickness. For details of μSR and LE-μSR we refer to (http://lmu.web.psi.ch/lem/index.html). We used a transverse field (TF) arrangement, i.e. the magnetic field is applied perpendicular to the direction of polarized muon spins which are in this way forced to precess around the field direction. The field used was 9.9 mT in plane of the film.

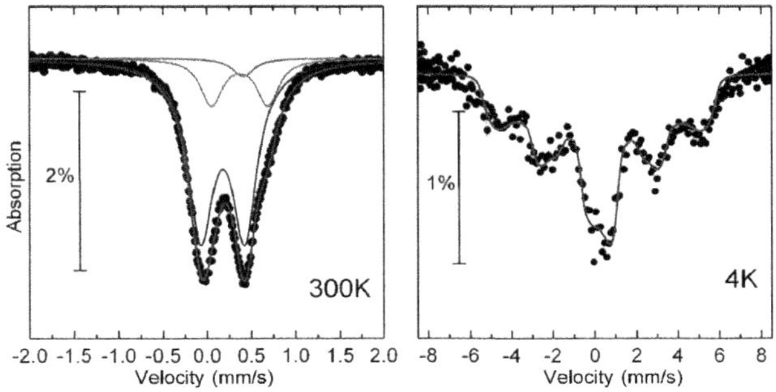

Fig. 1 ^{57}Fe Mössbauer absorption spectra of Fe clusters in a silver film (nominal conc. 1 *at*% Fe)

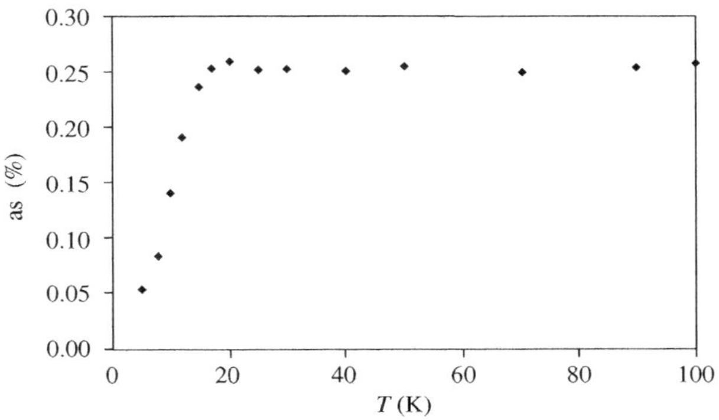

Fig. 2 Asymmetry of μSR signal in 9.9 mT TF

3 Results and discussion

Figure 1 shows Mössbauer spectra in the paramagnetic and the spin-frozen regime. The paramagnetic pattern reveals two quadrupole doublets assigned to the mentioned two types of clusters and a minor single line contribution from monomeric iron. For details of the fit procedure, see [2]. Fluctuation of cluster moments is fast compared to nuclear Larmor precession in locally acting fields so that no magnetic hyperfine splitting is observed. Below about 20 K the fluctuations are gradually frozen on the hyperfine time scale and magnetically split patterns appear. The broad patterns as shown in Fig. 1 (right side) are typical for fluctuation rates of cluster moments on the order of 0.01–1 GHz. A detailed study of the Mössbauer relaxation patterns will be reported elsewhere.

The temperature dependent asymmetry of the μSR spectra is shown in Fig. 2. Clearly visible is the loss of asymmetry below 20 K which is in agreement with the onset of spin freezing seen also from Mössbauer spectroscopy leading to a spin glass-like state of interacting particles.

The μSR pattern of pure silver in applied transverse field should reveal an undamped rotation pattern with a single frequency corresponding to the applied field. There is no damping since silver is diamagnetic and also contains no nuclear magnetic moments. In the case of our iron doped silver films, the influence of iron cluster moments causes damping. The damping function was assumed $\sim\exp(-(\lambda t)^{\beta})$

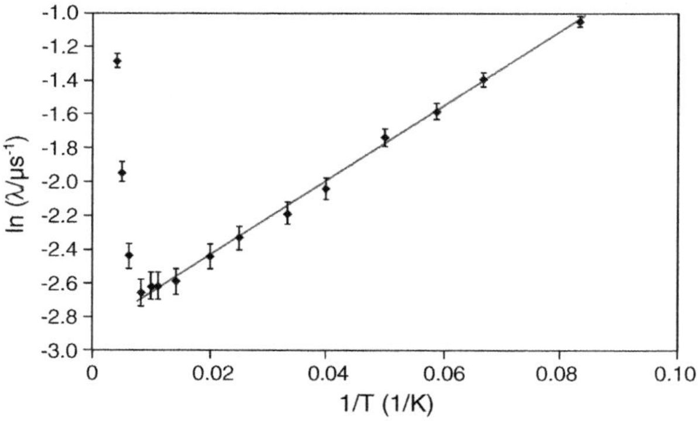

Fig. 3 TF μSR damping for clusters of Fe in a silver film

with $\beta = 1$. The damping parameter λ is proportional to the correlation time of the thermal "bath" (i.e. the fluctuating fields due to the superparamagnetic iron cluster moments) driving the muon spin. Figure 3 shows the temperature dependence of λ. Between about 15 and 100 K, λ reveals Arrhenius behavior with an activation energy $E_A \approx 22$ K $\cdot k_B$. An analysis using a stretched exponential damping function with $\beta = 1/2$ yields $E_A \approx 40$ K $\cdot k_B$. The steep increase of λ above 100 K is indicating the onset of muon diffusion and following trapping close to magnetic centers.

Comparison with results from mono-disperse Fe clusters of about 50 times bigger size reported in [4] shows that the damping parameters λ and also the activation energies range in the same order of magnitude. This can be understood, when taking into account that the sample used in [4] had bigger cluster moments, however, a smaller cluster concentration compared to our sample. In this way, both effects are compensating each other.

4 Conclusions

We have studied a silver film containing nano-sized iron clusters via LE-μSR and Mössbauer spectroscopy. The sample was thoroughly pre-characterized via XRD, magnetic and electric methods. Both μSR and Mössbauer data reveal an onset of spin glass freezing below about 20 K. The Mössbauer spectra exhibit partial magnetic dynamics even down to 4 K. Above 20 K, the iron clusters appear superparamagnetic on the Mössbauer time scale, revealing non-magnetic patterns.

In contrast, μSR is able to resolve the slowing down of superparamagnetic fluctuations of cluster moments over a wide temperature range from about 100 K down to 10 K. The correlation time of spin fluctuation follows an Arrhenius law.

The different sensitivities of both methods to magnetic fluctuations has to be sought in the different positions of probes: whereas μ^+ is at an interstitial position in silver and thus senses the magnetic fluctuations of an ensemble of relatively distant magnetic cluster moments, the hyperfine magnetic field at the Mössbauer nucleus is directly coupled to the moment of the cluster in which its atom is positioned. This means, the information on magnetic dynamics obtained from the two methods is complementary.

From the observed differences, it should be possible to distinguish various fluctuation scenarios (e.g. with axial or spherical anisotropy). Model calculations are under way.

Acknowledgements Essential support by the LEM crew at PSI, H. Luetkens, A. Suter and Th. Prokscha is gratefully acknowledged.

References

1. Steiner, P., Hüfner, S.: Phys. Rev. B **12**, 842 (1975)
2. Herrera, W.T., Xing, Y.T., Ramos, S.M., Munayco, P., Fontes, M.B., Baggio-Saitovitch, E.M., Litterst, F.J.: Phys. Rev. B **84**, 014430 (2011)
3. Morales, M.A., Passamani, E.C., Baggio-Saitovitch, E.: Phys. Rev. B **66**, 144422 (2002)
4. Jackson, T.J., Binns, C., Forgan, E.M., Morenzoni, E., Niedermayer, Ch., Glückler, H., Hofer, A., Luetkens, H., Prokscha, T., Riseman, T.M., Schatz, A., Birke, M., Litterst, J., Schatz, G., Weber, H.P.: J. Phys. Cond. Matt. **12**, 1399 (2000)

Author Index